£11·75

Vanier

D1234567

This book is dedicated to all those scientists who have contributed to the study and understanding of the channeling phenomenon. Foremost amongst these is the name of Professor Jens Lindhard whose contribution 'Influence of Crystal Lattice on Motion of Energetic Charged Particles' was a major advance in the theory of channeling.

Preface

The idea of the authors in writing this book was to encompass in a single volume a comprehensive discussion of the phenomenon of channeling and its application to physics and technology. This has been achieved by a coordinated co-authorship of nineteen scientists who have been engaged in various aspects of channeling research. In this way we sought to achieve a varied and balanced review of past and current work. One obvious danger in preparing a book of this kind is the possibility of producing sixteen independent reviews, with much overlap and many omissions. I am confident that we have surmounted this problem, not because of my own contribution as editor, but because of the extensive dialogue and constructive criticism between the co-authors, which has greatly simplified my task of giving cohesion to the volume.

The book covers all aspects of channeling, its theory, observation and applications. Chapter I introduces the concepts of directional effects of charged particles in crystals and provides a résumé of the historical development of the phenomenon, together with some elementary analytical concepts. An important message contained herein is that channeling is more than the simple transparency concept so obvious from a casual inspection of a crystal model. It is the *steering forces* and consequently the *stability* of channeled ions that make the phenomenon a major effect and not a minor one.

Chapter II develops the continuum model of directional effects along the lines of the classical 1965 paper of Professor Jens Lindhard. Lindhard's formulation of the continuum model is now accepted as a basis for the majority of current theoretical work. An alternative procedure for studying channeling is to use computer simulation. This, together with its uses and limitations, are discussed in Chapter III.

An important consequence of channeling is the striking reduction in the energy loss of such particles. Indeed, the crucial experiment which unambiguously established the existence of the channeling phenomenon was based upon the measurement of the energy loss of MeV protons in silicon. Energy loss and associated problems are discussed in Chapters IV, V and VI. Chapter VII is devoted to the physical processes which enable channeled particles to become unstable (i.e. become dechanneled), while Chapter VIII deals with range profiles—a parameter which is a sensitive function of both the channeling and dechanneling properties of the ion–crystal system. In Chapter IX the radiation damage problem is formulated, together with some discussion of the effect of radiation damage on channeling.

Thus far the discussion deals exclusively with charged particles which obey classical mechanics. The question of whether light particles such as electrons or positrons may be treated in this way is considered in Chapter X, and the transition from classical to quantum mechanics discussed.

Chapter XI represents a natural division in the book. Up to this point, directional effects are discussed for their own intrinsic interest; one may therefore ask the question 'can the phenomena be employed in branches of physics other than the rather specialized studies of the penetration of charged particles through crystals?' This question is answered in Chapters XI–XVI. Chapter XI considers the application to the study of crystal structures and Chapter XII to nuclear lifetime studies.

Chapters XIII, XIV and XV bring in three contributors who have been associated with some of the most successful and technologically important applications of channeling: (i) lattice site location studies, (ii) radiation damage studies and (iii) surface studies. Finally Chapter XVI deals with the applications to semiconductor technology.

The text is somewhat longer than originally envisaged; this, however, reflects the research situation which has existed over the last few years, where we have seen a remarkable growth in the number of channeling experiments and its application to an ever wider range of problems.

As editor, I thank most warmly all of the authors contributing to this book, and also our colleagues in the channeling field whose suggestions and criticism have played such an important role in unifying the sixteen contributions into a coordinated text on channeling. I am personally indebted to Professor J. A. Davies and Dr J. L. Whitton (Chalk River Nuclear Laboratories), to Dr D. Van Vliet (Cambridge), to Dr E. Bøgh (University of Aarhus), to Dr L. Feldman (Bell Telephone Laboratories) and to Professor J. W. Mayer (California Institute of Technology) for their advice and encouragement; to Mr P. Ashburn (Leeds) for his valuable assistance in compiling the Index. Finally, I should like to express my sincere thanks to the Leeds United Football Club, whose regular Saturday distractions have proved to be a valuable therapy during the compilation of this volume.

D. V. Morgan

Contributors

B. R. APPLETON   *Oak Ridge National Laboratory,* Oak Ridge, Tennessee, U.S.A.*

C. S. BARRETT   *Chemical Engineering and Metallurgy, University of Denver, Denver, Colorado, U.S.A.*

E. BØGH   *Institute of Physics, University of Aarhus, Aarhus C, Denmark.*

L. T. CHADDERTON   *Physical Laboratory II, H.C. Orsted Institute, University of Copenhagen, Copenhagen, Denmark.*

S. DATZ   *Oak Ridge National Laboratory,* Oak Ridge, Tennessee, U.S.A.*

G. DEARNALEY   *Nuclear Physics Division, Atomic Energy Research Establishment, Harwell, England.*

J. A. DAVIES   *Chemistry and Materials Division, Chalk River Nuclear Laboratories, Ontario, Canada.*

F. EISEN   *Science Center, North American Rockwell, Thousand Oaks, California, U.S.A.*

W. M. GIBSON   *Bell Laboratories, Murray Hill, New Jersey, U.S.A.*

F. GRASSO   *Istituto di Strattura della Materia, Universita di Catania, Catania, Italy.*

M. MARUYAMA   *Japan Atomic Energy Research Institute, Tokai-Mora, Ibaraki-ken, Japan.*

J. W. MAYER   *California Institute of Technology, Pasadena, California, U.S.A.*

C. D. MOAK   *Oak Ridge National Laboratory,* Oak Ridge, Tennessee, U.S.A.*

* Operated by Union Carbide Corporation for the U.S. Atomic Energy Commission.

D. V. MORGAN   *Department of Electrical and Electronic Engineering,*
                        The University of Leeds, Leeds, England.

R. S. NELSON   *Solid State Division, Atomic Energy Research*
                        Establishment, Harwell, England.

J. M. POATE   *Bell Laboratories, Murray Hill, New Jersey, U.S.A.*

M. W. THOMPSON   *School of Mathematical and Physical Sciences, The*
                        University of Sussex, Brighton, England.

D. VAN VLIET   *11 Willow Walk, Cambridge, England.*

J. L. WHITTON   *Chemistry and Materials Division, Chalk River*
                        Nuclear Laboratories, Ontario, Canada.

Contents

IV. Energy Loss of Channeled Ions: Low Velocity Ions
J. M. POATE

V. Energy Loss of Channeled Ions: High Velocity Ions
G. DEARNALEY

VI. Detailed Studies of Channeled Ion Trajectories and Associated Channeling Potentials and Stopping Powers
S. DATZ, B. R. APPLETON and C. D. MOAK

VII. Dechanneling
F. GRASSO

XII. Blocking Measurements of Short Nuclear Decay Times
W. M. GIBSON and M. MARUYAMA

XIII. Foreign Atom Location
J. A. DAVIES

XIV. Application to Radiation Damage
F. H. EISEN

XV. Application to Surface Studies
E. Bøgh

XVI. Applications to Semiconductor Technology
J. W. Mayer

CHAPTER I
An Introduction to Channeling

M. W. THOMPSON

1.1 INTRODUCTION

It is just sixty years since the existence of crystal lattices was conclusively demonstrated by the diffraction of X-rays. The first experiments showed that a single beam of X-rays is scattered by a crystal into many beams, which form the so-called Laue pattern on a photographic plate. At that time W. H. and W. L. Bragg were exploring the photon concept of X-radiation, thinking of it very much as particles. An historical account suggests that their first reaction to the Laue pattern was to try an explanation in terms of scattered photons passing out of the crystal along the open channels suspected to exist between the atomic rows.[1] They tested this idea by rotating a crystal through an angle and if their guess had been correct they expected the spots in the Laue pattern also to rotate by the same angle. In fact, this famous experiment showed rotations of *twice* this angle, leading to the picture of the crystal as a set of planes acting like mirrors, with Bragg's law giving the condition for reflection :

$$l\lambda = 2d \sin \psi \qquad (l = 1, 2, 3, \ldots)$$

Their first idea, although inapplicable in these circumstances, was later to form part of the channeling concept, whose validity for charged particles has been demonstrated in experiments during the past decade.

Figure 1.1 shows schematically what is meant by channeling. Three essential conditions must be met for a particle to be channeled. First it must find transparency in the form of an open channel between the rows of atoms, and this was seen by the Braggs. Secondly there must be a force acting which steers the particle towards the middle of the channel. Thirdly, for the channeled trajectory to be stable, it must not approach the rows of atoms too closely otherwise the gentle steering effect of many glancing collisions will be replaced by a wide angle deflection in one or more violent collisions with individual atoms. Thus one requires *transparency*, *steering* and *stability*.

At about the same time as the Bragg experiment, Stark wrote a paper[2] proposing an experiment with protons which bore a remarkable similarity to experiments which were performed in the early 1960's. Why was this idea

1

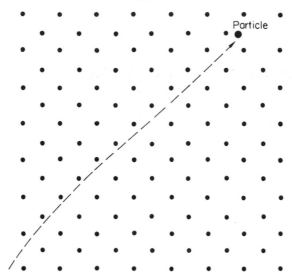

Figure 1.1. The channeling phenomenon, showing how
the particle is steered by glancing collisions along an
open passage in the crystal

not taken up immediately? Maybe it was the First World War which inter-
fered with scientific communication and swept away the possibilities of
experiment, or perhaps there were not suitable ion beams available? Whatever
the reason may be, the suggestion was neglected even though Rutherford,
and the other early nuclear physicists, must often have been close to a chance
discovery of channeling when they used thin mica crystals as absorbers and
scatterers in alpha particle experiments. Channeling had to re-emerge as an
independent suggestion by Robinson and Oen[3] who predicted that when a
copper crystal is bombarded with neutrons a proportion of the recoiling
copper atoms will become channeled within the crystal. This conclusion was
arrived at by a direct simulation of the situation in an electronic computer.[4]

Their computation followed a trajectory through thousands of collisions
and recorded its final position. By repeating this for several thousand
trajectories a statistical distribution of paths was generated for comparison
with experiments in which the range of heavy ions in solids were measured.[5]
In its earliest form the computer calculation assumed a structureless array
of atoms in the solid. Agreement between predicted and observed range
distributions was good, except that a few per cent of the ions in the experimental distribution lay in an 'exponential tail' which penetrated deeper
into the solid than had been predicted. A new computer program was written
in which the copper atoms lay on a crystal lattice. The first few runs were
abortive as the calculations failed to terminate within the time limit. In-

vestigation showed this was due to about one in a thousand trajectories becoming channeled. Since, in this early program, the central region of the channel was a region of zero potential energy, there was no means by which the particle, once channeled, could ever stop. The first two components of the channeling idea, transparency and steering, were thus identified and although it was realized that stability must also be considered, that item was not discussed much at this early stage.

The computer results were reported at a conference in Paris early in 1961 where some papers on sputtering were also presented. In one of these a Dutch group[6] reported that the sputtering rate of a single crystal in an ion beam was lowered whenever the direction of incidence came close to any simple crystallographic axis. They had previously[7] discussed the results in terms of the increased transparency of the crystal under such conditions but once the computer results were known the Paris conference saw a possible manifestation of channeling.

Immediately after the Paris conference several groups set out to find more direct evidence for channeling. The contrast with 1912, when Stark made his prediction, is most striking, for on this occasion the experimental means were at hand: in 1960 the successful exploitation of atomic energy demanded studies of radiation damage in solids and several research groups were working on the ion bombardment of crystals.

The range measurements have already been referred to, but as these had been conducted in polycrystalline solids the exponential tail could not be taken as more than a possible indication of channeling, with several other mechanism being equally likely. Experiments on the angular distribution of sputtered atoms from single crystals provided further inconclusive evidence. Here it was found[8] that when a copper crystal was bombarded with argon ions at energies of 25, 50 or 75 keV, a small but significant group of the copper atoms were ejected from the crystal parallel to the $\{110\}$ planes and the group became more numerous as the bombarding energy increased. This was consistent with copper recoils becoming channeled in the interplanar spaces. Subsequent investigations[9] showed that such atoms have energies far above the average in sputtering, thus confirming this earlier interpretation.

However the most striking evidence for channeling came from measurements of the penetration of ions into or through single crystals. Such experiments were conducted almost simultaneously by three groups: at Chalk River, Munich, and Harwell. These important investigations form the subject of the next section.

1.2 PENETRATION EXPERIMENTS

The experiments fall into two main categories. Either ions are transmitted through thin single crystals into a detector or radioactive ions are allowed to come to rest in a crystal and their range is measured by progressively stripping away the surface whilst measuring the net radioactivity. In both cases it was found that penetration is greatly enhanced when the ions are incident close to channel directions.

Figure 1.2 shows the Harwell penetration experiment in which protons accelerated to an energy of 50 keV passed through a thin single crystal into a

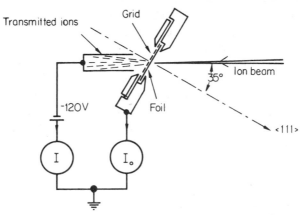

Figure 1.2. The Harwell penetration experiment[25] in which ions were transmitted through a single crystal into a detecting electrode

current-measuring electrode[25]. By rotating the crystal about an axis normal to its surface the lattice orientation could be varied without affecting the thickness to be penetrated. Thus any changes in current should indicate effects which depend on the crystal structure. Figure 1.3 shows how the transmitted current varied during the rotation of a gold crystal with the angle of incidence chosen to successively align the most closely packed atomic rows with the ion beam. It is between the most closely packed rows, of course, that the widest channels exist. Evidently the penetration was enhanced whenever the ion beam was within about 4° of the $\langle 110 \rangle$ axes. The effects were found to persist up to proton energies at least two orders of magnitude greater[10,11] although at such high energies the angular range over which the effect could be seen fell to a fraction of a degree. But these high energy experiments had the great advantage that the transmitted protons could be observed as single particles using silicon barrier detectors and the energy lost by each particle in transmission could be measured directly from the pulse height of the counter. Figure 1.4 shows energy spectra of transmitted protons coming through a thin silicon single crystal. It will be seen that whilst the particles in the non-aligned case lose a well defined, and relatively large, amount of energy, the particles in the aligned case fall into two groups: those which have an anomalously small energy loss and those which have a spread of energy losses

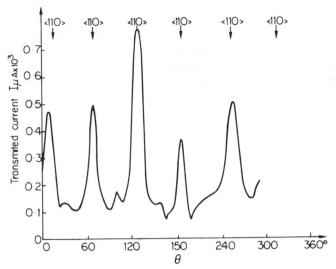

Figure 1.3. Showing how the transmitted ion current in Figure 1.2 varied with rotation of the crystal about the $\langle 111 \rangle$ axes[25]

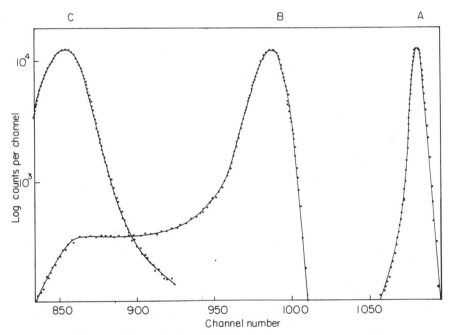

Figure 1.4. Showing the effect of crystal orientation on the energy spectrum of protons incident at 3 MeV on a silicon crystal 23 μm thick and transmitted through it into a small silicon barrier detector with an acceptance solid angle of 10^{-7} sr. A, the incident beam spectrum; B, beam aligned for channeling between $\{111\}$ planes; C, beam not aligned with simple crystallographic direction[35]

extending all the way from the anomalous peak to the normal peak in the nonaligned case. It is the first group which is associated with the channeling phenomena since such particles as these should lose energy predominantly by electron excitation in the solid. A channeled trajectory is thought to travel in a region of low electron density and therefore channeled particles should appear as a separate group in the energy distribution. The second group comes either from particles which fail to enter the crystal at a suitable point to become channeled, or which become dechanneled during the passage through the crystal.

The range measurements of the Chalk River and Munich groups showed the same enhancement of penetration for incidence close to channeled direction for heavy ions of kilovolt energies.[12,13] Increased range comes from a number of factors. First the channeled trajectory loses less energy in travelling a given distance, since fewer violent encounters occur with individual atoms. It also travels in a region of low electron density, although for such ions as these the dominant mechanism of energy loss is in elastic atomic collisions, in contrast to the light energetic ions where electronic loss dominates. Secondly, it travels more nearly in a straight line than a random trajectory which wanders through the crystal in a zig-zag path. Figure 1.5 shows how the distribution of krypton ions which came to rest in an aluminium crystal varied as the orientation of the crystal was changed relative to the direction

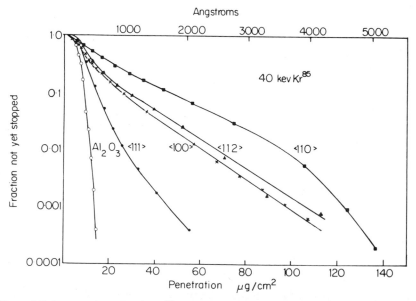

Figure 1.5. Penetration of 40 keV ^{86}Kr ions in the principal crystallographic directions of Al and in amorphous Al_2O_3.[57] The graphs show the fraction of incident ions not yet stopped at various depths below the surface

of ion incidence. This clearly illustrates the point made above that the most open channels are those between the most closely packed rows in the structure when one remembers that in fcc crystals the $\langle 110 \rangle$ direction is the most closely packed.

Although these experiments represented an impressive bulk of evidence in support of the channeling prediction there remained some important points to clear up. The enhanced penetration in thick single crystals might have been due to anomalously rapid diffusion of the implanted ions in some mobile defect configuration formed only when they were incident on the crystal along a simple crystal axis. For example interstitial defects in the lattice are well known to be very mobile in diffusion experiments. This point was finally answered by performing two range measurements in which the ions were incident along two equivalent channeling axes, one normal to the surface and the other at a wide angle to the normal.[14] It was found that the depth distribution in the latter case was shifted towards the surface thus proving that there is a genuine range effect and not an anomalous diffusion effect. However, there are some cases where the existence of an anomalous diffusion, dependent on channeled ion incidence, has been found to occur,[15] so the objection was not to be dismissed lightly.

The first transmission experiments with light ions were also open to an objection: that although an anomalously low energy loss had been demonstrated for certain directions of transmission it had not been proved that this was necessarily due to a spatial property of the trajectories such as channeling. Instead it was suggested as an alternative that the electron excitation process, responsible for the energy loss, might be a highly anisotropic function of direction in the crystal irrespective of the lateral position of the path with respect to the atomic rows. The answer to this objection came from experiments in which a nuclear reaction rate,[16,18] or a backscattered intensity[25] was studied as a function of crystal orientation.

1.3 EFFECT ON NUCLEAR AND ATOMIC PROCESSES

Figure 1.1 shows how a channeled trajectory always avoids the nuclei, which are the centres of repulsion for the positive ions and are producing the required steering action for channeling. One might expect the channeled particles to be less likely to initiate a nuclear reaction since such events require the particle to approach the nucleus to a distance roughly 10^{-4} times the channel width.[16,17] Several successful experiments were conducted using proton-initiated nuclear reactions such as:[18]

$$^{28}\text{Si} + \text{p} \rightarrow {}^{29}\text{P} + \gamma$$

or[16]

$$^{65}\text{Cu} + \text{p} \rightarrow {}^{65}\text{Zn} + \text{n}$$

Figure 1.6 shows, as an example, how the gamma-ray yield from an aluminium crystal changes as it is rotated in a proton beam. The minimum corresponds to incidence along a channeled axis, with clear evidence that the increased penetration is associated with avoidance of the nuclei, thus finally confirmed the channeling picture of Figure 1.1.

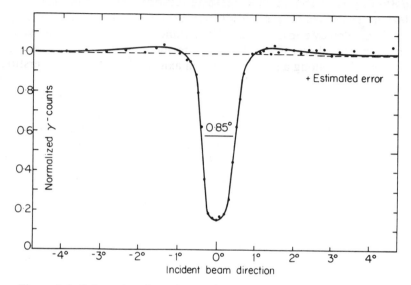

Figure 1.6. Orientation dependence of the (p, γ) resonant nuclear reaction yield in aluminium as the incident proton beam at 1·4 MeV is scanned through the ⟨110⟩ axis[19]

An important difference exists between the two cases given above, the (p, γ) reaction is a resonant reaction which only occurs if the proton has an energy within a narrow band a few tens of kilovolts wide. Therefore, if the proton can be made to stay channeled long enough for it to lose energy and pass through this sensitive resonance region it will fail to make any nuclear reactions. If the incident energy is chosen to be just above that of the resonance the proton will only have to stay channeled for the first hundred ångströms or so. On the other hand, if the energy is far above the resonance the particle will have to stay channeled for a much greater distance than this in order for an effect on the nuclear reaction rate to be seen. This provides an important means for studying the stability of channeled trajectories and it was observed that for energies close to the resonance the magnitude of the effect was larger than it was for energies well above the resonance.[19]

In contrast to the resonant case, the copper (p, n) reaction will only occur if the proton has an energy greater than a threshold energy of 2·1 MeV. In this case, in order to inhibit the nuclear reaction the proton must remain channeled until its energy has fallen below the threshold. In a typical case where the incident energy was 1 MeV above the threshold the particle would have to stay channeled for many thousands of ångströms. Consequently one might expect that the angular range over which this could be observed would be extremely small and this was indeed found to be the case as the angular width of the dip in nuclear reaction rate was only 0·3° at 3 MeV.[16]

Similar effects are seen when the proton-induced X-ray yield is measured, since this of course originates from collisions of the proton with electrons of the inner shell, close to the nucleus.[20] In fact a whole range of experiments, where some measured rate depends on close collisions between the bombarding ion and an atom in the crystal, show channeling effects. Examples are the rate of electron emission, the rate of erosion or sputtering, and the rate of radiation damage.[7,21,22] As mentioned in the previous section the second of these actually came before Robinson's theory in 1961 but was interpreted in terms of transparency, without the vital steering action being taken into account.

In an experiment closely related to the nuclear reaction, one may take a crystal containing alpha radioactive atoms as members of its lattice. The emitted α-particles will not be able to emerge from the crystal in the exact channel directions, since they would then have to pass through a row of intervening nuclei, and they are steered away as shown in Figure 1.7. This is often referred to as a *blocking* effect to distinguish it from the channeling effect. Thus if one sets a detector to collect α-particles emitted from a crystal in a specific direction, the counting rate should fall to a minimum when the

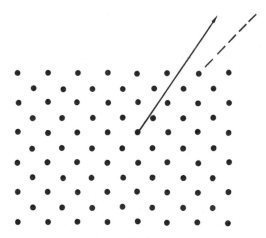

Figure 1.7. The blocking effect, showing that when particles are started from the region of an atomic nucleus they cannot emerge from the crystal in a simple crystallographic direction

detector is placed in the channel direction. Figure 1.8 shows the results of such an experiment.[23]

One can clearly see that there must be a reciprocal relationship between the emission from an atom which is observed in a particular direction and the inverse absorption process when the beam is incident along the reverse path.[24]

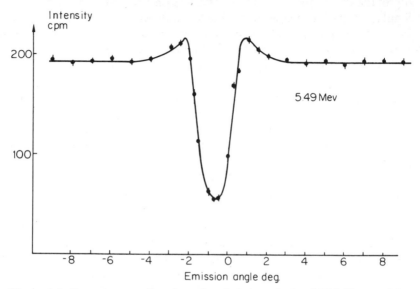

Figure 1.8. Intensity as a function of emission angle for 5·5 MeV α-particles from ^{222}Ra atoms embedded in a tungsten crystal. The dip is centred on a simple crystal axis. After reference 55

1.4 BACKSCATTERING EXPERIMENTS

Suppose we observe the reflection, or backscattering, of ions from a crystal as illustrated in Figure 1.9. We might expect that when channeling of the incident ion occurs, the likelihood of an ion being scattered through a large angle and returning to the surface will be greatly reduced. In the case of light ions at high energy, such as MeV protons, the reduction in scattering can be viewed in a similar way as the drop in nuclear reaction rate since the most important mechanism of reflections through large angles is the single close collision with the nucleus or Rutherford collision. We saw above how such collisions are absent from channeled trajectories. Figure 1.10 shows how the current of reflected ions changes as the direction of incidence is varied.[25] Minima are observed, not only for channels between atomic rows but also between atomic planes.[25,26] Planar channeling is an obvious extension of the idea of pipelike channels and both types of channel were envisaged in the early publications.[8,25] If a very small detector of light ions is used in the

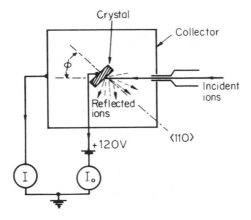

Figure 1.9. Experiment to measure backscattering[25]

experiment of Figure 1.8, and this is moved around whilst the crystal is held
fixed, minima appear in the current of detected ions whenever a chanelling
direction is encountered. This is a further example of the blocking effect of
Figure 1.7 if we view the scattering of a proton by a nucleus through a wide
angle as something similar to proton emission by the nucleus, since it demands
a rather close approach to the nucleus ($\sim 10^{-3}$ Å).

Perhaps the simplest method of observing blocking effects is to place a
fluorescent screen or photographic plate near to a crystal to record the
intensity of scattered ions at various angles.[27] Such a photograph is shown in
Figure 1.11. The dark bands indicate regions where the scattered intensity is
low and corresponds to the intersection of crystal planes with the detector
plate showing clearly the existence of blocking by crystal planes. The dark
blobs indicate regions of blocking by atomic rows. Such blocking patterns
have a clear application in crystallography and an account of this work will
be found in Chapter 11.

It is amusing to speculate on the likely course of events if Stark's proposed
experiment had been performed in 1914 and channeling discovered then. The
interpretation of blocking patterns is so much simpler in principle than the
interpretation of X-ray diffraction photographs, because one works with the
direct lattice in real space without having to transform into reciprocal space
and work with a reciprocal lattice. This removes many ambiguities in the
ensuing calculations of crystal structure from the observed intensities.
However, because blocking and channeling effects were not discovered until
X-ray crystallography had been in existence for nearly fifty years, when all
the simple structures had been solved, it is unlikely that much would be
gained by going over the old ground with a new technique. But for the future

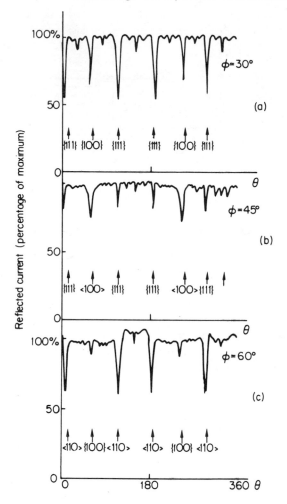

Figure 1.10. Showing how the current of back-scattered protons in the experiment of Figure 1.9 varied with the direction of incidence. Incident energy was 50 keV[25]

it may well be that in solving the more complex structures encountered in metallurgy the new technique will prove of some value.

A very important application of channeling and blocking has been in the study of defects in crystal lattices. In any experiment which measures the backscattered current of particles from a crystal we expect a large attenuation whenever the ion beam is channeling along a simple crystal direction. But if the channel contains atoms which are not at the normal lattice sites, that is,

Figure 1.11. A blocking pattern formed by protons back-scattered from a tungsten crystal onto a fluorescent screen

defects of the interstitial variety, then we must expect the backscattered yield to be increased above the level found for a perfect crystal. Since the back-scattered yield can be as little as 1 % of the normal value when the incident beam is channeled the presence of only 1 % of atoms not on lattice sites will roughly double the backscattered yield.

In the so called double alignment experiment, by setting the detector and crystal to give both channeling of the incident ion and blocking of the out-going ion, an enormous attenuation of intensity is obtained, sometimes by as dramatic a factor as 10^{-4}. This provides a sensitive means of locating defects in the crystal structure since each lattice atom is scattering on average at only 10^{-4} times its normal rate. Thus, if only one atom is 10^4 is off the lattice, and in a site from which it can scatter ions from the incident channel to the exit channel, the scattered intensity will be roughly doubled. Figure 1.12 illustrates the process.

Although the account here is oversimplified and neglects some important effects of spatial distribution of channeled trajectories, experiments of this type have managed to settle some outstanding problems of the defect solid state, such as whether certain impurity atoms occupy interstitial or substitutional sites on the crystal lattice and the nature of radiation damage.[28,29,30,31,32] Later chapters contain an account of such work in detail.

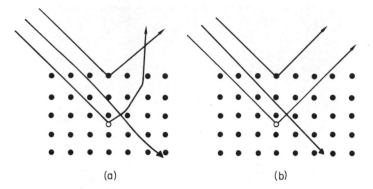

(a)   (b)

Figure 1.12. Showing the contribution to backscattering made by collisions with interstitial defects. (a) When the incident beam is aligned for channeling and ions are detected over a wide solid angle—the 'single alignment case'. (b) When the ions are incident along a channel direction and the detector select ions scattered in a channel direction—the 'double alignment case'. Note that in both cases scattering at the surface will contribute

1.5 STAR PATTERNS AND FORWARD-SCATTERING

A most striking method of studying channeling and blocking effects simultaneously is illustrated in Figure 1.13. The crystal is thin enough for most of the incident particles to pass through it and to be recorded on a photographic plate or fluorescent screen. This allows a rapid determination to be made of the angular distribution of particles which have travelled for some distance through the crystal. By using crystals of successively increasing thickness it is possible to build up a statistical picture of the particles' directions as they collide their way through the crystal.[33,34,35]

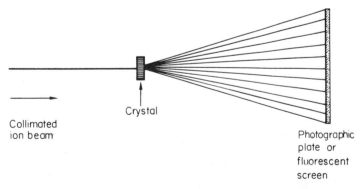

Collimated
ion beam

Crystal

Photographic
plate or
fluorescent
screen

Figure 1.13. Showing how a star pattern is formed by the forward-scattered ions from a thin crystal target

When the incident beam of ions is parallel to a simple axis of the crystal the plate develops an image such as that shown in Figure 1.14, whose appearance led to the obvious name 'star pattern' for such photographs. The pattern can be summarized as: (a) a smeared out, dark background, which falls off with increasing angle from the direction of incidence, (b) a pattern of dark streaks, which follow the intersection of crystal planes with the plate and whose intensity falls off at increasing angles to the direction of incidence, until eventually they become (c) light streaks in the dark background (a).

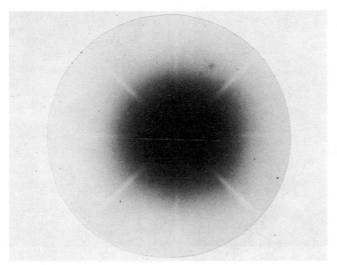

Figure 1.14. A star pattern formed by 4 MeV protons passing through a MgO crystal 5×10^{-5} m in thickness on to a photographic plate. The dark streaks show channeling, the light streaks the blocking effect and the dark smear is due to random scattering[35]

The background smear is attributed to nonchanneled particles which, because they travel in directions well away from the crystallographic planes, do not notice the lattice structure. The dark streaks are obviously from channeled particles which have been travelling between the lattice planes, but with their direction evidently fanning out from the direction of incidence parallel to the planes. The light streaks once again show the blocking effect whereby the nonchanneled particles are unlikely to leave the crystal close to the channeling direction. Using protons, the particle energy at any point in the star pattern may be measured with a small detector of the proportional type, and this confirms the above interpretation by showing that the dark streaks contain protons which have lost very much less energy in transit than those in the background. This is to be expected because the channeled particle travels in a region of low electron density. Figure 1.4 shows a typical energy spectrum in these circumstances.

Experiments such as these provide a means of studying the detailed shape of the trajectories of channeled ions within a crystal. As a simple example Figure 1.15 shows two star patterns which were made by a well collimated

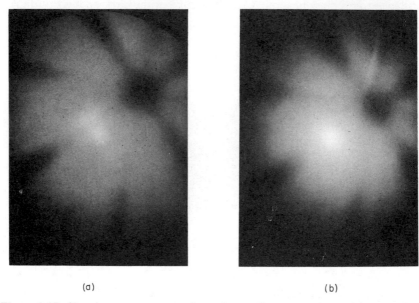

(a)   (b)

Figure 1.15. Showing a star pattern formed on a fluorescent screen by transmitting 250 keV protons through a very thin gold crystal (1200 Å). The direction of incidence was parallel to the {100} planes in (a) and 0·6° from the {100} planes in (b). The two bright spots in (b) show that the channeled trajectories are of a zigzag form[36]

beam of protons which were incident on a thin gold crystal parallel to or at an angle of 0·6° to a prominent planar channel.[36] The resulting star pattern contains two regions of high intensity, one almost in the direction of the incident beam and the other at an equal distance away from the plane on the opposite side. This can be interpreted in terms of trajectories which have a zigzag path and therefore emerge from the crystal symmetrically disposed on either side of the plane. From the angular separation of the intense spots and the thickness of the crystal one deduces that the trajectories are reflected from the channel walls six times during their passage through the crystal.

Experiments by the Oak Ridge group[37] have made detailed studies of energy loss in situations such as these from which they have been able to deduce information about the potential energy of ions moving in a crystal lattice. Their methods of analysis[38] are discussed elsewhere in this book.

1.6 SURFACE SCATTERING

Some of the earliest evidence for lattice correlated collision phenomena in crystals came from the experiments by Datz and Snoek[39] who used beams of heavy ions such as argon in the energy region around 30 keV. In scattering these from a single crystal of copper and measuring the energy spectrum of the scattered ions they discovered that a small fraction of the scattered intensity was in a sharp peak at an energy corresponding to a single collision between an argon ion and a copper beam. This peak was only seen clearly when the direction of incidence corresponded to a channel direction. It was suggested that those ions whose trajectories were able to pass directly into the channel were unlikely to be scattered back into the detector without a large loss of energy but the group which did not become channeled would be scattered back at the surface of the crystal by the atoms at the ends of atomic rows. Such ions would therefore suffer only the energy loss appropriate to the elastic collision between the ion and a single atom. The role of channeling here is simply to increase the signal to background ratio.

With light ions in the MeV energy range it was also found that a so called surface peak appeared in the backscattered energy spectrum. Again this was strongly enhanced whenever the direction of incidence corresponded to channeling and the interpretation was exactly that given above. Because the exact energy of the backscattered ion in the surface peak depends on the mass of the scattering centre some progress has been made in deducing the chemical composition of solid surfaces by determining the atomic mass of the atoms involved (e.g. Davies *et al.*[28]).

Such a picture only holds good if the scattering angles are large, in cases of grazing incidence the sharp peak corresponding to a single collision disappears and instead one obtains a smeared out energy distribution with some fine structure composed of several peaks. Extensive work has been done by the Russian groups using heavy ions such as argon in the 30 kV energy region. They attribute the smearing out of the energy distribution to a transition from single scattering events to plural scattering in which the ion interacts successively with several atoms. They found striking orientation effects whenever the plane of incidence came close to an atomic row in the surface.[40,41,42,43]

Experiments with light ions in the MeV energy range scattered at grazing incidence from very flat crystal surfaces show that direct reflection of an ion from a crystal is possible in a manner which must be rather similar to that responsible for the trapping of a trajectory by reflection between planes of atoms in the channeling process. This might therefore be called 'channeling with half a crystal'. The intensity patterns reproduced in Figure 1.16 show how, as the angle of incidence is decreased, the normal blocking pattern, which indicates that the ions have penetrated into the crystal lattice, changes to a

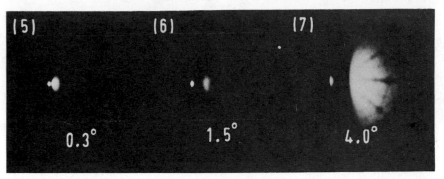

Figure 1.16. Three patterns from a fluorescent screen showing the scattered intensity of He ions incident at grazing angles at 1 MeV on a flat tungsten crystal. The small bright spot is formed by the residue of the incident beam. At grazing angles the ions are scattered as though by specular reflection. At 4° the normal blocking pattern is obtained, indicating that penetration of the lattice has occurred

single bright spot at angles less than 1°. This indicates the specular reflection process from the surface plane of atoms.

Experiments such as these may provide a powerful means of studying surface crystallography for the future and it has significant advantages over the alternative methods available, such as RHEED or LEED, all of which involve diffraction and demand very clean surfaces.

1.7 CLASSICAL THEORY OF CHANNELING

Robinson's theory of channeling suffered from the disadvantage that it required a computer calculation for each and every experimental situation. Analytical theories were therefore developed and the first of these were given, simultaneously, by Lehmann and Leibfried[44] and Nelson and Thompson.[25] Their methods were essentially similar and resulted in the crystal being replaced, for the purposes of calculation, by a continuum potential in which the channeled ion moved. The calculation leading up to this crystal potential is outlined in the following paragraphs.

Let us first consider, with reference to Figure 1.17, a glancing collision between a particle of mass M_1, energy E_1 incident upon an atom of mass M_2 initially at rest. By a glancing collision we mean one in which the *change* in momentum of the incident particle is very small in comparison to its *total* momentum. The collision is characterized by an impact parameter q which defines the distance between the undeflected trajectory and the atom's initial position, and by a potential energy of interaction which is a function $V(r)$ of the separation r between particle and atom at any instant. This function's derivative dV/dr gives the force which acts between particle and

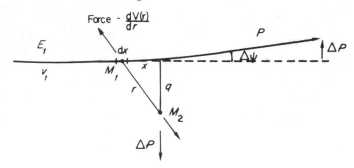

Figure 1.17. A glancing collision, defining the quantities used in the momentum approximation

atom during the interval dx along the particle's trajectory. This force has a component normal to the particle trajectory which, when multiplied by the time interval dx/v_1, gives the contribution to the momentum change from the interval dx at x. The total momentum change ΔP is then found by integration to be

$$\Delta P = \int_{-\infty}^{+\infty} - \frac{dV(r) \, q \, dx}{dr \, r \, v_1}$$

$$= \frac{2}{v_1} \int_{q}^{\infty} - \frac{dV(r) \, q \, dr}{dr(r^2 - q^2)^{\frac{1}{2}}}. \tag{1.1}$$

This integral will be referred to as $I(q)$ since for a given potential $V(r)$ it depends only on the impact parameter q. Knowing the momentum change ΔP it follows that the deflection of the particle's trajectory is simply

$$\Delta \psi = \Delta P / P$$

or

$$\Delta \psi = \frac{1}{E_1} I(q). \tag{1.2}$$

Thus the deflection depends on the energy of the particle.

Now consider the simplest case of channeling, illustrated in Figure 1.18,

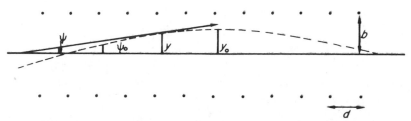

Figure 1.18. A channeled trajectory, between a pair of atomic rows

where the particle passes between a pair of rows of atoms. The particle trajectory is characterized at any instant by its displacement y from the channel axis and the small angle it makes to the channel axis. The interatomic spacing along the rows is d and the time taken to traverse a distance nd is just nd/v_1. During this interval the angle ψ changes by n times the deviation for a single collision, *provided that y and ψ change slowly for small angles*, hence the rate of change of ψ is given by:

$$\dot{\psi} = -\frac{v_1}{nd}\frac{n}{E_1}I(q)$$

or

$$\dot{\psi} = -\frac{2}{dM_1v_1}I(b-y). \tag{1.3}$$

Using equation (1.2) with $q = b - y$,

$$\dot{y} = v_1\psi$$

and

$$\ddot{y} = v_1\dot{\psi} \tag{1.4}$$

Hence using equations (1.3) and (1.4) the equation of motion in the y direction is

$$M_1\ddot{y} = -\frac{2}{d}I(b-y) = -F(y) \tag{1.5}$$

where $F(y)$ is the effective restoring force directed against the displacement of y. Note that this is independent of E_1, and we are thus dealing with a conservative force field. This is a consequence of using the approximation that the momentum change is small in a single collision compared with the total momentum and means that we can replace the crystal by an effective potential well $U_1(y)$, or *channel potential* defined as

$$U_1(y) = \int_0^v F(y)\,dy \tag{1.6}$$

$$= \frac{2}{d}\int_0^v I(b-y)\,dy$$

for the case of two rows illustrated in Figure 1.14. This will be valid provided that the impact parameters are always large enough for the above *momentum approximation* to be used and that the trajectory varies slowly with the number of collisions. We note that the potential depends inversely on the interatomic spacing d in the rows and is strongest, therefore, for the closely-packed rows.

A similar potential could also have been defined had we been considering a trajectory channeled between a pair of planes rather than a pair of rows, but in this case there would have been an inverse dependence on the density of atoms per unit area of plane, again showing the significance of close packing. In the case of energetic light particles like 100 keV protons the potential $V(r)$ can be approximated by the screened Coulomb form:

$$V(r) = \frac{Z_1 Z_2 e^2}{r} \exp(-r/a) \tag{1.7}$$

with the screening radius a given by

$$a = a_0 (Z_1^{\frac{2}{3}} + Z_2^{\frac{2}{3}})^{-\frac{1}{2}}$$

where a_0 is the Bohr radius, and this leads to an approximate form for the channel potential in the case of a pair of rows.

$$U_1(y) = \frac{2(2\pi)^{\frac{1}{2}} Z_1 Z_2 a_0 E_R}{d} \frac{\exp - (b - y)/a}{(b - y)^{\frac{1}{2}}/a^{\frac{1}{2}}} \tag{1.8}$$

where E_R is the Rydberg energy: 13·6 eV. The screening radius of the potential a is generally much smaller than the inter-row spacing $2b$. As an example this function is plotted in Figure 1.19 for the case of protons in copper travelling between a pair of (110) rows, which shows that for a large part of the channel very little force acts on the particle, but that as it approaches the channel wall

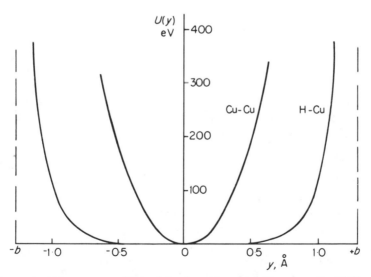

Figure 1.19. A comparison of the channel potential between the $\langle 110 \rangle$ rows of a copper crystal for protons and Cu atoms. Equation (1.8) is used for the proton case, the Cu case is based on an interatomic potential which accounts for the repulsion between electron shells[25]

a steeply rising potential is encountered. One might therefore approximate the situation by a *square well* potential for such cases and this leads to zig-zag trajectories, for which the star pattern of Figure 1.15 provides evidence.

The channel potential for the case of heavy ions, on the other hand, is quite different in the centre of the channel. Here the electron clouds, which such ions carry along with them, cause the interaction potential to extend out much further from the channel walls and the result is an almost *parabolic* potential well in which the particle is channeled. A typical case is illustrated in Figure 1.19, for copper atoms moving in a copper crystal between $\langle 110 \rangle$ rows and here we would expect a sinusoidal trajectory. Close consideration of the nature of the channel potential is given by Van Vliet in the next chapter and Datz *et al.* in Chapter 6.

We are now in a position to solve the equation of motion for the channeled particle, or at least to calculate the more important characteristics of its trajectory. For example, if one is given the angle ψ_0 at which the trajectory crosses the channel axis it is a simple matter to calculate the maximum deviation y_0 of the trajectory. As it crosses the axis the kinetic energy associated with the lateral movement is just $\psi_0^2 E_1$, when the trajectory amplitude y_0 is reached all of this kinetic energy has been converted into potential energy $U_1(y_0)$, thus:

$$\psi_0^2 E_1 = U_1(y_0)$$

or

$$\psi_0 = U_1(y_0)/E_1. \tag{1.9}$$

This gives us a simple relation between y_0 and ψ_0 in terms of the channel potential.

Thus far the theory concentrates on the behaviour of trajectories in the channel but an important characteristic of the experimental results is the angular width of the peaks and dips observed, which will clearly depend on the maximum permissible amplitude at which the particle trajectory can still remain channeled. This extreme trajectory will occur when the approximations developed above begin to break down. If the individual collisions become sufficiently violent so that the momentum changes are no longer small compared with the forward momentum, then wide angle scattering occurs and we can no longer treat the trajectory as a smoothly varying curve averaging out the effects of individual collisions. The result will be an unstable trajectory which eventually flies out through the channel walls.

Thus, having dealt with the first two facets of the channeling phenomenon, transparency and steering, we come to the third vital ingredient: *stability* of the channeled trajectory. It was by defining the criterion for stability that Lindhard[17,24] made his crucial contribution to channeling theory.

The atoms in the channel walls are in a normal state of lattice vibration and a mean square displacement x_{rms}^2. The distance of closest approach to the channel wall must not approach x_{rms} and it must also exceed the minimum q for which a single collision can be treated in the momentum approximation of Figure 1.17. In the case of screened Coulomb potentials it is known that the momentum approximation breaks down when q is roughly equal to the screening radius a. Now a^2 and x_{rms}^2 are of the same magnitude, and as a basis for rough calculation it was suggested that the distance of closest approach to a channel wall for a stable channeling to occur might be taken in the convenient form $m(a^2 + x_{rms}^2)^{\frac{1}{2}}$ with m a numerical constant of order unity. Writing $\hat{y}_0$ for the maximum amplitude of a stable trajectory

$$\hat{y}_0 = b - m(a^2 + x_{rms}^2)^{\frac{1}{2}} \tag{1.10}$$

and hence the maximum angle $\hat{\psi}_0$ is found by inserting this into equation (1.9). Taken with the channel potential appropriate for rows of atoms with the screened Coulomb potential, this leads to

$$\hat{\psi}_0^2 = \frac{2(2\pi)^{\frac{1}{2}} Z_1 Z_2 a_0 E_R}{d E_1} \frac{\exp\{-m(1 + x_{rms}^2/a^2)^{\frac{1}{2}}\}}{(1 + x_{rms}^2/a^2)^{\frac{1}{2}}} \tag{1.11}$$

Since x_{rms}^2 and a^2 are of the same magnitude, the order of magnitude for $\hat{\psi}_0$ is given by

$$\hat{\psi}_0 \approx \left(\frac{Z_1 Z_2 a_0 E_R}{E_1 d} \right)^{\frac{1}{2}} \tag{1.12}$$

Thus by introducing a stability criterion into the continuum potential models of Lehmann and Leibfried or Thompson and Nelson, Lindhard deduced a critical angle that could be easily compared with experimental observations. Taking the example of 50 keV protons in the $\langle 110 \rangle$ channels in Cu this predicts $\hat{\psi}_0 \simeq 3°$, in clear agreement with the experiments of Figure 1.10. The prediction that the angular width of channeling effects decreases with increasing energy or less closely-packed atomic rows is also in good agreement with the experiments.

More precise expressions, such as equation (1.11), give better agreement with experiments, using values of m between 1 and 3. The value used depends on the nature of the experiment and whether it demands that the channeling shall persist over long distances, as in a transmission experiment, or for short distances as in the case of the α-particle emission. In the first case m has to be fairly large in order to keep the trajectory out of trouble, whereas in the second it can be quite small. Expressions like equation (1.11) also allow temperature effects to be introduced via x_{rms}^2. A more detailed discussion of these points will be found in the next chapter by Van Vliet.

Lindhard went on to show how to calculate the reduction in an atomic collision process brought about by channeling, using the old transparency ideas. Suppose we consider the fraction of particles which are channeled between rows when a beam is incident on a crystal plane normal to a channeling axis. Then for each atomic row an area $\pi m^2(a^2 + x_{rms}^2)$ must be crossed if an ion is to contribute to the collision process concerned. Otherwise the ion becomes channeled and is lost to the process. Then if each row contributes an average area $A(\simeq d^2)$ to the crystal plane the probability of an incident ion contributing to the collision process is reduced by a factor χ_{min} given by:

$$\chi_{min} = \pi m^2(a^2 + x_{rms}^2)/A.$$

Then, since only A is a function of the crystal direction, we see that the most transparent direction, with the largest value of A, has the least value of χ_{min} and the greatest reduction in the yield of the atomic collision process. For a close packed direction, where A is typically of the order $100\ \pi a^2$, the predicted reduction is by a factor roughly $1/30$. This is found in experiments with carefully prepared and aligned crystals.

In planar channeling one finds χ_{min} from the ratio of the thickness of the planar layer in which channeling is unstable to the interplanar separation, that is

$$\chi_{min} \simeq m(a^2 + x_{rms}^2)^{\frac{1}{2}}/d.$$

These generally give a less dramatic reduction in yield than do the axial channels between rows, since $a/d \simeq 1/10$. This is clear in Figure 1.10 where dips owing to planar channels are not so pronounced as those owing to axial channels.

A model of the crystal which is particularly useful for understanding star patterns, and all planar channeling effects, is illustrated in Figure 1.20. Planes are replaced by a square potential barrier of height U_0 approximately

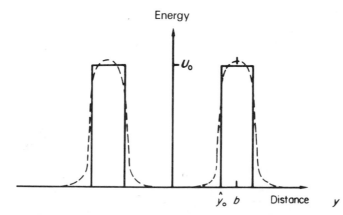

Figure 1.20. An approximate potential for describing planar channeling effects with light ions

equal to $U_1(\hat{y}_0)$, and the channels of width $2\hat{y}_0$ are the potential wells in between.[45,35] The crystal is thus divided into two types of region: (1) where the potential is zero and (2) where the potential is U_0. Blocking is now easily understood for particles emitted or scattered from nuclei, since their trajectories always originate in a region of potential U_0. Thus in region 1 they always have a transverse kinetic energy of at least U_0 and cannot travel parallel to planes, or rows, in region 2. Particles emerging from the crystal do so from either region 1 or 2. Those coming from region 1 can have any direction, but the rest will have the minimum transverse energy and hence be excluded from the angular region $\pm \hat{\psi}_0$.

An optical analogue of a crystal, along these lines, can be constructed from a stack of glass plates with spaces in between, immersed in a bath of dense liquid with a refractive index slightly greater than the glass plates. Total internal reflection will prevent rays of light from entering the glass if they make less than the critical angle to the interface of the glass and there will therefore be a set of channeled rays. At wider angles rays pass into the glass, suffering refraction as they cross the interfaces. It is this *refraction* which leads to the small maxima on either side of the blocking minima seen in Figures 1.6, 1.9 and 1.14 owing to the bunching of rays close to $\hat{\psi}_0$ in region 1. The bunching effect is illustrated in Figure 1.21 and will be familiar to all

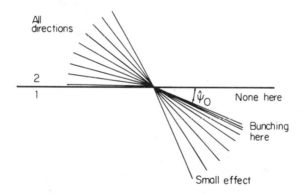

Figure 1.21. Showing how trajectories randomly directed in the region of high potential are bunched near to $\psi = \hat{\psi}_0$ in the region of lowest potential

students of optics. Because this optical analogy shows how channeling can apply to wavelike phenomena with wavelengths small in comparison with channel width, it serves as a useful link to Section 1.8 where particle waves, rather than light waves, are considered.

All of this explains the first order effects observed in channeling but it cannot explain, for example, why in the star pattern the streaks fan out from the central channeling direction in which the ion beam was incident, nor does it provide an adequate description of the dechanneling of particles which are injected into a channel within the critical angular range. This requires a consideration of the possible perturbations which can act on a channeled trajectory. The first of these will be the presence of defects in the structure,[46,28,47,48] the second will be the inability of a continuum potential to describe the lattice of atoms, particularly when this is in vibration,[35,45] and the third will be the scattering of the particle by electrons in the solid whose position is not uniquely determined by the lattice structure.[25,49] The last two can be treated by the theory of multiple scattering and it is found that in the case of heavy elements, such as gold or tungsten, the most important perturbation on the trajectory is the multiple scattering by the nuclei, whilst in the light elements, such as silicon, the most important contribution is usually the multiple scattering by electrons.[35,49] The effect will be that a channeled beam, after being injected into the crystal, will lose its memory of the starting conditions and diffuse in angle around these to give streaks in the star pattern. The perturbation of the trajectory may become so extreme that dechanneling occurs. An account of these calculations will be found in the following chapter by Van Vliet.

1.8 THE LIMITS OF CLASSICAL MECHANICS

So far the discussion has been carried out in classical terms with little apology or explanation, but the scene has been set by the optical model for a wave-mechanical treatment. It will now be shown that the classical model appears adequate to explain any experiment so far conducted with protons or heavier ions, and can even account for some observations with electrons and positrons. However, a wave-mechanical treatment is really necessary for these lighter particles, as we know from electron diffraction experiments, and the dynamical theory of electron diffraction provides a reliable, though complicated, method of solving problems.[50] At some point in the future when more precise experiments are performed it may be necessary to use such methods for the proton case and this point is considered in Chapter 10 by Chadderton. The discussion is introduced here at an elementary level using the WKB approximation.[51] The wavelength of the particle is given by the de Broglie relation,

$$\lambda = \frac{h}{(2M_1 E_1)^{\frac{1}{2}}} \tag{1.13}$$

In the case of protons, even at energies as low as 10 keV, this wavelength is of the order 10^{-3} Å (10^{-7} μm) and this is so small that one might feel justified in using a classical treatment. However, the channeled particle moves at a small angle to the crystal planes, and its transverse motion may therefore have associated with it a large enough wavelength that departures from the classical theory become important. This is the point which will now be investigated and a simple example will illustrate some of the more important principles involved.

An idealized channel potential can be taken as a square well of width $2\hat{y}_0$ and depth U_0, as illustrated in Figure 1.20 and 1.22. In this type of potential

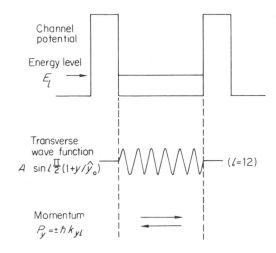

Figure 1.22. Particles waves in a channel

the particle is described by a wavefunction like $A(y)\exp(ik_x x)$, where $A(y)$ is the amplitude which varies according to the transverse position and $\exp(ik_x x)$ describes the longitudinal motion. The wavenumber k_x is related to the wavelength of the resolved component of the motion along the channel and hence to the momentum P_x.

$$\hbar k_x = P_x.$$

For the moment assume that penetration of the barrier by quantum tunnelling can be neglected; this will be examined below. Boundary conditions then require that $A(y)$ must disappear at $\pm\hat{y}_0$, but no restrictions apply to $\exp(ik_x x)$ since the channel is supposed to be very long. A suitable form for $A(y)$ is the standing wave $A \sin l(\pi/2)(1 + y/\hat{y}_0)$ which has a wavelength λ_y that fits the channel width according to

$$2\hat{y}_0 = l\lambda_y/2$$

where $l = 1, 2, 3 \ldots \hat{l}$. Thus there is a wavefunction for each value of l. The wavenumber of the transverse motion is

$$k_{yl} = l\pi/2\hat{y}_0$$

and the transverse momentum is

$$P_{yl} = l\pi h/2\hat{y}_0 \qquad (1.14)$$

but if ψ_0 is the angle that total momentum P makes with the channel axis, the limits on P_y mean that because

$$P_y = \psi_0 (2M_1 E_1)^{\frac{1}{2}}$$

ψ_0 can only take values given by

$$\psi_{0l} = l \frac{\pi h}{2y_0 (2M_1 E_1)^{\frac{1}{2}}}. \qquad (1.15)$$

The angular spacing between these permitted values is just ψ_{0l}/l, and from equation (1.15) we see that the spacing decreases as either $M_1^{\frac{1}{2}}$ or $E_1^{\frac{1}{2}}$ increases but is independent of U_0 and is related to the crystal only through the channel width $2\hat{y}_0$.

A set of wavefunctions thus exists, the lth of which corresponds to the particle momentum making angles $\pm\psi_{0l}$ to the channel axis, and the momenta for successive l values are equally spaced in angle. The total number of wavefunctions $\hat{l}$ is found approximately by putting

$$\psi_{0i} = \hat{\psi}_0 = (U_0/E_1)^{\frac{1}{2}}$$

when the transverse kinetic energy is enough to surmount the barrier U_0, that is,

$$\hat{l} = \frac{2\hat{y}_0}{\pi h}(2M_1 U_0)^{\frac{1}{2}}. \qquad (1.16)$$

In the case of protons it is calculated that this is of order 100, and in general the number increases with both $M_1^{\frac{1}{2}}$ and $U_0^{\frac{1}{2}}$ but is independent of E_1. The large number of wavefunctions, most of which have wavelengths much smaller than the well dimensions and energies well below U_0, justifies both the assumption that proton tunnelling of the barrier can be neglected for most of the functions, and our previous use of the classical approximation.

The probability that the particle is in the lth state depends on the experimental conditions. For example, one might expect an angular distribution of particles within the channel to look like Figure 1.23(a) with a fine structure superimposed on the classical curve. We notice that no particles should travel at $\psi = 0$ corresponding to the fact that in equation (1.14) $l \neq 0$ which is the well known requirement for a zero-point motion. Furthermore, if the wavelength from equation (1.13) is introduced into equation (1.15), one obtains

$$l\lambda = 4\hat{y}_0\psi_{0l}. \qquad (1.17)$$

This is almost the small angle version of Bragg's law from Section 1.1,

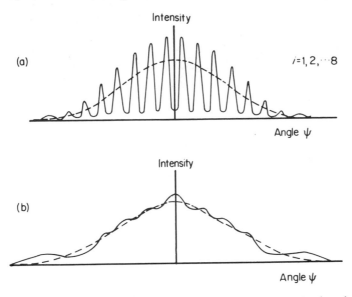

Figure 1.23. Schematic diagrams showing fine structure introduced into the angular distribution of channeled particles by wave mechanics when $l = 10$. (a) Within the crystal, idealized to the case of no energy spread. (b) Outside the crystal, allowing for the effects of aperture diffraction by the channel exit, a small energy spread and a realistic potential exists. Dashed curves indicate the classical result

but with $2\hat{y}_0$ the channel width, substituted for the interplanar spacing d. The link is thus established between channeling and diffraction.

Turning now to another aspect, each wavefunction corresponds to a sharp energy level for each value of l and these are defined by

$$E_l = l^2 \frac{\pi^2 \hbar^2}{8 \hat{y}_0^2 M_1} \tag{1.18}$$

It is clear that with 100 or so levels, for protons, the spacing is narrow enough for many different processes to cause transitions between them. For example, phonon creation or annihilation will result from collisions with the atoms in the channel walls, plasmons will be excited and it might even be possible to find electromagnetic radiation in the infrared region as a result of transitions between levels.

Such distribution as Figure 1.23(a) illustrates have never been seen with protons or heavier particles, but for experimental reasons alone this is not surprising since $\psi_0/\Delta\psi \simeq \hat{l} \simeq 100$, and one would require extremely fine angular resolution (better than 0·01°) and a highly perfect crystal. Note that $\Delta\psi \simeq \theta_B$, the first Bragg angle for diffraction (i.e. $\theta_B = \lambda/4b$ and $l \simeq \hat{\psi}_0/\theta_B$).

Another smearing effect will arise when the true shape of the potential well is introduced, for this will make it impossible to define a wavefunction having a single value of wavelength λ_y and also removing sharpness from $k_{yl}P_{yl}$ and ψ_{0l}. The peaks at ψ_{0l} inside the crystal will then be spread out and all values of angle found between them. In the classical case this just corresponds to the change from zigzag to more curved trajectories. But even if the experimental conditions could be achieved, a more fundamental reason would prevent the fine structure from being so marked since in any experiment the particle must both enter and leave the channel.

The particles in a beam approach the crystal as plane waves and see the channel entrance as an aperture of width $2y_0$. Just as in optics, there will be diffraction by the aperture which smears out the particles' directions into a distribution of the type sketched in Figure 1.24. The angular width of the central peak will be roughly $\lambda/\hat{y}_0$. Hence the particles will have almost equal probability of being found in one of several wavefunction states inside the channel.

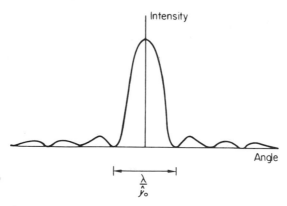

Figure 1.24. The intensity of initially plane waves after diffraction by passing through an aperture of width $2y_0$

Similarly when the particle leaves the crystal, from the lth state, aperture diffraction will again occur and give peaks centred roughly on $\pm\psi_{0l}$ with oscillating side bands. Since several adjacent valves of l will contribute in any experiment, the resulting angular distribution outside the crystal will be a superposition of several peaks whose width is roughly the same as their spacing. One cannot therefore expect sharply defined peaks outside the crystal which correspond to the individual states inside the crystal, though there may be slight departures from the classical predictions, perhaps of the form sketched in Figure 1.23(b). The best chance of seeing such departures would be for cases where barrier tunnelling becomes appreciable and effectively increases the aperture width thus decreasing the angular broadening owing to aperture diffraction.

A further smearing effect will arise if all the particles do not have the same energy, because equation (1.15) implies that the position of the lth peak depends on $l/E_l^{\frac{1}{2}}$; thus if there is a slight spread of energy amonst the protons, of only about one per cent, the peaks inside the crystal will merge with each other except for a few in the centre. Beating effects amongst the outer peaks will then give oscillations with longer periodicity, however, and make experimental detection easier, though unless the energy spread could be controlled, the results would be highly nonreproducible and perhaps make the experimenter distrust any effects that he saw! Altogether it appears that it will not be easy to find departures from the classical results and may even prove impossible. Further consideration of the point will be found in the chapter by Chadderton.

1.9 ELECTRONS AND POSITRONS

The case of electrons or positrons is quite different because of their smaller mass. First of all the model above suggests $\hat{l}$ between 1 and 10, but perhaps more important, the transverse particle wavelength λ_y is much closer to the barrier width and quantum tunnelling will allow the particle to be found in the classically forbidden region allowing coherence to exist between waves in adjacent channels. Thus the wavefunction does not describe states within a single channel. Not only does this alter the wavefunctions inside the crystal, it also makes aperture diffraction less serious at the entrance and exit. This problem has been solved by the electron diffraction theorists, particularly in the context of the electron microscope.[50] There it has been found that the angle of incidence can be sufficiently defined that only one l level is excited, corresponding to a single Bragg reflection. For $l = 1$ two wavefunctions are needed then to describe the observations adequately. One of them has its maxima in the channels, the other in the channel walls (these are similar to the two standing wave solutions of the Schrödinger equation in the Kronig–Penney model of metal). In the jargon of the electron microscopist this is a two-wave or two-beam approximation.

It is important to distinguish between the positron and the electron. The positron will find the regions of low potential in the same places as a proton would, but the electron's negative charge means that its waves with lowest potential energy travel within the atomic rows or planes. There will be a tendency therefore for such electron waves to interact strongly with the atoms, whereas their counterparts, tunnelling in the potential barrier, will not. There will be strong absorption of the first wave type and the slight absorption of the second, causing the two waves for a given value to have slightly different velocities, which leads to the beating or *thickness extinction* effects familiar to electron microscopists and quite inexplicable in classical terms.

If we turn now to the case of positrons or electrons at extreme relativistic speeds where $E_1 \gg m_0 c^2$ (i.e. $E_1 \gg 0.51$ MeV) the relativistic increase in mass will reduce the barrier tunnelling effect and lead to a more nearly classical solution to the problem.[52] For $l < \hat{l}$ the particle again becomes bound within the channel. Also, the number of states $\hat{l}$ will increase owing to the smaller de Broglie wavelength and we might expect the relativistic positron to behave rather like a proton with a fine structure showing $2\hat{l}$ (i.e. $8\hat{y}_0\hat{\psi}_0/y$) peaks within the angular range $\pm\hat{\psi}_0$. For MeV positrons one expects $\hat{l} \simeq 10$ and hence some 20 subsidiary peaks.

The observation of electron and positron channeling at low relativistic energies in experimental situations similar to these used for heavy particles has been the subject of many recent papers.[53] For example, experiments with radioactive electron emitters, where the particles are only just relativistic, similar to the α-emission experiments of Section 1.3, have been performed. Provided the emitting nuclei are near the surface, a *peak* of channeled electrons appear at the row directions. The contrast with the α-particle case, where a minimum is observed, is most striking and arises because the electron finds its lowest potential in the same regions as its trajectory starts out.[54] In cases where the emitting nuclei are distributed throughout a thick crystal a very complex structure of peaks and dips is seen, which so far defies analysis because so many wavefunctions must be involved.[55]

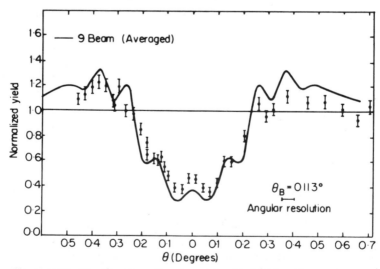

Figure 1.25. The blocking dip observed with MeV positrons scattered parallel to a {110} plane by a thin silicon crystal. The observations are compared with a prediction of a 9 beam dynamical diffraction theory[55]

Positrons are more like light protons, their lowest potential energy being in the regions of low atomic density. Relatively few experiments have been performed with them, because they are more difficult to work with experimentally. In those experiments which do not require the positron to travel far in the crystal, the light proton picture seems adequate, and both blocking and channeling have been seen, with the classical equations predicting roughly the correct angular width.[53] But where the thickness is sufficient for two or more wavefunctions to be set up, the experiments present much greater complexity. This appears a profitable area for future study. Figure 1.25 shows the angular distribution of MeV positrons scattered from a thin gold crystal in a blocking direction. The results show evidence of fine structure and are compared with dynamical diffraction theory assuming nine wave types, or beams, are present. Agreement appears to be excellent.[56]

1.10 CONCLUSIONS

The present state of channeling theory presents some interesting challenges. Classical theory works well enough for experiments with protons or heavy ions, though an adequate description of multiple scattering introduces great complexity. Wave mechanics are essential for all but the most rudimentary descriptions of electrons or positrons but the methods of calculation appear cumbersome at present and it is to be hoped that some simpler solution to the problem will be found.

Fundamental experimental studies of particle trajectories and energy losses will continue to provide insight into the fields which exist within crystals and the interaction of ions with the electronic systems of atoms. Advances can be expected in the general field of atomic collision physics as a result of channeling.

Amongst the applications of ion channeling, the crystallography of defect structures, and surfaces are perhaps the most promising though its direct use in ion implantation may be of greater technical importance. For electron channeling the most important applications for the immediate future will be in the MeV electron microscopes for which contrast analysis theories are now being developed.

References

1. von Laue, M., *Historical Introduction to 'International Tables for X-ray Crystallography'* Vol. 1. Ed. N. F. M. Henry and K. Lonsdale, Kynoch Press, (1952).
2. Stark, J., *Phys. Z.*, **13**, 973 (1912).
3. Robinson, M. T. and Oen, O. S., *Conf. Proc.: Le Bombardement Ionique*, Ed. J. J. Trillat. C.N.R.S., Paris. (1962).
4. Robinson, M. T. and Oen, O. S., *Phys. Rev.*, **132**, 1385 (1963).
5. Davies, J. A. and Sims G. A., *Can. J. Chem.*, **39**, 601 (1961).

6. Fluit, J., *Conf. Proc.*: *Le Bombardement Ionique*, Ed. J. J. Trillat, C.N.R.S. Paris, p. 119 (1962).
7. Rol, P. K., Fluit, J. M., Viehbock, F. P. and de Jong, M., *Proc. 4th Int. Conf. on Ionisation Phenomena in Gases*, Amsterdam, North Holland, Publishing Co. (1959).
8. Nelson, R. S. and Thompson, M. W., *Phys. Lett.*, **2**, 124 (1962).
9a. Thompson, M. W. *Defects and Radiation Damage in Metals*, Cambridge University Press (1968).
9b. Thompson, M. W., 1968, *Phil. Mag.*, **18**, 377 (1968).
10. Dearnaley, G., *I.E.E.E. Trans. Nuc. Sci.*, **11**, 249 (1964).
11. Erginsoy, C., Wegner, H. E. and Gibson, W. M., *Phys. Rev. Lett.*, **13**, 530 (1964).
12. Piercy, G. R., Brown, F., Davies, J. A. and McCargo, M., *Phys. Rev. Letters*, **10**, 399 (1963).
13. Lutz, H. and Sizmann, R., *Phys. Lett.*, **5**, 113 (1963).
14. Davies, J. A. and Jespersgaard, P., *Can. J. Phys.*, **44**, 1631 (1966).
15. Kornelsen, E., Brown, F., Davies, J. A. and Piercy, G. R., *Phys. Rev.*, **136A**, 849 (1964).
16. Thompson, M. W., Harwell Conference on Collision Cascades; *Unclassified U.K.A.E.A.*, *report A.E.R.E.-R-4694* (from HMSO); *Phys. Rev. Lett.*, **13**, 756 (1964).
17. Lindhard, J., *Phys. Lett.*, **12**, 126 (1964).
18. Bogh, E., Davies, J. A. and Nielsen, K. O., Harwell Conference on Collision Cascades: *Unclassified U.K.A.E.A. Report A.E.R.E.-R-4694* (from HMSO) *Phys. Lett.*, **12**, 129 (1964).
19. Andersen, J. U., Davies, J. A., Nielsen, K. O. and Andersen, S. L., 1965 *Nuc. Instrum. and Meth.*, **38**, 210 (1965).
20. Brandt, W., Khan, J. M., Potter, D. L., Worley, R. D. and Smith, H. P., *Phys. Rev. Lett.*, **14**, 42 (1965).
21. Mashkova, E. S., Molchanov, V. A. and Odintsov, D. D., *Fiz. Tverdogo Tela*, **5**, 3426 (1963) or *Sov. Phys. Solid St.*, **5**, 2516 (1964).
22. Noggle, T. S. and Oen, O. S., *Phys. Rev. Lett.*, **16**, 395 (1966).
23. Domeij, B. and Bjorqvist, K., *Phys. Lett.*, **14**, 127 (1965).
24. Lindhard, J., *Dansk. Vid. Selsk. Mat. Fys. Medd.*, **34**, 14 (1965).
25. Nelson, R. S. and Thompson, M. W., *Phil. Mag.*, **8**, 1677 (1963).
26. Bøgh, E. and Uggerhoj, E., *Nucl. Instrum. and Methods*, **38**, 216 (1965).
27. Tulinov, A. F., Kulikauskas, V. S. and Malov, M. M., *Phys. Lett.*, **18**, 304 (1965).
28. Davies, J. A., Denhartog, J., Eriksson, L. and Mayer, J. W., *Can. J. Phys.* **45**, 4053 (1967).
29. Matzke, H. and Davies, J. A., *J. app. Phys.*, **38**, 805 (1967).
30. Andersen, J. U., Andreasen, O., Davies, J. A. and Uggerhoj, E., *Rad. Effects*, **7**, 25 (1971).
31. Alexander, R. B. and Poate, J., 1972, to be published in *Rad. Effects*.
32. Pabst H. and Palmer D. W., 1972, to be published in *Rad. Effects*.
33. Gemmell, D. S. and Holland, R. E., *Phys. Rev. Lett.*, **14**, 945 (1965).
34. Appleton, B. R., Erginsoy, C. and Gibson, W. M., *Phys. Rev.*, **161**, 330 (1967).
35. Dearnaley, G., Farmery, B. W., Mitchell, I. V., Nelson, R. S. and Thompson, M. W., *Phil. Mag.*, **18**, 155, 985 (1968).
36. Marwick, A. D., *Thesis*, University of Sussex (1971).
37. Datz, S., Noggle, T. S., Appleton, B. R. and Lutz, H. O., *Phys. Rev.*, **179**, 315 (1969).
38. Robinson, M. T., *Phys. Rev.*, **179**, 327 (1969).
39. Datz, S. and Snoek, C., *Phys. Rev.*, **134A**, 347 (1964).

40. Mashkova, E. S., Molchanov, V. A., Parilis, E. S. and Turaev, N. Y., *Phys. Lett.*, **18**, 7 (1965).
41. Mashkova, E. S., Molchanov, V. A. Skripka, Y. G. Snisa, V. A. and Soshka, W., *Conf. Proc. Atomic Collisions Phenomena in Solids*, Ed. Palmer, Townsend and Thompson, Amsterdam, North Holland Publishing Co. p. 608 (1970).
42. Yurasova, V. E., Brzhezinskij, V. A. and Ivanov, G. M., *JETP (1964) Z.E.T.*, **47**, 473 (1964).
43. Yurasova, V. E., Shulga, V. I. and Karpuzov, D. S., *Can. J. Phys.*, **46**, 729 (1968).
44. Lehmann, C. and Leibfried, G., *J. appl. Phys.*, **34**, 2821 (1963).
45. Thompson, M. W., Proc. Brookhaven Solid State Conf., Ed. Goland USAEC Unclassified BNL report—BNL 50083 (1967).
46. Farmery, B. W., Nelson, R. S., Sizmann, R. and Thompson, M. W., *Nuc. Instrum. and Meth.*, **38**, 231 (1965).
47. Morgan, D. V. and Van Vliet, D., *Conf. Proc.: Atomic Collision Phenomena in Solids*, Ed. Palmer, Townsend and Thompson, Amsterdam, North Holland Publishing Co. p 476 (1970).
48. Delsarte, G., Jousset, J. C., Mory, J. and Quere, Y., 1970, *Conf. Proc.: Atomic Collision Phenomena in Solids*, Ed. Palmer, Townsend and Thompson, Amsterdam, North Holland Publishing Co. p. 501 (1970).
49. Appleton, B. R., Feldman, L. and Brown, W., Proc. Brookhaven Solid State Conf., Ed. Goland, USAEC unclassified report BNL 50083 (1967).
50. Hirsch, P. B., Howie, A., Nicholson, R. B. and Pashley, D. W. and Whelan, M. J., 1965 *Electron Microscopy of Thin Crystals*, London, Butterworth (1965).
51. Thompson, M. W., *Contemp. Phys.*, **9**, 375 (1968).
52. Lindhard, J., *Conf. Proc.: Atomic Collision Phenomena Solids*, Ed. Palmer, Townsend and Thompson, Amsterdam, North Holland Publishing Co. p. 1 (1970).
53. Andersen, J. U., Augustyniak and Uggerhoj, E., *Phys. Rev. B*, **3**, 705 (1971).
54. Uggerhoj, E. and Andersen, J. U., *Can. J. Phys.*, **46**, 543 (1968).
55. Domeij, B., 1965, *Nuc. Instrum. and Meth.*, **38**, 207 (1965).
56. Pedersen, M. J., Andersen, J. U. and Augustyniak, W. M., 1972, *Rad. Effects*, to be published.
57. Piercy, G. R., McCargo, M., Brown, F. and Davies, J. A., *Can. J. Phys.*, **42**, 1116 (1964).
58. de Wames, R. E. Hall, W. F. and Chadderton, L. T., Proc. Brookhaven Solid State Conf. Ed. Goland, USAEC unclassified BNL report, 50083 (1967).
59. Lutz, H., Datz, S., Moak, C. D. and Noggle, T. S., *Phys. Rev. Lett.*, **17**, 285 (1966).

CHAPTER II

The Continuum Model of Directional Effects

D. Van Vliet

2.1 INTRODUCTION

The passage of energetic,* positively charged ions through a crystal lattice is normally a complicated process. The number of possible collision sequences with the crystalline atoms is infinite and, if the ion undergoes a number of large-angle deflections, is essentially unrelated to the precise structure of the medium, depending only on its density. In these cases collisions with the lattice atoms occur with the same statistical frequency as they would in a completely structureless medium and one can determine properties as averaged over all possible trajectories, for example the range, by applying the appropriate statistical considerations.

On the other hand experiments over the last ten years, as discussed in Chapter 1, have clearly shown that ion beams which are nearly parallel to major axes or planes in a single crystal exhibit quite distinct directional effects such as increased ranges or reduced nuclear reaction yields. The regular atomic rows and planes steer these ions into stable orbits down the open channels between them, the 'channeling effect', Such trajectories are examples of 'governed' motion,† where the regular structure of the crystal is a crucial factor in determining both their path and properties. Another example is that of ions which originate from an atomic site, e.g. by nuclear disintegration, but are consistently steered away or 'blocked' from orbits parallel to major axes.

---

* In this context typically between 1 keV and 10 MeV.
† We may draw a distinction with directional effects associated with 'ungoverned' motion, i.e. a beam of particles whose properties are in some way influenced by the lattice structure but whose motion is not. Imagine, for example, a neutron beam striking a thin crystal at a low angle to a major axis. Those entering near the middle of the open channels can traverse the entire crystal in a straight line possibly without ever coming close to an atomic nucleus, statistically an unlikely situation in a random medium but here very likely. However, in this instance there is no question of the lattice atoms steering the neutrons away from the nuclei. A more complete discussion of the relationship between directional effects and governed and ungoverned motion is given by Lindhard.[1]

The continuum model shows how such governed motion is related to the basic interactions between the ion and the crystalline atoms and to the regular lattice structure. The first comprehensive treatment of its principles was given by Lindhard[1] in 1965 although previously Lehmann and Leibfried[2] had used continuum ideas to describe proper channeling (see Section 2.4.2). The essential simplification in the continuum model is to average the ion–atom interaction potential along the major axes or planes so that the scattering is reduced from three to two or one dimensions respectively. In principle this means that individual trajectories may now be calculated with relatively small uncertainties, although exact solutions present considerable mathematical difficulties and are not normally attempted. However the great advantage of the continuum model is, as we shall see, that it enables one to describe the average properties of a given trajectory or, alternatively, to describe the average properties of an entire beam of ions.

In this chapter we shall develop the continuum model from first principles for both axes and planes, describe how and why it leads to different types of governed trajectories and discuss their different characteristics. The limitations of the model both in terms of angular limits and of multiple scattering are also described. Section 2.6 shows how the twin phenomena of channeling and blocking are related by reversibility while Sections 2.7 and 2.8 illustrate this principle as well as providing a quantitative application of the continuum model. A glossary of the symbols used appears at the end of the chapter.

First, however, let us briefly review the basic interactions between an ion and a crystal.

2.1.1 The Ion–Crystal Interaction

The interaction between an ion and a crystal is normally divided[3] into elastic interactions with the atomic nuclei and inelastic interactions involving the electrons.* Because of the large mass differences between an ion and an electron, inelastic collisions lead to relatively small angular deflections so that the ion's path is essentially controlled by the elastic component. On the other hand, at high energies (as a rule of thumb for $E > Z_1$ (keV)) the rate of energy loss is primarily determined by the inelastic component.

The elastic interaction between an ion and an atom is described by an interaction potential $V(r)$ (for a recent review of this topic see Reference 4). At low separations $V(r)$ is normally described by the coulombic repulsion of the bare nuclei modified by some function of the separation which de-

---

* In this context elastic implies conservation of total kinetic energy between the collision partners; in an inelastic event some might be given up in, say, the ionization of a bound electron.

scribes the electronic screening. The following approximations are commonly used:

(i) The Lindhard standard potential[5]:

$$V(r) = \frac{Z_1 Z_2 e^2}{r} \left\{ 1 - \frac{r}{(r^2 + C^2 a^2)^{\frac{1}{2}}} \right\} \tag{2.1}$$

where the adjustable parameter C is normally set equal to $\sqrt{3}$.

(ii) The Moliere potential[6]:

$$V(r) = \frac{Z_1 Z_2 e^2}{r} (0 \cdot 1 \, e^{-6r/a} + 0 \cdot 35 \, e^{-0 \cdot 3r/a} + 0 \cdot 55 \, e^{-1 \cdot 2r/a}). \tag{2.2}$$

(iii) And (less frequently used) the Bohr potential[3]:

$$V(r) = \frac{Z_1 Z_2 e^2}{r} e^{-r/a} \tag{2.3}$$

In all three the screening radius a is, in theory, an adjustable parameter, but in practice it is usually determined by an analytical expression, the most common being:

$$a = 0 \cdot 8853 a_H (Z_1^{\frac{2}{3}} + Z_2^{\frac{2}{3}})^{-\frac{1}{2}} \tag{2.4}$$

where $a_H = 0 \cdot 528$ Å is the Bohr radius.

At relatively large separations the interatomic potential is often described by a Born–Mayer potential:[7]

$$V(r) = A_{BM} \, e^{-Br} \tag{2.5}$$

where A_{BM} and B are arbitrary parameters which are generally fitted to experimental data.

The inelastic interaction is usually described by a continuous energy loss rate $(-\partial E/\partial z)_{inel}$ which is either determined by an analytical expression or by experimental data.

Our entire treatment here, both of the trajectory of an ion in a crystal and of the continuum model, assumes that each ion may be treated as a localized wave packet to which the laws of classical mechanics may be satisfactorily applied and that any quantum-mechanical corrections are relatively small. The justification of this rather important point has been the subject of considerable discussion and will be treated in Chapter 10. For the moment our primary function will be to discuss the precise implications of the classical approach which, although we do not mention comparison with experimental results here, has proved extremely successful.

2.2 DIRECTIONAL EFFECTS NEAR ATOMIC STRINGS

2.2.1 The Continuum Approximation

Consider an ion moving at an angle ψ to a row (or string) of atoms of spacing d along the z-axis and in a plane containing them. Let $\rho(z)$ be its distance from the atomic string. The average interatomic potential experienced by an ion at ρ is given by:[1]*

$$U_1(\rho) = \frac{1}{d} \int_{-\infty}^{\infty} dz V(\sqrt{z^2 + \rho^2}). \qquad (2.6)$$

If ψ is sufficiently small that ρ changes relatively slowly the so called 'continuum potential' $U_1(\rho)$ becomes a well defined property of the trajectory. In this case the ion should experience an average repulsive force $-U_1'(\rho)$ perpendicular to the row, and its trajectory would approach a continuous curve as illustrated in Figure 2.1. To examine the validity of such an approach consider that section of the trajectory between $-d/2$ and $+d/2$ of an atom located at $z = 0$. The net change in momentum perpendicular to the atomic row is given by

$$\Delta p_\perp = -\int_{-d/2}^{d/2} U_1'(\rho) \frac{dz}{v} \qquad (2.7)$$

and therefore the net angular deflection by

$$\Delta\psi_c = \Delta p_\perp/p \simeq -\frac{d}{2E} U_1'(\rho') \qquad (2.8)$$

where, if $\Delta\psi_c$ is small, we approximate $U_1'(\rho)$ by its value at $z = 0$, $\rho = \rho'$.

In reality, however, owing to the short range of the interatomic forces ($a \ll d$) the ion is given an almost instantaneous deflection, θ, as it passes an atom so that its trajectory between $-d/2$ and $+d/2$ should, very nearly, be given by two straight line segments, the impulse model illustrated in Figure 2.1. If θ is small it may be estimated by the first-order momentum approximation[8] which, effectively, integrates the net change of momentum normal to the ion's velocity:

$$\Delta\psi_i = \theta = -\frac{1}{2E} \int_{-\infty}^{\infty} dz \frac{\partial}{\partial s} V(\sqrt{z^2 + s^2}) \qquad (2.9)$$

where s is the impact parameter. Comparing equations (2.9) and (2.8) with equation (2.6) shows that $\Delta\psi_c$ and $\Delta\psi_i$ are equivalent so long as $\rho' \simeq s$. (The major source of difference between the two arises from the curvature in the continuum trajectory. Between $z = -d/2$ and $z = 0$, $\rho(z)$ diverges from

---

* Changes in the definition of $U_1(\rho)$ for vibrating strings are discussed in Appendix 2B.

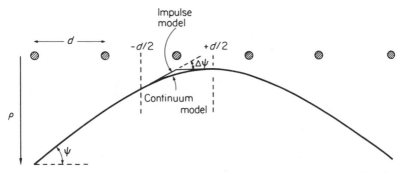

Figure 2.1. The trajectory of an ion almost parallel to a regular atomic string as predicted by the continuum model. The comparable trajectory given by the impulse model is illustrated for one atom at $z = 0$

the straight line impulse trajectory by an amount:

$$\delta p' \simeq -\frac{d^2}{16E} U'_1(\rho') \qquad (2.10)$$

Since $|s - \rho'| \simeq \delta \rho'$ the discrepancy in the angle of deflection is given by:

$$\delta \Delta \psi \simeq \delta \rho' \frac{\partial \Delta \psi_c}{\partial \rho} \simeq \frac{d^2}{16E} U'_1(\rho') \frac{d}{2E} U''_1(\rho'). \qquad (2.11)$$

The equivalence of the two models requires that $\delta \Delta \psi \ll \Delta \psi_c$ or, by substitution,

$$U''_1(\rho') \ll 16E/d^2 \qquad (2.12)$$

Since U''_1 is a monotonically decreasing function of ρ, equation (2.12) is in fact satisfied at large values of ρ' and hence at relatively small scattering angles.*)

Similarly it may be easily demonstrated that both the continuum model and the impulse model yield, to a first approximation, the same value of $\rho(d/2)$. By extending the argument to the next and to all subsequent atoms in

---

* If equation (2.12) is satisfied, or in fact if the more accurate condition $\rho' > \rho_{min}$ (2.28) is satisfied, then one may demonstrate that two other implicit assumptions of the continuum approximation are also satisfied. First, higher order terms in the momentum approximation are negligible if $s > \rho_{min}$; hence equation (2.9) is accurate. Second, although there is a further error of approximately $\frac{1}{2}\rho'\psi(0)^2$ in equating s to ρ', it may be shown that for the maximum transverse angle ψ_2 equation (2.35) permitted by equation (2.12) and near the critical point $\rho \simeq \rho_{min}\frac{1}{2}\rho_{min}\psi_2 \ll \delta\rho'$ given by equation (2.10). The only other necessary assumption to be made in deriving the continuum approximation is that the atoms are at fixed locations along the string axis and as we see later, Section 2.2.4, thermal displacements establish the other limit to the continuum approximation. N.B. In point of fact we need not have assumed that a large number of scattering events are required to define $U_1(\rho)$; the continuum approximation can be applied to as few as two atoms.

the string it may be easily appreciated that both the continuum and the impulse model continue to give the same values of $\rho(z)$ and $\psi(z)$ at $z = 3d/2$, $5d/2$, etc. The same conclusions also hold if the path of the ion does not lie in the same plane as the atomic string. Hence, within the limits discussed below, the scattering of an ion from an individual atomic string may be adequately represented by a continuously varying trajectory and the string itself, by its continuum potential field $U_1(\rho)$. It is perhaps instructive to think of an atomic row as a perfectly elastic rod which grows harder as one penetrates nearer its core.

2.2.2 Continuum Potentials

A relatively simple analytical expression for $U_1(\rho)$ may be obtained from the Lindhard standard potential by substituting equation (2.1) into equation (2.6). This we shall refer to as the 'standard continuum potential':[1]

$$U_1(\rho) = \frac{2Z_1 Z_2 e^2}{d}\zeta(\rho/a)   \tag{2.13}$$

where*

$$\zeta(\rho/a) = \tfrac{1}{2}\ln\{(Ca/\rho)^2 + 1\}.$$

In the limits of small and large ρ/a respectively ζ may be approximated by

$$\zeta(\rho/a) \simeq \ln(Ca/\rho)   \tag{2.14}$$

and

$$\zeta(\rho/a) \simeq \tfrac{1}{2}(Ca/\rho)^2.   \tag{2.15}$$

Comparable expressions derived from the Moliere, Bohr and Born–Mayer potentials are listed in Appendix 2A and compared with the standard potential in Figure 2.2.

In a full crystal lattice it may be easily appreciated that when viewed parallel to a low index direction all the atoms are positioned along regular atomic strings, the positions of which form a regular array in the 'transverse plane' normal to the strings. If the interaction between an ion and the ith atomic string may be represented by its continuum potential $U_1(\rho_i)$ then the total potential field in which the ion moves is found by summing over all the strings:†

$$U(\boldsymbol{\rho}) = \sum_i U_1(\rho_i) - U_{\min}   \tag{2.16}$$

---

* Note that our definition of $\zeta(\rho/a)$ is one half that given by Lindhard.[1]

† There is evidence that, under some circumstances, when an ion is interacting strongly with two or more atomic strings simultaneously its interaction cannot be separated into independent collisions with individual strings. For example, Morgan and Van Vliet[10] have found that the continuum approximation breaks down more easily for a collection of strings than for individual strings when, at low energies, the critical approach distance (equation (2.29)) becomes comparable to the separation of the strings.

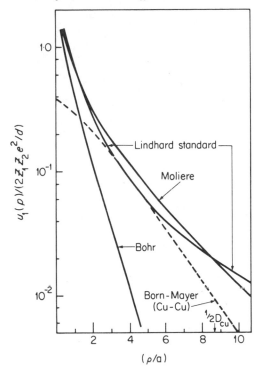

Figure 2.2. The continuum string potentials for the Lindhard standard, Moliere, Bohr and Born–Mayer potentials, the latter being plotted for the Gibson II Cu–Cu parameters with $a = 0.1073 \, \text{Å}$

where $\boldsymbol{\rho}$ is its vector coordinate in the transverse plane and $U_{\min}$ is the minimum potential energy at any point in the transverse plane. The choice of $U_{\min}$ is in a sense arbitrary since, being constant, it does not affect the equations of motion. We choose it out of convenience so that $U(\boldsymbol{\rho})$ goes to zero at its minimum.

Two examples of $U(\boldsymbol{\rho})$ are illustrated in Figures 2.3 and 2.4 for α-particles in the $\langle 001 \rangle$ and $\langle 101 \rangle$ transverse planes of copper. In both cases $U(\boldsymbol{\rho}) \gtrsim U_1(\rho_i)$ when $\boldsymbol{\rho}$ lies near the ith row and hence $U(\boldsymbol{\rho})$ is rotationally symmetric about the rows. In the $\langle 001 \rangle$ plane $U(\boldsymbol{\rho})$ is also symmetric about the mid-channel axis for $U(\boldsymbol{\rho}) \lesssim 5 \, \text{eV}$ and it is sometimes useful to compare this region of the potential to a harmonic bowl[9]. On the other hand the $\langle 101 \rangle$ channel has two equivalent minima but only shows a circular symmetry for extremely low values of $U(\boldsymbol{\rho})$, i.e. $\ll 0.1 \, \text{eV}$.

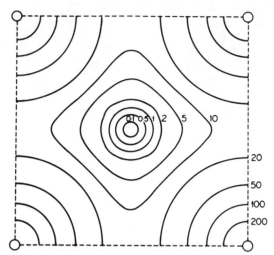

Figure 2.3. The net continuum potential $U(\rho)$ for α-particles in the {001} transverse plane of copper in eV (Van Vliet[9])

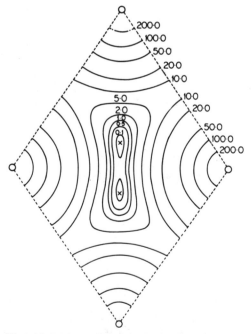

Figure 2.4. The net continuum potential $U(\rho)$ for α-particles in the {101} transverse plane of copper in eV (Van Vliet[9])

2.2.3 Axial Motion In The Continuum Model

Consider the motion of an ion moving at a relatively low angle ψ to a low index direction in a crystal. Its Hamiltonian may be written:

$$H = U(\rho) + (p_x^2 + p_y^2 + p_z^2)/2m_1. \tag{2.17}$$

Two important consequences of the continuum approximation are now evident from equation (2.17):

(i) Since $U(\rho)$ is independent of z the momentum component parallel to the string, p_z, and the parallel velocity, v_z, are constants of the motion and $z = v_z t$. The trajectory can therefore be completely described by solving the projected or 'transverse' motion in the xy plane. In this transverse motion the ion moves with a velocity $v_\perp = v \sin \psi \simeq v\psi$ in the potential field $U(\rho)$.

(ii) Since $U(\rho)$ is velocity independent the total energy of the projected motion is also a constant, the 'transverse energy':

$$E_\perp = U(\rho) + (p_x^2 + p_y^2)/2m_1$$

$$\simeq U(\rho) + E\psi^2. \tag{2.18}$$

Several illustrations of projected motion are given in Figure 2.10; we defer to Section 2.4.2 a discussion of the characteristics of trajectories for which the continuum description is valid (or nearly so).

Although we have reduced the equations of motion from three dimensions to two it is still extremely complex to solve them explicitly. Nevertheless the principle of the conservation of transverse energy represents a powerful tool in deriving averaged properties of the motion. For example, one knows that ions of transverse energy $E_\perp$ can only reach that portion of the transverse plane where $U(\rho) < E_\perp$, an accessible area which we define as $A(E_\perp)$. Lindhard[1] has shown that once statistical equilibrium is reached there is an equal probability of finding an ion anywhere within its accessible transverse space so that we may write:

$$P_0(E_\perp, \rho) = \begin{cases} 1/A(E_\perp) & U(\rho) \leqslant E_\perp \\ 0 & U(\rho) > E_\perp \end{cases} \tag{2.19}$$

There is obviously a minimum approach distance to the centre of a string, ρ_c, determined by:

$$E_\perp = U(\rho_c) \simeq U_1(\rho_c) \tag{2.20}$$

Hence, discounting thermal vibrations, all collisions of impact parameter $s < \rho_c$ are screened out. This leads to the important anomalies characteristic of channeling, e.g. the marked reduction in yield from any process which requires a near head-on collision with an atomic nucleus, e.g. large-angle Rutherford scattering.

We may also use equation (2.19) to determine the average of any property f of an ion trajectory, such as its elastic or inelastic energy loss rate, which is a function of its position in the transverse plane:

$$f(E_\perp) = \int P_0(E_\perp, \rho) f(\rho) \, dA$$

$$= \int_{U(\rho) < E_\perp} f(\rho) \frac{dA}{A(E_\perp)} \tag{2.21}$$

If $f(\rho)$ decreases rapidly with increasing separation from the strings and $E_\perp$ is large it is very often convenient to replace equation (2.21) by an integral about a single string. Writing $f(\rho) = f(\rho)$:

$$f(E_\perp) = \int_{\rho_c}^{\rho_0} \frac{2\rho \, d\rho}{(\rho_0^2 - \rho_c^2)} f(\rho) \tag{2.22}$$

where the upper limit of integration ρ_0 is chosen such that $\pi \rho_0^2 = (Nd)^{-1}$, the area per string in the transverse plane. An example of equation (2.22) is the calculation of thermal multiple scattering in equation (2.56).

Furthermore, if we know the distribution in transverse energy, $n(E_\perp)$, the average of f over all ions is given by:

$$\langle f \rangle = \int n(E_\perp) f(E_\perp) \, dE_\perp. \tag{2.23}$$

The most common situation where $n(E_\perp)$ is known is that of a collimated beam of ions striking a crystal with a transverse angle ψ_{in}. The transverse energy of an ion which enters at ρ_{in} is:

$$E_\perp = U(\rho_{in}) + E\psi_{in}^2 \tag{2.24}$$

and equation (2.23) may be replaced by an integral over a representative area of the entrance surface:

$$\langle f \rangle = \frac{1}{A_0} \int dA \, f(U(\rho_{in}) + E\psi_{in}^2). \tag{2.25}$$

To illustrate equation (2.25) consider the case where $f(E_\perp)$ is itself $P_0(E_\perp, \rho)$, the flux at a fixed point ρ in the transverse plane. The total flux through ρ averaged over all incident ions and taken relative to the random flux, $1/A_0$, is given by:

$$F(\rho) = \int_{U(\rho_{in}) + E\psi_{in}^2 \geq U(\rho)} \frac{dA}{A(U(\rho_{in}) + E\psi_{in}^2)}. \tag{2.26}$$

Solutions of equation (2.26) have been examined in detail by the author.[9]

2.2.4 The Limitations Of The Continuum Approximation

The continuum model of scattering is only valid for ions which remain relatively far from any atomic string and at a low transverse angle to them. The approximation breaks down when an ion has sufficient transverse energy to penetrate into the core of the strings where it may be scattered at large angles from individual atoms. The limiting conditions may be expressed in two equivalent ways: either as the 'critical approach distance', ρ_{crit}, within which noncontinuum scattering is possible; or, alternatively, as the 'critical angle', ψ_{crit}, or 'critical transverse energy', E_{crit}, required to penetrate that far.

$$U(\rho_{crit}) = E\psi^2_{crit} = E_{crit} \tag{2.27}$$

We have already derived in equation (2.12) a condition for the validity of the continuum approximation in a single collision with an atom in a rigid string. A similar condition has also been found by Morgan and Van Vliet[10] who have used computer simulation to study the breakdown of the continuum approximation for two parallel atomic strings. They found a rather sharp transition between channeled trajectories at low transverse angles and nonchanneled trajectories at higher angles, as illustrated in Figure 2.5

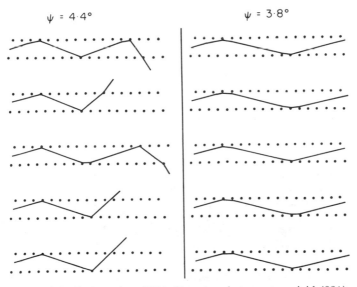

Figure 2.5. Trajectories of 20 keV protons between two rigid ⟨001⟩ copper rows for initial transverse angles just above and just below the critical angle of 4·0° calculated with the Bohr potential. The vertical axis is magnified four times for clarity. Each trajectory corresponds to a different initial position along the mid-channel axis (after Morgan and Van Vliet[27])

for a simple two-dimensional lattice. At the transition the point of closest approach ρ_{min} to the strings is quite accurately given by:

$$U_1''(\rho_{min}) \simeq 5E/d^2. \tag{2.28}$$

While equation (2.28) cannot in general be solved explicitly for ρ_{min} a good numerical solution for the Moliere continuum potential (equation 2.89) is given by:

$$\rho_{min}/a \simeq \tfrac{2}{3}\sqrt{\alpha}(1 - \sqrt{\alpha}/19 + \alpha/700) \tag{2.29}$$

where $\alpha = Z_1 Z_2 e^2 d/a^2 E$.

ρ_{min}/a is plotted in Figure 2.6 as a function of α and compared with an exact numerical solution of equation (2.28) for the Moliere potential and

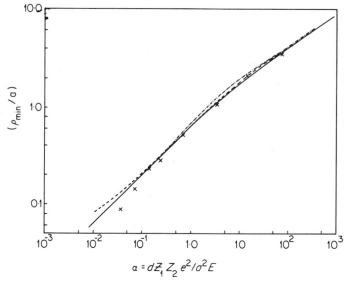

Figure 2.6. The critical approach distance for rigid rows, ρ_{min}, found by numerical solutions of equation (2.28) for the Lindhard standard (dot-dash) and Moliere continuum potentials (dashed line) and given by the approximation (2.29) (solid line). The experimental points are those determined by Morgan and Van Vliet[10] for protons in $\langle 001 \rangle$ copper

also with the computed results of Morgan and Van Vliet. Also shown is the solution of equation (2.28) for the standard potential (equation 2.13), and the only significant differences occur in the limit of large ρ_{min}/a, low energy.

Apart from the intrinsic limit (equation 2.28) to continuum scattering there is a further error introduced by the fact that the atoms are not located precisely along the string but are in fact slightly displaced due to their constant thermal vibrations.* This introduces a further discrepancy between ρ' and s which is of the order of x_{rms} and we must therefore require

$$\rho' \gg x_{rms} \qquad (2.30)$$

From equation (2.29) it is apparent that equation (2.30) becomes important at high energies when ρ_{min} becomes small. Computer simulation experiments by Barrett[11] at relatively high energies may be interpreted† as implying that the continuum approximation holds as long as

$$\rho' \geqslant 1{\cdot}2x_{rms} \qquad (2.31)$$

There is further evidence both from the calculations of Barrett† and of Morgan and Van Vliet that the intrinsic limit (2.28) and the extrinsic limit (2.31) add quadratically so that the critical approach distance may be expressed as:

$$\rho_{crit}^2 \simeq \rho_{min}^2 + (1{\cdot}2x_{rms})^2 \qquad (2.32)$$

for all energies.

2.2.5 Critical Angles

The maximum allowed transverse angle, ψ_{crit}, may now be obtained by substituting equation (2.32) into equation (2.27). In the limit of high energies

$$\psi_{crit} \rightarrow (U(1{\cdot}2x_{rms})/E)^{\frac{1}{2}}$$

$$\simeq \left[\frac{2Z_1 Z_2 e^2}{dE} \right]^{\frac{1}{2}} \zeta (1{\cdot}2x_{rms}/a)^{\frac{1}{2}} \qquad (2.33)$$

using equation (2.13). Since $\zeta (1{\cdot}2x_{rms}/a)^{\frac{1}{2}}$ is generally of the order of unity ψ_{crit} is of the order of the characteristic angle ψ_1 initially introduced by Lindhard:[1]

$$\psi_1 = \left(\frac{2Z_1 Z_2 e^2}{dE} \right)^{\frac{1}{2}}. \qquad (2.34)$$

In the limit of low energies substituting equation (2.29) into equation (2.27) leads to rather complicated expressions, although a rather simple

---

* We may neglect the actual motion of the atom during the collision since the thermal vibrational period ($\sim 10^{-13}$ s) is very much greater than the typical collision times, for example, $t_{coll} \simeq 2a/v \simeq 10^{-16}$ s for a 1 keV proton.
† See Chapter 3 or, for a more detailed discussion, Chapters 8 and 9 of reference 12.

approximation, also due to Lindhard[1], is found to hold over a wide range of values:

$$\psi_{\text{crit}} \simeq \psi_2 = \left\{ \frac{Ca}{d\sqrt{2}} \psi_1 \right\}^{\frac{1}{2}}$$

$$= \left\{ \frac{aC}{d} \left(\frac{Z_1 Z_2 e^2}{dE} \right)^{\frac{1}{2}} \right\}^{\frac{1}{2}}. \tag{2.35}$$

Note that in the high energy limit $\psi_{\text{crit}} \propto E^{-\frac{1}{2}}$ whereas in the low energy limit $\psi_{\text{crit}} \propto E^{-\frac{1}{4}}$. The point of transition between the two is not sharply defined and depends to a certain extent on x_{rms}. However it is customary to distinguish between a high energy regime, $E > E'$, and a low energy regime, $E < E'$, where ψ_1 and ψ_2 are applicable respectively, with:

$$E' = \frac{2Z_1 Z_2 e^2 d}{a^2 E}. \tag{2.36}$$

It should be emphasized that ψ_1 and ψ_2 are only characteristic of the angular extent of continuum string effects and are not intended to have a precise experimental interpretation; Section 2.8 briefly discusses the relationship between the angular half-width, $\psi_{\frac{1}{2}}$, in a particular experiment and ψ_{crit}. Nonetheless ψ_1 and ψ_2 are found to be generally in good qualitative agreement with experimentally measured half-widths and to predict very accurately the dependence upon, say, energy or interatomic spacing.

2.3 PLANAR EFFECTS

2.3.1 The Continuum Approximation For Planes

An analagous and, in many respects, simpler treatment than that applied to atomic strings may be applied to the scattering of an ion from a dense plane of atoms. Here the average or continuum potential at a distance y from an xz plane is given by:

$$Y_1(y) = (Nd_p) \int_0^\infty 2\pi r \, dr V(\sqrt{r^2 + y^2}) \tag{2.37}$$

where Nd_p is the number of atoms per unit area in the plane. That $Y_1(y)$ reproduces the *average* scattering of an ion from an atomic plane is demonstrated in Appendix 2C.

The Lindhard standard potential yields the following expression for $Y_1(y)$ by substituting equation (2.1) into (2.37):[1]

$$Y_1(y) = 2\pi Z_1 Z_2 e^2 (Nd_p)a[\{(y/a)^2 + C^2\}^{\frac{1}{2}} - y/a]. \tag{2.38}$$

The comparable Moliere and Bohr continuum potentials are listed in Appendix 2A and compared with the standard potential (equation 2.38) in Figure 2.7.

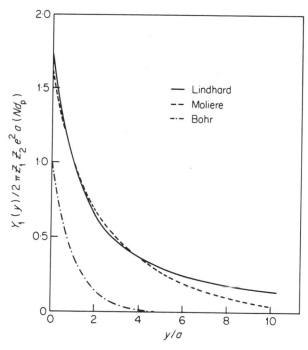

Figure 2.7. The planar continuum potentials as calculated by the Lindhard standard, Moliere and Bohr potentials

2.3.2 Continuum Motion Near Planes

Since a crystal lattice may be thought of as a succession of atomic planes let us consider the motion of an ion in an interplanar channel of width d_p. Its Hamiltonian may be written:

$$H = Y(y) + (p_x^2 + p_y^2 + p_z^2)/2m_1 \qquad (2.39)$$

where

$$Y(y) = Y_1(y) + Y_1(d_p - y) - 2Y_1(0.5d_p) \qquad (2.40)$$

is the net continuum potential within the channel. Once again (cf. equation (2.16)) an arbitrary constant term, $-2Y_1(0.5d_p)$, is selected so that $Y(y)$ goes to zero along the mid-channel minimum. The extent of the overlap is illustrated in Figure 2.8 for protons in the {010} copper channels.

From equation (2.39) the momentum components parallel to the plane, p_x and p_z, are seen to be constants of the motion so that we need only consider the projected motion along the transverse y-direction. If the angle which the ion makes with the planes is ϕ then its transverse velocity is $v_\perp = v \sin \phi \simeq v\phi$.

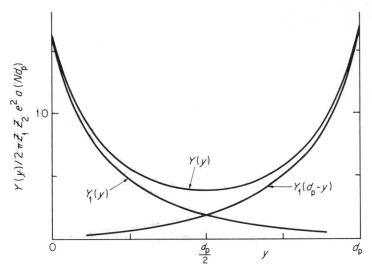

Figure 2.8. The net planar continuum potential for protons in the (010) interplanar channel as given by the Moliere potential (after Morgan and Van Vliet[15])

The transverse energy associated with the projected motion is also a constant:

$$E_\perp = Y(y) + E\phi^2. \qquad (2.41)$$

Just as atomic strings could be compared with elastic rods so may two planes of atoms be thought of as infinite elastic sheets.

While in principle it is easy to solve the one-dimensional equation of motion to give $y(z)$, where $z(=vt)$ is the penetration depth, (see Chapter 6) owing to the complex form of $Y(y)$ it is normally only possible to obtain analytical expressions for $y(z)$ by introducing certain approximations. For example, a series of solutions have been obtained for the Moliere continuum potential (equation 2.92) by Robinson.[13] However, one may still say quite a lot about the solutions.

For example, we may note from equation (2.38) that $Y_1(y)$ and hence $Y(y)$ approach finite values at $y = 0$ as opposed to $U_1(\rho)$ which approaches an infinite value, and that there is thus a finite potential barrier $E_B = Y(0)$ which an ion must overcome in order to penetrate through one atomic plane into the next channel. It is evident from Figures 2.7 and 2.8 that E_B depends both on the interatomic potential chosen and on d_p but as a good general approximation:

$$E_B \simeq 2\pi Z_1 Z_2 e^2 (N d_p) a \qquad (2.42)$$

and a measure of the transverse angle necessary to overcome the barrier is:

$$\phi_r = (E_B/E)^{\frac{1}{2}} = \left(\frac{2\pi Z_1 Z_2 e^2 (Nd_p)a}{E}\right)^{\frac{1}{2}} \tag{2.43}$$

This leads to an obvious division between ions with $E_\perp$ above and below the barrier. If $E_\perp < E_B$ the ion is trapped within a single interplanar channel where it oscillates from side to side as it passes down the channel. This is, of course, the channeling effect illustrated by case (1) in Figure 2.9. On the

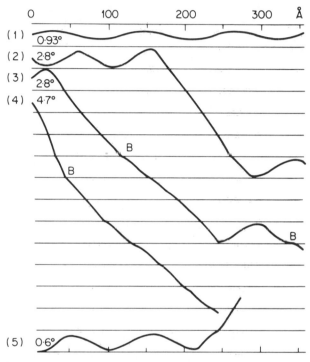

Figure 2.9. Trajectories of 20 keV protons along (010) inter-planar channels in copper. Cases (1) through (4) were initiated from a point on the mid-channel axis at the transverse angle indicated; case (5) was initiated from a lattice site. $\phi_r = 3.1°$

other hand an ion with $E_\perp > E_B$ is free to skip from one interplanar channel to the next, being bent toward the planes as it passes through them and bent away again as it moves past. Thompson[14] has drawn an analogy with an optical beam passing through a medium of periodically varying refractive index.

In either case one may estimate the equilibrium spatial distribution of ions with a given $E_\perp$ since the fraction of time which an ion spends at any point y (provided that $E_\perp > Y(y)$) is proportional to its $v_\perp$ at that point. Hence:

$$P_0(E_\perp, y) = \begin{cases} 0 & E_\perp < Y(y) \\ K(E_\perp)\left(\dfrac{E_\perp}{E_\perp - Y(y)}\right)^{\frac{1}{2}} & E_\perp \geqslant Y(y) \end{cases} \tag{2.44}$$

where $K(E_\perp)$ is a normalization constant.[15]

In contrast with the uniform spatial distribution (see equation 2.19) in the case of axial motion the ions are now more likely to be found where their transverse velocity is lowest, i.e. in those regions of the highest accessible transverse energy. In fact an ion with $E_\perp \geqslant Y(0)$ should spend a greater than normal fraction of its trajectory in the vicinity of the atomic planes where the atomic density is greatest and therefore, on average, it experiences a higher than normal atomic density. This is one of the characteristic properties of 'quasi-channeled' ions, to be discussed in Section 2.4.1.

2.3.3 The Limitations of Continuum Scattering From Planes

The conditions under which the interaction between an ion and an atomic plane may be adequately represented by a continuum potential are less easy to establish than those for atomic strings. Appendix 2C shows that for a trajectory parallel to the plane the uniform scattering rate implied by the continuum field is equivalent to the *average* rate of deflection derived from an impulse model. However, there are large deviations about this average for two reasons: (i) the distance between scattering centres is random, not fixed; and (ii) there is no correlation in the lateral x-component of the impact parameter in successive deflections, only in the transverse component. This produces intrinsic fluctuations in the transverse energy (see equation 2.41) as well as in the lateral momentum components, p_x and p_z. To minimize this effect a large number of scattering events must be included implying that a slow variation in y, i.e. low ϕ, is the essential condition for the application of continuum scattering to planes.

If the maximum permitted transverse angle is ϕ_{crit} it follows that there is also a critical approach distance given by:

$$Y_1(y_{\text{crit}}) = E\phi_{\text{crit}}^2 \tag{2.45}$$

such that for $y < y_{\text{crit}}$ the scattering cannot be represented by a continuum potential.

Lindhard[1] proposes that the continuum scattering remains valid so long as the maximum angle of deflection in any single collision at a distance y from the plane does not exceed the transverse angle required to reach y, so that y_{crit} is found from:

$$\theta(y_{\text{crit}}) = \phi_{\text{crit}} = (Y_1(y_{\text{crit}})/E)^{\frac{1}{2}}. \tag{2.46}$$

This leads to slowly varying values of y_{crit}, generally of the order of the screening radius a at high energies, and increasing with decreasing energy. Due to the relatively slow variation of $Y(y_{crit})$ with y_{crit} the comparable critical angles vary roughly between approximately $0.5\phi_r$ at very low energies up to approximately ϕ_r at high energies.

As was the case with atomic strings thermal vibrations also set an extrinsic limit to the continuum scattering since the distance of closest approach must be large compared with the thermal displacements of atoms out of the plane. Computer simulation calculations by Barrett[11] indicate that in the limit of high energies:

$$y_{crit} \simeq 1.6x_{rms} \tag{2.47}$$

is an approximate relationship. This too leads to critical angles which are of the order of ϕ_r.

2.4 THE CLASSIFICATION OF TRAJECTORIES

Our discussion so far has shown that within certain angular and spatial limits the elastic interaction between an ion and a crystal lattice may be considerably simplified by the introduction of continuum scattering and we have seen how the transverse energy of an ion governs many of the properties of its trajectories. Since the influence of the crystal structures on the actual trajectories is in fact the central issue we now discuss it in detail, beginning with the simpler and perhaps more instructive case of ions near major planes.

2.4.1 Planar Trajectories

Figure 2.9 shows several trajectories calculated by computer simulation which illustrate both governed and ungoverned motion almost parallel to major planes. Cases (1) to (4) were for ions commenced from a point in the middle of the channel with a transverse angle ϕ_i as indicated and a random lateral angle within the plane; case (5) was initiated from a lattice site at a small transverse angle. ϕ_r (equation 2.43) in this instance is $3.1°$.

(1) *Channeled Trajectories*

At relatively low transverse angles as in case (1), $\phi_i \ll \phi_r$ and $E_\perp = E\phi_i^2 \ll Y(0)$, the ions oscillate from side to side in a single interplanar channel as they pass between the planes. In each collision with a channel wall the angle of reflection equals (or very nearly so) the angle of incidence so that the oscillation is stable: the transverse energy and longitudinal momentum are conserved with a relatively small tolerance. Such a trajectory we term a 'channeled' trajectory and is the most simple example of how correlations in the elastic scattering can govern the motion of an ion within a crystal.

Since, from Figure 2.9, a channeled trajectory represents a stable wavelike oscillation one may defined its wavelength or repeat distance by

$$\lambda = 4 \int_{y_c}^{0\cdot5d_p} \left(\frac{E}{E_\perp - Y(y)} \right)^{\frac{1}{2}} dy. \tag{2.48}$$

Hence λ is proportional to $E^{\frac{1}{2}}$ but otherwise depends only the transverse energy and the definition of $Y(y)$. For relatively high values of $E_\perp$ one finds as a reasonable approximation:

$$\lambda \simeq \pi d_p/\phi_r = \left(\frac{\pi E d_p}{2Z_1 Z_2 e^2 Na} \right)^{\frac{1}{2}}. \tag{2.49}$$

Alternatively for very low $E_\perp$ one has:

$$\lambda \simeq 2\pi\{E/\tfrac{1}{2}Y''(0\cdot5d_p)\}^{\frac{1}{2}} = 2\pi\{E/Y_1''(0\cdot5d_p)\}^{\frac{1}{2}} \tag{2.50}$$

$$\simeq \left(\frac{\pi E d_p}{2Z_1 Z_2 e^2 Na} \right)^{\frac{1}{2}} \left(\frac{d_p}{2C^2 a} \right)^{\frac{1}{2}} \tag{2.51}$$

for the Lindhard potential (equation 2.38). Since, in most cases $d_p > 6a$, ions of high transverse energy have shorter wavelengths, but the variation over the range of channeled trajectories is generally less than a factor of 3.[16]

A channeled trajectory has a relatively well-defined distance of closest approach to either wall given by $Y(y_c) = E_\perp = E\phi_i^2$, so that, discounting thermal vibrations, all collisions of impact parameter less than y_c are screened out. A very important property of channeled trajectories is that the probability of a close encounter with an atomic nucleus is virtually nil, the only possibility being that an atom is thermally displaced a distance greater than y_c from a channel wall.

This has several consequences. On average, a channeled ion imparts far less energy to the crystal via elastic collisions than it would in a random medium; the reduction in the elastic stopping power may be orders of magnitude. It also follows that the rate of radiation damage produced by violent elastic collisions with the host lattice atoms is also considerably reduced.

The inelastic stopping power is also reduced relative to the random medium but by a much smaller amount, generally of the order of 50%. The reasons for this are discussed fully in Chapters 4 and 5; briefly, the inelastic stopping is sensitive both to interactions with the evenly distributed conduction electrons and to long range 'resonance' interactions, neither of which is strongly dependent on position within the crystal.

The maximum transverse angle at which channeled trajectories can occur is, by reason of their stability, synonymous with the upper angular limit ϕ_{crit} for which the continuum approximation remains valid. This, as discussed in Section 2.3.3, varies from approximately $0\cdot5\phi_r$ at low energies up to approximately ϕ_r at high energies, but is never precisely defined.

(2) *Random Trajectories*

Consider, on the other hand, the opposite extreme: case (4) in Figure 2.9, where ϕ_i is large compared with ϕ_r, $E_\perp \gg Y(0)$. According to the continuum model such an ion should pass straight through each plane, suffering only a slight deviation as it does so. However, clearly in this event the ion interacts with only relatively few atoms per plane and the impact parameters in these collisions are essentially random and uncorrelated. The average scattering implied by the continuum approximation breaks down completely and the regular crystal structure plays no direct role in governing the trajectory. The properties of the ion are in almost all respects equivalent to those in a perfectly random medium of the same density; e.g. the same stopping powers and the same probability $Nd\sigma$ of a close collision. Such trajectories we refer to as 'random'.

(3) *Quasi-channeled and/or Blocked Trajectories*

In the intermediate region $\phi_i \simeq \phi_r$ (cases (2) and (3) in Figure 2.9) the projected motion often bears some resemblance to the channeled trajectories. However, such ions easily penetrate into the planar core where they 'see' the atoms as individual scattering centres rather than as part of a smeared-out continuum. Nonetheless, an element of correlation does remain in the scattering; thus while approaching a plane each deflection tends to bring the ions parallel to the plane, while on leaving a plane the deflections consistently push them away. The deviations from the continuum scattering, however, are quite significant so that neither the transverse energy nor the lateral momentum are conserved for long and the apparent oscillatory motion is quite unstable. Such trajectories we refer to as 'quasi-channeled'.[17,18]

The lifetime of the quasi-channeled state is controlled by chance. Thus in case (2) the proton suffers a large angle collision on its fourth reflection sending it into an essentially random trajectory, whereas in case (3) with the same value of ϕ_i a similar transition occurs on the first reflection.

Closely associated with the concept of quasi-channeling is that of 'blocking'. In its most strict sense blocking refers to trajectories which originate from a lattice site, case (5) in Figure 2.9; those ions initially directed parallel to the plane are consistently deflected away from it by the first and subsequent atoms and hence 'blocked' from low transverse angles. Since $E_\perp > Y(0)$ for all ions originating from the top of the potential barrier, blocked trajectories have similar transverse energies and hence similar properties to quasi-channeled ions which originate from the centre of the channel with $\phi_i \gtrsim \phi_r$.

Strictly speaking, a quasi-channeled trajectory might be defined as a below-the-barrier trajectory which is reflected back from a plane and a blocked trajectory, as one which passes through. However, owing to the fluctuations in the transverse energy on each reflection the distinction is never precise and transitions from one to the other as well as to and from the random are frequent. Thus case (3) in Figure 2.9 goes from quasi-channeled to a mixture of random and blocked, back to quasi-channeled and finally to blocked again. Since both quasi-channeled and blocked ions have common characteristics, apart possibly from their origin, it is generally convenient to refer to them as a single class.

The most important property of a quasi-channeled/blocked trajectory is that, owing to the partially correlated scattering experienced at the atomic planes, the ions tend to spend a greater-than-normal fraction of their motion within the planar cores and therefore to sample a higher-than-random atomic density. The 'refractive' effect at the planes is clearly illustrated at the points marked B in Figure 2.9.

Thus, in contrast to the channeled ions, a quasi-channeled ion has a *higher*-than-normal chance of a close collision with an atomic nucleus, generally of the order of 50 % above random. While the presence of channeled orbits leads to the characteristic dip in the angular dependence of a close-collision yield it is the quasi-channeled orbits which lead to the equally characteristic shoulders (see Figure 2.15). It also follows that the rates of elastic energy loss and of radiation damage for quasi-channeled orbits exceed the random, again by the order of 50 %. The inelastic energy loss rate is also higher than random but since the inelastic processes are far less sensitive than the elastic losses to small impact parameters the increase is considerably less, of the order of a few per cent.

The upper angular limit at which quasi-channeled states degenerate into random is not well defined but as a general rule is of the order of $1 \cdot 5\phi_r$.

2.4.2 Axial Trajectories

Similarly trajectories which lie nearly parallel to atomic strings range from the completely governed, channeled trajectories at low transverse angles, through a partially governed, quasi-channeled state at intermediate angles, and on to the completely ungoverned or random trajectories at large angles. Representative trajectories, plotted as the projected motion on the $\langle 001 \rangle$ transverse planes, are illustrated in Figure 2.10 for 20 keV protons in copper.

(1) *Proper Channeling*

A subclass of the channeling behaviour may also be distinguished at very low transverse angles, *viz.* 'proper channeling', Figure 2.10(a). The ion is trapped within the central region of a single channel where it executes essentially a helical orbit as it passes down the channel. Since the potential barrier for escape into a neighbouring channel is small (see Figures 2.3 and 2.4), generally of the order of $5Z_1$ eV,[1] and the maximum transverse angle is

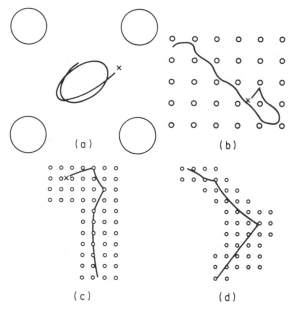

Figure 2.10. Trajectories of 20 keV protons almost parallel to the [001] axis in copper and viewed as projections onto the transverse plane. The circles are the projected positions of the strings. Four different categories are represented: (a) proper channeling, (b) normal channeling, (c) quasi-channeling, and (d) random

approximately $(5Z_1/E)^{\frac{1}{2}} \ll \psi_1$ proper channeling is only likely to be important for near perfect alignment. It may, in fact, be quite influential in determining the maximum flux down the centre of axial channels.[9]

(2) *Channeling*

At higher values of $E_\perp$ the ion is free to wander in the transverse plane, the normal channeled trajectory in Figure 2.10(b). Its orbit is extremely difficult to calculate analytically but its equilibrium spatial distribution has already been given (equation 2.19). The most important features of channeled orbits are the complete correlation of successive scattering centres, stability and conservation of the transverse energy and the longitudinal momentum, and a very low probability for close collisions. As was the case with planes the latter leads to substantial reductions in the rates of elastic energy loss and radiation damage* and, to a much lesser extent, in the rate of inelastic energy loss.

* It is still, however, possible that a stable channeled ion can displace an atom from an atomic string; see Kumakhov[19] and Chapter 11, reference 12 for a discussion of this point.

The maximum transverse angle ψ_{crit} for which the continuum approxima-tion is valid and stable channeled orbits occur has already been discussed in Section 2.4 and is given approximately by the Lindhard characteristic angles ψ_1 (equation 2.34) and ψ_2 (equation 2.35) at high and low energies respectively.

(3) *Quasi-channeled and/or Blocked Trajectories*

Quasi-channeled or blocked trajectories* occur for $E_\perp$ just in excess of $E\psi_1^2$ (or $E\psi_2^2$), Figure 2.10(c). Superficially there is very little difference between the projected motions just above and just below the transition point. In both cases the continuum approximation is valid for motion everywhere in the transverse plane except within the core of the strings; the distinction rests on the fact that the quasi-channeled orbits eventually penetrate the cores and are there-fore unstable while the channeled orbits cannot.

A refractive effect, analogous to that found at atomic planes, also operates within the core of the atomic strings. Thus axial quasi-channeled ions spend a greater-than-normal fraction of their motion in close proximity to the atomic nuclei; their cross section for close collisions is of the order of 50% above random.

Both experiment and computer simulation calculations indicate that quasi-channeling characteristics persist for transverse angles up to approxi-mately twice ψ_1, i.e. $E_\perp \leqslant 4E\psi_1^2$.

(4) *Random*

At transverse angles $\psi \gg \psi_1$ an ion can pass straight through a string without suffering a significant deflection, see Figure 2.10(d), and the regular lattice structure has no direct influence on its trajectory.

2.5 MULTIPLE SCATTERING

Our previous discussions of channeling have always assumed that, within certain spatial and angular limitations, the transverse energy associated with both axial and planar motion is a constant. However, any deviation from a perfect lattice structure, such as thermal vibrations or lattice defects, tends to violate this rule and leads to fluctuations in the transverse energy. Processes of this nature we refer to as 'multiple scattering' and can have important consequences. For example, they control the transition rate from channeled

---

* In the high energy regime $E > E'$ (equation 2.36) blocked and quasi-channeled trajectories are essentially the same but at low energies a distinction may be drawn. If $E\psi_2^2 < E_\perp < E\psi_1^2$ the projected motion is unstable but the ions are unable to penetrate right to the core of the strings where the transverse energy barrier is greater than $E\psi_1^2$. By definition a blocked trajectory is equivalent to one which originates from an atomic nucleus so that over this intermediate regime the trajectories may be considered as quasi-channeled yet not equivalent to blocked.

to quasi-channeled or random trajectories, i.e. 'dechanneling', to be discussed in Chapter 7.

An exact treatment of multiple scattering is extremely complex since both increases and decreases in $E_\perp$ must be allowed for and diffusion-type equations may be required.[1] However, for most cases of interest it appears to be valid to consider that the mean transverse energy increases uniformly and to neglect fluctuations about the mean.

Within the crystal itself multiple scattering arises from several sources which we discuss from the point of view of axial channeling:

2.5.1 Atomic Thermal Displacements*

In any collision between an ion and a lattice atom that is displaced by an amount δ from its equilibrium site the scattering angle, θ, differs slightly from that predicted by the continuum model because of:

(a) A change in the magnitude of θ owing to a change $\delta_{//}$ in the impact parameter s, where $s \simeq \rho$:

$$\delta\theta \simeq \frac{\partial\theta}{\partial s}\delta_{//} \tag{2.52}$$

(b) A change in the direction of the deflection owing to the component of δ normal to the impact parameter:

$$\delta\psi \simeq \theta\delta_\perp/\rho. \tag{2.53}$$

These processes lead, on average, to a change in the transverse energy of

$$\delta\langle E_\perp\rangle(\rho) \simeq \frac{\rho_{\rm rms}^2}{2}E\left\{\left(\frac{\theta}{\rho}\right)^2 + \left(\frac{\partial\theta}{\partial\rho}\right)^2\right\} \tag{2.54}$$

since $\langle\delta_{//}^2\rangle = \langle\delta_\perp^2\rangle = \rho_{\rm rms}^2/2$. Since $\theta(\rho) \simeq -dU_1'(\rho)/2E$ we have for the standard continuum potential (see equation 2.13):

$$\delta\langle E_\perp\rangle(\rho) \simeq \frac{(Z_1Z_2e^2)^2}{E}\frac{\rho_{\rm rms}^2}{2}\left(\frac{\zeta'(\rho/a)^2}{a^2\rho^2} + \frac{\zeta''(\rho/a)^2}{a^4}\right). \tag{2.55}$$

The mean rate of increase of $E_\perp$ with depth may now be evaluated as a function of $E_\perp$ by averaging over the available space in the transverse plane using equation (2.19)†. Lindhard[1] gives:

$$\frac{\partial\langle E_\perp\rangle}{\partial z} \simeq \frac{2\pi Z_1^2Z_2^2e^4}{E}N\frac{\rho_{\rm rms}^2}{C^2a^2}\exp(2E_\perp/E\psi_1)^2(1 - \exp(-2E_\perp/E\psi_1^2)^3 \tag{2.56}$$

* The effect of thermal vibrations on the mean scattering potential $U_1(\rho)$ is discussed in Appendix 2B.
† It should be noted that equation (2.56) represents the rate of increase of $E_\perp$ averaged over relatively long penetration depths. For relatively short δz the deviations can be very significant especially when an ion is near its point of closest approach to the row (see, for example, Section 2.8).

This expression does not hold at very low values of $E_\perp$; for a best channeled ion confined near the centre of an axial channel we use (2.55) with $\rho = \rho_{ch}$, the channel radius, to obtain:

$$\frac{\partial \langle E_\perp \rangle}{\partial z} \simeq n \frac{Z_1^2 Z_2^2 e^4}{dE} \frac{\rho_{rms}^2}{2} \left(\frac{10 C^4 a^4}{\rho_{ch}^8} \right) \tag{2.57}$$

where n is the number of strings surrounding the channel.

2.5.2 Inelastic Multiple Scattering

Since the angular deflections involved in inelastic collisions are considerably smaller than those in elastic collisions the scattering caused by the former only becomes significant under channeling conditions where the primary effect of the elastic steering forces is to conserve the transverse energy.

In a free Rutherford collision between an ion and an electron the transverse deflection ψ_e is related to the energy transfer ΔE_{inel} by:[3,12]

$$\psi_e^2 \simeq \frac{m_e}{m_1} \frac{\Delta E_{inel}}{E}. \tag{2.58}$$

This would lead to a change in transverse energy of $E\psi_e^2 \simeq \Delta E_{inel} m_e/m_1$; hence:

$$\frac{\partial \langle E_\perp \rangle}{\partial z} \simeq \frac{m_e}{m_1} \left(\frac{-\partial E}{\partial z} \right)_{free} \tag{2.59}$$

where we only include the energy lost via free collisions where equation (2.58) applies as opposed to the resonance collisions where it does not. We can allow for resonance effects empirically by writing:

$$\frac{\partial \langle E_\perp \rangle}{\partial z} \simeq c \frac{m_e}{m_1} \left(\frac{-\partial E}{\partial z} \right)_{inel} \tag{2.60}$$

where $c \simeq \frac{1}{2}$ and using experimentally determined stopping powers for $(-\partial E/\partial z)_{inel}$. Equation (2.60) does not include any specific dependence of the energy loss rate on the transverse energy; Lindhard[1] has estimated this by:

$$\frac{\partial \langle E_\perp \rangle}{\partial z} \simeq \frac{1}{2} \frac{m_e}{m_1} \left(\frac{-\partial E}{\partial z} \right)_{inel} \left\{ 1 - \exp \left(\frac{-2E_\perp}{E\psi_1^2} \right) \right\}. \tag{2.61}$$

2.5.3 Intrinsic Errors in the Continuum Approximation

While there are slight deviations in $E_\perp$ arising purely from the elastic scattering from a rigid atomic string (and which, in fact, limit the continuum model at low energies, Section 2.2.4) they are, for all practical purposes, less than the thermal fluctuations (given in equation 2.56) and may in general be ignored for axial channeling.* A full discussion of this point is given in Appendix A of Lindhard.[1]

2.5.4 Crystal Defects

We include in this category all defects which distort or disrupt the atomic strings such as dislocations, stacking faults, point defects, etc. The multiple scattering from such defects has never been examined in detail, although estimates of the total dechanneling probabilities from a variety of defects have been made.[16,18,20]

In a somewhat different category we might include defects such as grain boundaries or, in some circumstances, twin boundaries where the preferred orientation is completely lost after the defect and dechanneling is total.

2.5.5 External Multiple Scattering

We should also mention two sources which limit the definition of transverse energy in any actual experiment but which are external to the crystal: the angular spread of an incident beam or the angular width of a detector, and the scattering from amorphous surface layers, e.g. an oxide or contamination layer. In the former case the mean fluctuation in transverse energy is:

$$\delta \langle E_\perp \rangle \simeq E \langle (\psi_{in} - \bar{\psi}_{in})^2 \rangle \qquad (2.62)$$

In the latter cases it may be expressed in the general form:

$$\delta \langle E_\perp \rangle = E \langle \psi_s^2 \rangle \simeq \gamma E \left(\frac{Z_1 Z_s e^2}{aE} \right)^2 \qquad (2.63)$$

where $\langle \psi_s^2 \rangle$ is the mean square scattering angle in the surface layers. The parameter γ depends on the thickness of the amorphous layers, Δz_s, and is determined by either single, plural or multiple scattering as Δz_s and the number of collisions per ion increases.[†]

The energy dependence and relative magnitudes of the various sources of multiple scattering are illustrated in Figure 2.11 which plots the mean increase in transverse energy for α-particles channeled along $\langle 001 \rangle$ copper and evaluated, arbitrarily, at a depth of 200 Å. The following conditions are represented:

    (i) 0°C, $\rho_{rms} = 0.104$ Å; (The increase due to elastic collisions is given for a best-channeled ion using equation (2.57); $\delta E'_\perp$, for an intermediate transverse energy of $E\psi_1^2/2$ using equation (2.56); and for a random

---

* This does not hold true for interplanar channeling where the intrinsic fluctuations discussed in Section 2.3.3 are comparable with or greater than those induced by thermal vibrations.
† A similar expression may also be applied to isolated impurities within the crystal with $\Delta z_s = c_s \delta z$.

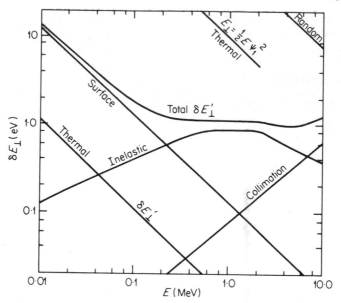

Figure 2.11. The different contributions to the multiple
scattering for α-particles along the ⟨001⟩ channels in
copper as a function of energy. The exact conditions
depicted are given in the text (after Van Vliet[9])

trajectory. All the other sources are taken to be independent of transverse energy.)

(ii) inelastic scattering evaluated from equation (2.60) with $c = \frac{1}{2}$ and
using experimental stopping powers;

(iii) an rms beam collimation of $\pm 0.017°$;

(iv) 4 atomic planes (7·2 Å) of amorphous copper on the surface.*

Practically speaking these represent near optimum channeling conditions
but the curves could be scaled for a less stringent set of conditions. Several
general conclusions are worthy of note:

(1) For large $E_\perp$ the elastic multiple scattering caused by thermal displacements is almost always the most important single factor, although this
also depends on the depth considered relative to the degree of surface scattering and beam collimation. Equally the multiple scattering in a random
medium is dominated by the elastic collisions.

(2) For low $E_\perp$, best-channeled ions, thermal multiple scattering is almost
always of secondary importance except, possibly, at very low energies or

---

* We assume single particle scattering to obtain $\gamma \simeq 0.061^2$. There is a 50% probability of at
least one collision of impact parameter less than $\bar{s}$ where $N\Delta z_s \pi \bar{s}^2 \simeq 0.5$. In this particular
instance $\bar{s} \simeq 0.5$ Å and for the Moliere potential (2.2) $\theta\,(0.5 \text{ Å}) \simeq 0.061\,Z_1 Z_s\, e^2/aE$.

over large penetration depths. $\delta E'_{\perp}$ is controlled by surface scattering at low energies, by the beam collimation at high energies and, if both are relatively good, by inelastic scattering at intermediate energies. For the particular choice of conditions in Figure 2.11 this intermediate regime coincides with the peak in the stopping power so that $\delta E'_{\perp}$ is virtually energy independent over a wide spread of energies; this is not, however, a general occurrence.

2.6 THE RULE OF REVERSIBILITY

Directional effects in crystals as observed experimentally are in general dominated by either channeling or blocking phenomena. For example, in the former an external ion beam is aligned with either a low-index row or plane in a single crystal and, say, the backscattered yield monitored along some random direction; in the latter, see Chapter 9, the ion beam is incident at a random angle and the backscattered yield is detected about the low-index direction. These two modes are closely related by the 'rule of reversibility'.[1]

In its most strict sense reversibility is based on the fact that, according to the laws of classical mechanics, particle trajectories in a fixed potential field are completely reversible as long as the energy losses are negligible. Imagine an ion originating from a point A outside a rigid crystal lattice and reaching point B on the inside; reverse its velocity at this point and, if no energy is lost, it will retrace its steps to A again. In a somewhat looser sense reversibility requires that the probability of an ion going from A to B is the same as that of going from B to A, so that any randomness introduced by thermal vibrations is accounted for.

The rule of reversibility therefore states that the relative intensity of ions originating from a point B inside the crystal and observed at an external angle ψ is equal to the relative flux of ions at B from an external beam incident along ψ. In the particular case where B is an atomic site and ψ lies along an aligned direction we see that the probability of an aligned beam undergoing a close collision with a lattice atom is the same as the probability of an ion originating (e.g. backscattered) from that site emerging along the aligned direction. Hence, in the above example, the angular variation observed in both channeling and blocking experiments are equivalent; experimentally this has been confirmed by Bøgh and Whitton.[21]

Reversibility starts to break down when the energy lost, whether by inelastic or elastic processes, becomes significant. Obviously an ion losing energy between A and B would not regain it going from B to A; equally it would not find the atoms en route from B to A recoiling toward their equilibrium sites prior to the collision. It follows that the rule is best applied over relatively short penetration depths and for light ions where the recoil of lattice atoms is minimized.

It is common practice to use reversibility to simplify the analysis of directional effects; for example, to calculate the number of close collisions from an aligned beam it is easiest to follow the ions backwards from a close collision to the exterior. This particular situation is illustrated in the following two sections: Section 2.7 calculates the distribution in transverse energy of ions emitted from a string relative to a random medium and hence, by reversibility, the relative probability of a close collision as a function of transverse energy: Section 2.8 thereby transforms the distribution in $E_\perp$ for an external beam into a mean collision probability. Note that we are using the concept of transverse energy as the common denominator in relating the properties of ions which originate both internally and externally.

2.7 THE EMISSION OF PARTICLES FROM ATOMS IN A STRING

Consider a large number of ions emitted isotropically from an atomic nucleus which is part of a regular string.[1,22] We wish to calculate $\Pi_{out}(E_\perp)\,dE_\perp$, the probability that the ions emerge from the string with a transverse energy between $E_\perp$ and $E_\perp + dE_\perp$; or, more appropriately, the ratio between Π_{out} for emission from a string and Π_{out} for emission in a random medium. In the latter case $\Pi_{out}(E_\perp) \simeq 1/4E$ for small $E_\perp$. Therefore we define

$$\Pi(E_\perp) = 4E\Pi_{out}(E_\perp) \tag{2.64}$$

as the relative change due to string effects.

Such a calculation is important for two reasons:
(1) The transverse angle at a large distance from the emitting string, ψ_e, is given by:

$$E\psi_e^2 = E_\perp \tag{2.65}$$

so that the probability of a transverse angle ψ_e occurring relative to that in a random medium (i.e. isotropic) is also given by $\Pi(\psi_e) = \Pi(E\psi_e^2)$. Ignoring the interaction with any subsequent strings $\Pi(\psi_e)$ represents to a first approximation the angular distribution of the emitted ions emerging from the crystal.
(2) By reversibility $\Pi(E_\perp)$ is also the relative probability that an ion of transverse energy $E_\perp$ can penetrate near enough the centre of an atomic string to undergo a close collision. (A slight correction must be incorporated when we go from a single string into a lattice of many strings.)

Consider a single string and the emission geometry shown in Figure 2.12. Ion A is emitted from the atom B with energy E, transverse angle ψ and azimuthal angle v. Its transverse energy may be measured at the plane halfway to the next atom:

$$E_\perp = E\psi^2 + U_1(\rho^*) \tag{2.66}$$

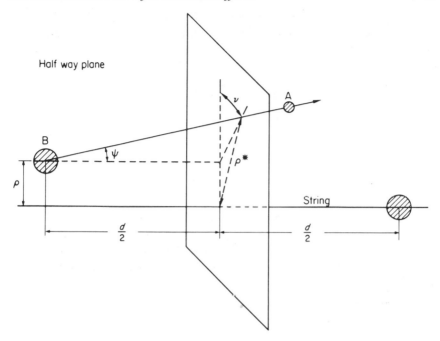

Figure 2.12. Emission of an ion from a string atom and the geometry for the halfway plane model (after Andersen[22])

where

$$\rho^{*2} = \rho^2 + (\psi d/2)^2 + \rho\psi d\cos v$$

and ρ is the thermal displacement of the emitting atom in the transverse plane normal to the string. The distribution in ρ is normally represented by a Gaussian:

$$dP(\rho) = \exp(-\rho^2/\rho_{rms}^2)d(\rho^2)/\rho_{rms}^2 \qquad (2.67)$$

where $\rho_{rms}^2 = 2x_{rms}^2$.

The distribution in transverse energy is given by:

$$\Pi_{out}(E_\perp) = \int_0^\infty dP(\rho)\frac{1}{4\pi}\int_0^\pi\int_0^{2\pi}\sin\psi\,d\psi\,dv\,\delta(E_\perp - U_1(\rho^*) - E\psi^2) \qquad (2.68)$$

and at small transverse energies:

$$\Pi(E_\perp) = \int_0^\infty dP(\rho)\int_0^E d(E\psi^2)\int_0^{2\pi}\frac{dv}{2\pi}\delta(E_\perp - U_1(\rho^*) - E\psi^2). \qquad (2.69)$$

This approach we term the 'halfway plane model'.[22]

A simpler expression may be obtained if one approximates ρ^* to ρ in equation (2.66); i.e. evaluate $E_\perp$ at the point of emission rather than at the first halfway plane:

$$\Pi(E_\perp) \simeq \int dP(\rho)\Big|_{U_1(\rho) < E_\perp} \tag{2.70}$$

For the gaussian distribution this reduces to:

$$\Pi(E_\perp) = \Pi(\psi_e) = \exp(-\rho_c^2/\rho_{rms}^2) \tag{2.71}$$

where

$$U_1(\rho_c) = E_\perp = E\psi_e^2.$$

Typical angular distributions as obtained from the halfway plane model (equation 2.69) and the simplified continuum model (equation 2.71) are shown in Figure 2.13. An inherent error in the latter is that there is no angular compensation for the dip.* However, it does lead to a very simple rule for the angular half width, $\psi_{\frac{1}{2}}$, the transverse angle at which the yield reaches 50% normal:

$$\psi_{\frac{1}{2}} \simeq [U_1\{\rho_{rms}(\ln 2)^{\frac{1}{2}}\}/E]^{\frac{1}{2}}$$
$$= [U_1(1\cdot18x_{rms})/E]^{\frac{1}{2}}. \tag{2.72}$$

The very close correspondence with the critical angle (see equation 2.33) and approach distance (2.31) derived in the limit of high energies by Barrett's[11] computer simulation of channeling is to be noted. The critical approach distance

$$\rho_{crit} = (\ln 2)^{\frac{1}{2}}\rho_{rms} = 1\cdot18x_{rms} \tag{2.73}$$

is therefore a fundamental property of a vibrating string. In the limit of high energies the stability of channeled trajectories is synonymous with a reduced probability for close collisions.

We may set a lower energy limit to the simplified continuum model (equation 2.71). Since setting $\rho^* = \rho$ in equation (2.74) neglects terms in ρ^* of relative magnitude $\psi d/\rho_{rms}$; a reasonable criterion for the validity of equation (2.71), and hence (2.72), is that:

$$\rho_{rms}/\psi_1 d > 1 \tag{2.74}$$

or

$$E > 2Z_1Z_2e^2d/\rho_{rms}^2.$$

---

* Obviously the ions which do not emerge near low ψ_e must emerge elsewhere so that within some shoulder regime $\Pi(\psi_e)$ must exceed unity in order to compensate. Equivalently quasi-channeled ions have a greater-than-normal cross section for close collisions as discussed in Section 2.4.

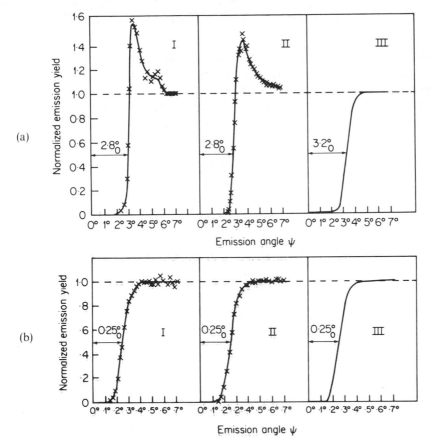

Figure 2.13. The angular distribution $\Pi(\psi_e)$ for string emission as calculated by: (I) computer simulation with the emitting atom only allowed to vibrate; (II) the halfway plane model (equation 2.69); and (III) the simplified continuum model (equation 2.71) for (a) 400 keV protons from [001] string in tungsten at 0 K and (b) for 7 MeV protons from a [101] string in silicon at 300 K (Andersen and Feldman[23])

This corresponds closely with the proposed transition energy E' (equation 2.36) between intrinsically and extrinsically limited channeling.

This limit has been verified by Andersen and Feldman[23] who used computer simulation to generate exact emission profiles $\Pi(\psi_e)$; these are compared with the two analytical expressions (2.69) and (2.71) in Figure 2.13. At the higher energy tested, 7 MeV protons emitted from $\langle 110 \rangle$ silicon strings, the agreement between all three is excellent. However, at the lower energy, 400 keV protons emitted from $\langle 100 \rangle$ tungsten, the halfway plane results are still in good agreement with the 'exact' results, but the simplified continuum model overestimates the angular halfwidth and entirely misses the shoulder.

2.8 THE CLOSE-COLLISION YIELD FOR AN EXTERNAL ION BEAM

Consider a collimated beam of ions entering a crystal at an angle ψ_{in} relative to a set of atomic strings; we ask how the yield χ from a process which requires a close collision varies with ψ_{in}.

On entering the crystal the beam may be divided into channeled, quasi-channeled and random components, their relative populations being determined by the fraction of ions whose transverse energy (equation (2.24)) lies in the regimes: $E_\perp < E\psi^2_{crit} = U(\rho_{crit})$; $U(\rho_{crit}) < E_\perp < 4E\psi^2_1$; and $E_\perp > 4E\psi^2_1$ respectively. For example, Figure 2.14 illustrates the division of the $\{001\}$

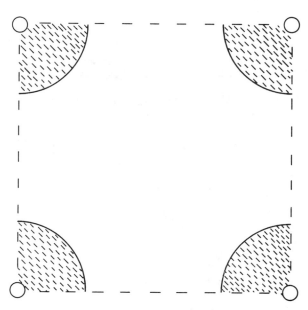

Figure 2.14. The division of the entrance $\{001\}$ transverse plane in copper into regions of channeling and quasi-channeling, the shaded region, for an angle of alignment of $0.95\psi_1$. The random region centred about each string is negligibly small

entrance plane in copper for $\psi_{in} = 0.95\psi_{crit}$ where, for simplicity, we take $\psi_{crit} = \psi_1$. For $\psi_{in} < \psi_{crit}$ the quasi-channeled region is virtually a circle about each string of radius $\hat{\rho}$ where:

$$U(\hat{\rho}) = U(\rho_{crit}) - E\psi^2_{in}$$

or

$$U_1(\hat{\rho}) \simeq U_1(\rho_{crit}) - E\psi^2_{in}. \tag{2.75}$$

Figure 2.15 plots the fraction in each category as a function of ψ_{in} where, again, $\psi_{crit} = \psi_1$. The initial random component is seen to be small until $\psi_{in} \rightarrow 2\psi_1$.

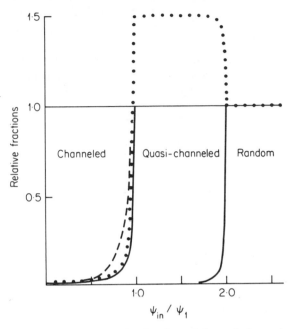

Figure 2.15. The relative fraction of channeled, quasi-channeled and random trajectories as a function of the incident alignment angle, illustrated here for α-particles in $\langle 001 \rangle$ copper. The dots represent the relative probability for a close collision χ from both the quasi-channeled and random ions; the dashed line includes the additional contribution from the channeled ions

From our discussion of the characteristics of quasi-channeled and random trajectories in Section 2.4 and using equation (2.71) for channeled ions the relative probability of a close collision as a function of $E_\perp$ is:

$$\chi(E_\perp) = \begin{cases} 1 \cdot 0 & E_\perp > 4E\psi_1^2 \\ 1 \cdot 5 & U(\rho_{crit}) < E_\perp < 4E\psi_1^2 \\ \exp(-\rho_c^2/\rho_{rms}^2) & E_\perp < U(\rho_{crit}) \end{cases} \qquad (2.76)$$

where $U(\rho_c) = E_\perp$. We may now use equation (2.25) to evaluate the mean value of χ averaged over all incident ions as a function of ψ_{in}. This is also

illustrated in Figure 2.15 and we note the significant contribution from
channeled ions for $\psi_{in} \leqslant \psi_{crit}$.

It is apparent: (1) that the shoulder in the yield is a direct consequence
of quasi-channeling and may be used to study its properties; and (2) that
the angular half width $\psi_{\frac{1}{2}}$ at which a 50% yield is reached is less than ψ_{crit},
here by approximately 5%, and corresponds to a nonchanneled component
of just under one third*.

However, let us turn our attention to the minimum yield, χ_{min}, obtained
at $\psi_{in} = 0$. Here $\hat{\rho} = \rho_{crit}$ and, since the random fraction is negligible, the
contribution to χ_{min} from all nonchanneled ions is:†

$$\chi_2 = 1 \cdot 5 N d\pi \rho_{crit}^2 \tag{2.77}$$

so that, in the limit of high energies as $\rho_{crit} \rightarrow 1 \cdot 2\, x_{rms}$:

$$\chi_2 \simeq 1 \cdot 1\, N d\pi \rho_{rms}^2. \tag{2.78}$$

The channeled contribution arises primarily from ions incident in the
region just beyond $\hat{\rho}$ so we may evaluate it by an integral around a single
string:

$$\chi_1 = N d \int_{\hat{\rho}}^{\rho_0} 2\pi \rho_{in}\, d\rho_{in} \exp(-\rho_c^2/\rho_{rms}^2)$$

$$= N d\pi \rho_{rms}^2 \exp(-\rho_{crit}^2/\rho_{rms}^2) \tag{2.79}$$

since $\rho_c = \rho_{in}$ and $\hat{\rho} = \rho_{crit}$. Writing $\rho_{crit} = 1 \cdot 2 x_{rms}$ gives:

$$\chi_1 \simeq 0 \cdot 5\, N d\pi \rho_{rms}^2. \tag{2.80}$$

We have, however overlooked the fact that those channeled ions for
which $\rho_{in} \gtrsim \hat{\rho}$ are subject to a good deal of multiple scattering so that, on
average, $\rho_c < \rho_{in}$. Since these ions are automatically at their distance of
closest approach, i.e. ρ_{in}, not equally distributed in the transverse plane,
we cannot use equation (2.56) to estimate the fluctuation in $E_{\perp}$. However,
we know from equation (2.55) the mean variation in $E_{\perp}$ per collision, and
also that the number of important collisions near the surface must be of the
order of $\rho_{in}/d\theta(\rho_{in})$. Therefore, the net fluctuation in $E_{\perp}$ from the surface
layers is approximately:

$$\delta\langle E_{\perp}\rangle_s \simeq \frac{\rho_{in}}{d\theta(\rho_{in})} \delta\langle E_{\perp}\rangle(\rho_{in}). \tag{2.81}$$

The new estimate of ρ_c is approximately:

$$\rho_c \simeq \rho_{in} + \delta\langle E_{\perp}\rangle_s/U_1'(\rho_{in}). \tag{2.82}$$

---

* In fact Barrett[11] has shown that $\psi_{\frac{1}{2}}$ may be written as $\psi_{\frac{1}{2}} \simeq k\psi_{crit}$ where $k \simeq 0 \cdot 80$.
† We adopt here the notation of Lindhard[1] who gives $\chi_2 \simeq N d\pi a^2$ and $\chi_1 \simeq N d\pi \rho_{rms}^2$.

Substituting equations (2.82) and (2.81) into (2.79) and using the low limit of the standard potential (equation 2.14) we now obtain twice our previous result:

$$\chi_{1(s)} \simeq 2 N d\pi \rho_{rms}^2 \exp(-\rho_{crit}^2/\rho_{rms}^2) \qquad (2.83)$$

$$\simeq N d\pi \rho_{rms}^2 \qquad (2.84)$$

for $\rho_{crit} = 1\cdot 2 x_{rms}$. This illustrates how sensitive χ_{min} is to deviations from perfect crystallinity in the first few layers and equally how sensitive it would be to amorphous surface layers.*

We may now combine equations (2.78) and (2.84) to estimate χ_{min} as:

$$\chi_{min} = \chi_{1(s)} + \chi_2 \simeq 2\cdot 1 \, N d\pi \rho_{rms}^2. \qquad (2.85)$$

In the limit of high energies the channeled and nonchanneled components contribute approximately equally to χ_{min} and both are proportional to the mean squared thermal displacement amplitude, hence to the temperature at high values. This behaviour has been confirmed qualitatively by Barrett[11] using computer simulation; he, however, obtains a value of $3\cdot 0 \pm 0\cdot 2$ instead of $2\cdot 1$ for the proportionality constant in equation (2.85), probably due to measuring the yield over large penetration depths where significant multiple scattering has occurred†.

In the low energy limit where $\rho_{crit} \to \rho_{min} > 1\cdot 2\, x_{rms}$ the nonchanneled contribution χ_2 (equation 2.77) increases while the channeled contribution χ_1 (equation 2.79) decreases. Hence the minimum yield approaches an intrinsically determined, nonchanneled value:

$$\chi_{min} \to \chi_2 \simeq 1\cdot 5 \, N d\pi \rho_{min}^2 \qquad (2.86)$$

$$\simeq \tfrac{2}{3} N d\pi Z_1 Z_2 e^2 d/E. \qquad (2.87)$$

using equation (2.29).

---

* Lindhard[1] gives the extra contribution to the minimum yield from surface layers as:

$$\chi_3 \simeq \sum_s v_s (Z_1 Z_s e^2/E\psi_{\frac{1}{2}}^2)^2$$

where v_s is the number of atoms per unit surface area of type s.

† We estimate the increase in χ_1 with depth z owing to the equilibrium thermal multiple scattering by using equation (2.56) instead of (2.81) in (2.82). This gives:

$$\chi_1(z) = N d\pi \rho_{rms}^2 \exp(-\rho_{crit}^2/\rho_{rms}^2) \exp[v(z)] \qquad (2.88)$$

where

$$v(z) = 2\pi Z_1 Z_2 e^2 N d z/E$$

valid for $v(z) < 1$. The increase at shallow depths is approximately linear although this may be largely overshadowed by oscillations before equilibrium is attained.

Appendix 2A—Continuum Potentials

The following expressions for the string potential $U_1(\rho)$ (equation 2.6) are derived from the Moliere (equation 2.2), Bohr (equation 2.3) and Born-Mayer (equation 2.5) potentials:

Moliere:[24] $\quad U_1(\rho) = \dfrac{2Z_1Z_2e^2}{d}\{0\!\cdot\!1K_0(6\rho/a) + 0\!\cdot\!55K_0(1\!\cdot\!2\rho/a)$

$$+0\!\cdot\!35K_0(0\!\cdot\!3\rho/a)\}. \tag{2.89}$$

Bohr:[2] $$U_1(\rho) = \frac{2Z_1Z_2e^2}{d}K_0(\rho/a). \tag{2.90}$$

Born–Mayer:[2] $$U_1(\rho) = \frac{2A_{BM}}{d}K_1(B\rho) \tag{2.91}$$

where K_0 and K_1 are the reduced Hankel functions of order 0 and 1 respectively.

These are compared with the Lindhard standard potential (equation 2.13) in Figure 2.2. Since the Born–Mayer potential is defined in a different way from the other three it has been plotted here for the specific case of copper in copper using the Gibson II parameters[25] in $V_{BM}(r)$. ($A_{BM} = 22\!\cdot\!5$ keV and $B = 5 \times 10$ Å^{-1}) and taking $a = 0\!\cdot\!1073$ Å.

The Moliere and Lindhard potentials are in quite close agreement over the entire range of ρ up to approximately $10a$, and may, in general, be used interchangeably. The Bohr string potential fits well at low ρ but falls off much too rapidly at large ρ. The Born–Mayer continuum potential on the other hand is only applicable for large ρ since it approaches an unrealistically finite value as $\rho \to 0$.

Planar Continuum Potentials

The planar continuum potentials $Y_1(y)$ (equation 2.37) for the Moliere and Bohr interatomic potentials are given respectively by:[24]

$$Y_1(y) = 2\pi Z_1Z_2e^2a(Nd_p)\left[\frac{0\!\cdot\!1}{6}e^{-6y/a} + \frac{0\!\cdot\!55}{1\!\cdot\!2}e^{-1\!\cdot\!2y/a} + \frac{0\!\cdot\!35}{0\!\cdot\!3}e^{-0\!\cdot\!3y/a}\right] \tag{2.92}$$

$$Y_1(y) = 2\pi Z_1Z_2e^2a(Nd_p)e^{-y/a} \tag{2.93}$$

A planar continuum potential is not strictly defined for the Born–Mayer potential since the integral in equation (2.37) diverges due to the slow decrease of $V_{BM}(r)$ at large r.

The Moliere and Bohr potentials are compared with the Lindhard standard potential (equation 2.38) in Figure 2.7. The Lindhard and Moliere potentials are once again in good agreement for $y \lesssim 5a$, but at larger values the Lindhard potential decreases much less rapidly. The simpler Bohr potential drops off too rapidly at large y and also approaches a distinctly different maximum value as $y \to 0$.

Appendix 2B—Thermally Modified Continuum Potentials

Erginsoy[24] has pointed out that the average continuum potentials as defined by equations (2.6) and (2.37) should be modified to account for the random displacements of atoms away from their equilibrium sites. Thus one should really define effective continuum potentials by:

$$U_{1T}(\rho) = \int_0^\infty dP(\mathbf{r})U_1(|\boldsymbol{\rho} - \mathbf{r}|) \tag{2.94}$$

and

$$Y_{1T}(y) = \int_0^\infty dP(y)Y_1(y) \tag{2.95}$$

where $dP(\mathbf{r})$ and $dP(y)$ represent the distribution of thermal displacements in the axial transverse plane and the planar transverse direction. For the gaussian distribution (2.67), (2.94) becomes:

$$U_{1T}(\rho, \rho_{\text{rms}}) = \int_0^\infty \frac{dr^2}{\rho_{\text{rms}}^2} \exp(-r^2/\rho_{\text{rms}}^2) \int_0^\pi \frac{d\theta}{\pi} U_1\{(\rho^2 + r^2 - 2r\rho \cos \theta)^{\frac{1}{2}}\} \tag{2.96}$$

Different analytical representations have been given by a number of authors for both U_{1T}[11,26] and Y_{1T}[15,26]. An example of thermally modified row continuum potentials is shown in Figure 2.16. The important changes are that

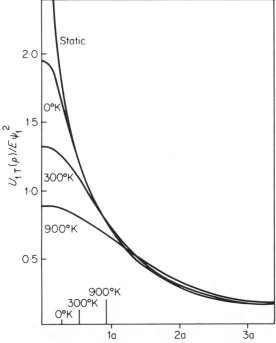

Figure 2.16. Thermally modified continuum potentials for $\langle 001 \rangle$ atomic strings in tungsten (after Andersen and Feldman[23])

$U_{1T}(\rho)$ becomes finite as $\rho \to 0$ but slightly exceeds $U_1(\rho)$ at large separations. Similarly planar continuum potentials become reduced at small y and increased at large y.

Despite their conceptual appeal thermally modified potentials are rarely used in practice. First, equations (2.94) and (2.95) lead to extremely complicated expressions which are difficult to handle. Second, the differences only become significant for $\rho < a$ or $y < a$ and thus do not strongly alter the potential in the range of interest for channeled trajectories. Moreover, at low separations changes in the mean scattering potential are generally very much less important physically than the multiple scattering (equation 2.55).

Appendix 2C—Planar Continuum Scattering

Consider an ion a distance y from an atomic plane and moving a distance Δz parallel to the z-axis in the plane. According to the continuum scattering implied by equation (2.37) the ion is deflected by:

$$\Delta\phi_c = -\frac{Y'_1(y)}{2E}\Delta z \qquad (2.97)$$

Consider, on the other hand, only instantaneous deflections as the ion passes closest to the atoms in the plane. The probability of finding an atom at $(x, x + dx)$ is $(Nd_p) dx \, \Delta z$; the component of the deflection along y is $(y/\sqrt{x^2 + y^2})\theta_i(\sqrt{x^2 + y^2})$ where θ is given by the momentum approximation (2.6); and the average deflection $\overline{\Delta\phi}$ is given by:

$$
\begin{aligned}
\overline{\Delta\phi} &= -(Nd_p)\,\Delta z \int_{-\infty}^{\infty} dx \frac{y}{\sqrt{x^2 + y^2}}\frac{1}{2E}\int_{-\infty}^{\infty} dz \frac{\partial V(\sqrt{z^2 + x^2 + y^2})}{\partial\sqrt{x^2 + y^2}} \\
&= -\frac{\Delta z}{2E}(Nd_p)\int_{-\infty}^{\infty} dx \int_{-\infty}^{\infty} dz \frac{\partial V(\sqrt{z^2 + x^2 + y^2})}{\partial y} \\
&= -\frac{\Delta z}{2E}\frac{\partial}{\partial y}(Nd_p)\int_{-\infty}^{\infty} dx \int_{-\infty}^{\infty} dz V(\sqrt{z^2 + x^2 + y^2}) \\
&= -\frac{\Delta z}{2E}Y'_1(y)
\end{aligned}
\qquad (2.98)
$$

according to our definition (2.37). Hence $\Delta\phi_c \equiv \overline{\Delta\phi}$.

GLOSSARY OF SYMBOLS

a	Screening radius
$C = \sqrt{3}$	Parameter in the Lindhard standard potential (2.1)
d	Distance between atoms in a string

d_p	Distance between planes
e	Electronic charge (in esu)
E	Kinetic energy of particle
$E_\perp$	Transverse energy of particle
m_e	Mass of electron
m_1	Mass of incident ion
m_2	Mass of lattice atom
N	Atomic density
p	Momentum of ion
r	Distance between ion and atom
s	Impact parameter
v	Velocity of particle
$v_\perp$	Transverse component of velocity
x_{rms}	Root-mean-square thermal displacement in one direction
y	Distance of particle from plane
z	Penetration depth
Z_1	Atomic number of incident ion
Z_2	Atomic number of lattice atom
δ	Thermal displacement of lattice atom
λ	'Wavelength' of channeled particle
ρ	Distance of ion from string
$\mathbf{\rho}$	Vector co-ordinate of ion in transverse plane
ρ_c	Distance of closest approach to a string
ρ_{rms}	Root-mean-square thermal displacement in a plane
ν	Azimuthal angle
ϕ	Transverse angle relative to an atomic plane
ψ	Transverse angle relative to an atomic string
θ	Scattering angle in a single collision

References

1. Lindhard, J., *Dansk. Vid. Selsk., Mat. Fys. Medd.*, **34**, No. 14 (1965).
2. Lehmann, C. and Leibfried, G., *J. appl. Phys.*, **34**, 2821 (1963).
3. Bohr, N., *Dansk. Vid. Selsk., Mat. Fys. Medd.*, **18**, No. 8 (1948).
4. Torrens, I. M., '*Interatomic Potentials*', Academic Press, New York, in press.
5. Lindhard, J., Nielsen, V. and Scharff, M., *Dansk. Vid. Selsk., Mat. Fys. Medd.*, **36**, No. 10 (1968).
6. Moliere, G., *Zeits. fur Nat.*, **2a**, 133 (1947).
7. Born, M. and Mayer, L., *Z. Physik*, **75**, 1 (1932).
8. Lehmann, C. and Leibfried, G., *Z. Physik*, **172**, 465 (1963).
9. Van Vliet, D., *Rad. Effects*, **10**, 137 (1971).
10. Morgan, D. V. and Van Vliet, D., *Rad. Effects*, **8**, 51 (1971).
11. Barrett, J. H., *Phys. Rev. B*, **3**, 1527 (1971).
12. Van Vliet, D., 'The Computer Simulation of Ion Channelling', Vol. 5, *Radiation Effects in Solids*, Gordon and Breach, New York, in press.

13. Robinson, M. T., *Phys. Rev.*, **179**, 327 (1969).
14. Thompson, M. W., *Contemp. Phys.* **9**, 375 (1970).
15. Morgan, D. V. and Van Vliet, D., *Rad. Effects*, **5**, 157 (1970).
16. Quéré, Y., *Ann. de Physique*, **5**, 105 (1970).
17. Beeler, Jr J. R. and Besco, D. G., *Phys. Rev.* A, **134**, 530 (1964).
18. Chadderton, L. T. and Krajenbrink, F. G., *Atomic Collision Phenomena in Solids*, Amsterdam, North-Holland, p. 456 (1970).
19. Kumakhov, M. A., *Phys. stat. sol.*, **30**, 379 (1968).
20. Van Vliet, D., *Phys. stat. sol.* (a), **2**, 521 (1970).
21. Bøgh, E. and Whitton, J. L., *Phys. Rev. Lett.*, **19**, 553 (1967).
22. Andersen, J. U., *Dansk. Vid. Selsk.*, *Mat. Fys. Medd.*, **36**, No. 7 (1967).
23. Andersen, J. U. and Feldman, L. C., *Phys. Rev.* B, **1**, 2063 (1970).
24. Erginsoy, C., *Phys. Rev. Lett.*, **15**, 360 (1965).
25. Gibson, J. B., Goland, A. N., Milgram, M. and Vineyard, G. H., *Phys. Rev.*, **120**, 1229 (1960).
26. Appleton, B. R., Erginsoy, C. and Gibson, W. M., *Phys. Rev.*, **161**, 330 (1967).
27. Morgan, D. V. and Van Vliet, D., *Can. J. Phys.* **46**, 503 (1968).

CHAPTER III
Computer Simulation of Channeling

D. V. Morgan

3.1 INTRODUCTION

The techniques of computer simulation (or modelling) have been utilized extensively in scientific research. Digital computers have been used to simulate problems in disciplines as far apart as urban planning and solid state physics. This chapter, will consider the application of computer simulation to the study of the phenomenon of channeling. Channeling, as we have seen, is concerned with the anomalous penetration of energetic ions along the open avenues which exist inside the crystal structure, and involves the interaction of an incident ion with a large number of host atoms. As we saw in Chapter 2, one way of handling the problem is to replace the individual interactions by a continuum potential. An alternative method, to be discussed here, is to consider each interaction individually with a high speed digital computer.

There are two main objectives of computer simulations. Firstly, we have the direct simulation of experiments, here the object is to produce a 'theoretical' analogue for comparison with experiments. Thus the variation of computed parameters with, for example, beam energy or crystal temperature, may be estimated, allowing the physical phenomena causing the experimental variations to be isolated and studied. In the second category, computer simulation is used as an *ideal experiment* on which to test and develop analytical models. Computer simulation closely resembles the nature of an idealized experiment, the results are subject to statistical uncertainties and also must be understood inductively.

3.2 COMPUTER MODELS

To simulate the passage of an ion through a crystal lattice on a computer, we proceed as follows: the three-dimensional structure of the crystal lattice is stored in the computer memory, together with the interaction forces which bind the atoms together. Subsequently, the response of this lattice to the motion of some incident ion is followed. Computer simulation models can be divided into two categories; the N-body (or Vineyard) model, and the binary collision model. We will now discuss these models briefly.

3.2.1 The N-Body Model

Consider a system of N interacting atoms. Let the ith atomic coordinate at time t be $\bar{r}_i(t)$ and the associated velocity be $v_i(t)$ ($i = 1,2,\ldots N$). We assume that the force on the ith atom can be expressed as the sum of pair-wise interaction between the ith atom and all other atoms taken separately.

$$\mathbf{F}_i = \sum_{n \neq i}^{N} \mathbf{F}_{i,n}(\mathbf{r}_{i,n}) \tag{3.1}$$

Letting m be the mass of the atom i, the classical equations of motion for the system are

$$\ddot{\mathbf{r}}_i = m^{-1}\mathbf{F}_i \tag{3.2}$$

$$\dot{\mathbf{r}}_i = \mathbf{v}_i(t) \tag{3.3}$$

One procedure[1] is to replace the time derivations by finite differences with arbitrary interval Δt

$$\ddot{\mathbf{r}}_i(t) \simeq \{\mathbf{v}_i(t + \Delta t/2) - \mathbf{v}_i(t - \Delta t/2)\}/\Delta t \tag{3.4}$$

$$\dot{\mathbf{r}}_i(t + \Delta t/2) \simeq \{\mathbf{r}_i(t + \Delta t) - \mathbf{r}_i(t)\}/\Delta t \tag{3.5}$$

Replacing t by $t + \Delta t/2$ in equation (3.3), inserting (3.4) and (3.5) into (3.2) and (3.3) gives

$$\mathbf{v}_i(t + \Delta t/2) \simeq \mathbf{v}_i(t - \Delta t/2) + \Delta t m^{-1}\mathbf{F}_i \tag{3.6}$$

$$\mathbf{r}_i(t + \Delta t) \simeq \mathbf{r}_i(t) + \Delta t \mathbf{v}_i(t + \Delta t/2) \tag{3.7}$$

Starting with a crystal containing N atoms at positions $\mathbf{r}_i(t)$ at arbitrary time t and corresponding velocities $\mathbf{v}_i(t - \Delta t/2)$, the computer determines the new coordinates $\mathbf{r}_i(t + \Delta t)$ and velocities $\mathbf{v}_i(t + \Delta t/2)$. The procedure is repeated for $(t + l\Delta t)$ ($l = 1, 2, \ldots$, etc.). The accuracy of the procedure depends on the magnitude of Δt, this is set by the criterion that during one iteration no atom moves a long way relative to the range of the interatomic forces. For equilibrium simulations $\Delta t \simeq 10^{-14}$ s while for radiation damage cascades $\Delta t \leqslant 10^{-15}$ s. In practice a compromise must be reached between the desired accuracy and available computer time. The limit to N, the number of atoms, is set by the available store of the computer. A 'time-saving' procedure is to consider only the atoms in the immediate neighbourhood of the event being studied. Since most interatomic potentials fall away rapidly with distance $\mathbf{r}$, generally this means, at most, a sphere with a radius of 4–5 atomic spacings.

Due to the time consuming nature of this model it is generally used for studies where the relaxation of the surrounding lattice drastically alters the damage, displacement energies and atom relaxation at lattice defects (see the review in Reference 3). In channeling studies we are primarily interested with the trajectory of one swiftly moving ion with an energy greater than tens of keV. In these circumstances the ion is un-influenced by the relaxation of a struck atom enabling the binary collision model to be employed.

3.2.2 Binary Collision Model

In this model, originally used for channeling studies by Robinson and Oen[2], an ion is assumed to be moving with fixed velocity and energy in a lattice, and the computer finds the first lattice atom that the ion will strike with an impact parameter less than some cut-off value. The collision is then treated as though it were an isolated two-body collision. The centre of mass scattering angle θ and the time integral τ are determined by solving the standard scattering integrals:

$$\theta = \pi - 2s \int_R^\infty \frac{dr}{\{r^2 f(r)\}} \tag{3.8}$$

$$\tau = (R^2 - s^2)^{\frac{1}{2}} - \int_R^\infty \left\{ \frac{1}{f(r)} - \left(1 - \frac{s^2}{r^2}\right)^{-\frac{1}{2}} \right\} dr \tag{3.9}$$

where

$$f(r) = (1 - s^2/r^2 - V(r)/E_r)^{\frac{1}{2}} \tag{3.10}$$

and s is the impact parameter, R the distance of closest approach, $V(r)$ the interatomic potential, and $E_r = \{m_2/(m_1 + m_2)\}E$; m_1 and m_2 are the masses of the incident and target atom respectively. The outgoing asymptote of the scattered ion is found, and the ion is now considered to move along its asymptote from the point of impact (if the trajectory of the recoiling atom is to be followed its asymptote and acquired energy are also determined.) The next atom is found and the procedure is repeated. The final trajectory consists of a series of straight line segments, any localized curvature being neglected.

Treating the ion–atom interactions as isolated, two-body events involves two assumptions:

(i) The ion must interact strongly with only one atom at a time.
(ii) The interaction of the scattering atom with the rest of the crystal during the actual collision must be unimportant.

In channeling studies, which are concerned with relatively high energy ions, all these criteria are satisfied. A detailed discussion of these points is given in reference (3).

3.2.3 Thermal Vibrations

Thermal vibrations may be introduced into the model by allowing the lattice atoms to vibrate independently about their mean lattice position. The distributions $D(x)$ of the displacements is chosen from the gaussian distribution:

$$D(x) = (2\pi x_{rms}^2)^{-\frac{1}{2}} \exp(-x^2/2x_{rms}^2) \qquad (3.11)$$

or the triangular approximation:

$$D(x) = (\sqrt{6}x_{rms} - |x|)/6x_{rms}^2 \qquad (3.12)$$

where x_{rms} is the root mean displacement in one dimension. The total displacement consists of three such components along the three orthogonal axes.

3.2.4 Interatomic Potentials

The interatomic potential $V(\mathbf{r})$ is central to the calculation of scattering angles and hence the trajectory of a channeled ion. A detailed account of interatomic potentials is given by Torrens[4] and the relevant potentials summarized in Chapter 2.1. The potentials used for channeling calculations take into account only the short range repulsive component, since the high interaction energies concerned ensure that the effect of the attractive component is negligible.

3.2.5 Inelastic Multiple Scattering

The inelastic interactions between the ion and the crystal electrons may be accounted for by a continuous energy loss rate using known experimental values of $(\partial E/\partial z)_{inel}$. In order to account (to a first approximation) for the effects of the dispersion of the channeled beam by the very small deflections arising from the inelastic interactions we proceed as follows:[5] in an inelastic collision in which ΔE_{inel} is lost, the momentum transfer is

$$\Delta p \simeq (2m_e \Delta E_{inel})^{\frac{1}{2}}$$

where m_e is mass of the electron. If the momentum transfer is normal to the ion's velocity then it is deflected by an angle

$$\Delta\psi = \frac{\Delta p}{p} \simeq \left(\frac{m_e}{m_1 E}\right)^{\frac{1}{2}} (\Delta E_{inel})^{\frac{1}{2}} \qquad (3.13)$$

where m_1 is the mass of the ion. In the computed trajectories, each time the ion moves a distance Δz between two elastic collisions it is given an inelastic loss $\Delta E_{inel} = \{(-\partial E/\partial z)_{inel}\Delta z\}$ plus a random deflection normal to its velocity determined by equation (3.13). Strictly speaking ΔE_{inel} in equation (3.13) must be multiplied by some parameter c, where $0 < c < 1$ to correct for the negligible contribution of resonance excitations where the momentum transfer $\Delta p \simeq 0$. The precise value of c could be important in relating computed curves to experimental ones.

We will now consider a selected number of examples chosen to illustrate how computer simulation may be used for channeling studies. The discussion will be selective rather than exhaustive, since the object is to discuss the basic principles of the technique and its flexibility as an auxiliary tool for studying particle–solid interactions.

3.3 CRITICAL ANGLES AND CRITICAL APPROACH DISTANCES

In Chapter 2, we saw that in the development of the continuum model of channeling two fundamental parameters which define the region between channeled and nonchanneled trajectories are the critical approach distance ρ_{crit} and the associated critical angle ψ_{crit}. Continuum scattering breaks down when ions have sufficient transverse energy to penetrate deep into the cores of the rows and 'see' the atoms as individual scattering centres rather than smeared-out rows. Transverse energy is no longer conserved, the projected motion is no longer stable and the particle becomes dechanneled.

Several concepts may be linked with the onset of this breakdown. ρ_{crit} is the critical approach distance necessary for an ion to undergo individual rather than continuum scattering. This is an intrinsic property of each individual row. Secondly, there must be a maximum transverse energy E_{crit} for those ions which cannot penetrate closer than ρ_{crit}, i.e.

$$E_{crit} = U_1(\rho_{crit}) = E\psi_{crit}^2 \qquad (3.14)$$

Thirdly, we see from equation (3.14) that from the value of E_{crit} we define a critical angle ψ_{crit}, which is the maximum transverse angle relative to any single atomic string for which channeling can occur.

3.3.1 Evaluation of ψ_{crit} and ρ_{crit}

Computer simulation has been used to evaluate ψ_{crit} and ρ_{crit} directly, and to compare the results with those obtained from the continuum model. The central feature of the continuum model is the conservation of transverse energy for channeled trajectories. Thus if we fire a large number of ions at an angle ψ to a string of atoms (Figure 2.5), and determine the number of atoms which conserve transverse energy as a function of ψ, the point of transition between channeled and nonchanneled (ψ_{crit}) may be calculated. Ions travelling at the critical angle will have a transverse energy E_{crit} and a critical approach distance ρ_{crit} given by equation (3.14). This procedure for calculating ψ_{crit} is particularly suitable for two-dimensional lattices. Alternatively, ψ_{crit} can be defined as the incident angle for a beam entering the crystal such that 50% of the incident beam is dechanneled after some given distance in the solid. Generally, the stability condition is applied after about 10 oscillations of the incident beam.[6,7]

Table 3.1. Critical limits for axial channeling in a two-dimensional rigid lattice

Z_1	Z_2	E (keV)	d (Å)	ψ_{crit}				ρ_{crit}/a	
				Computed	Theory	ψ_2	ψ_1	Computed	Theory
H$^+$	Cu	1	3·60	7·1°	6·4°	8·8°	—	3·50	3·70
—	—	5	3·60	5·6°	5·7°	5·9°	—	2·11	2·07
—	—	20	3·60	4·5°	4·4°	4·2°	—	1·07	1·13
—	—	100	3·60	2·8°	2·7°	2·8°	—	0·51	0·54
—	—	300	3·60	1·97°	1·90°	—	1·59°	0·28	0·32
—	—	500	3·60	1·62°	1·59	—	1·23°	0·23	0·24
—	—	1000	3·60	1·28°	1·23	—	0·87°	0·14	0·17
—	—	2000	3·60	0·98°	0·94	—	0·62°	0·09	0·12
H$^+$	Au	20	3·60	4·8°	4·9°	4·6°	—	2·31	2·26
—	—	20	4·07	4·6°	4·5°	4·2°	—	2·31	2·38
—	—	20	4·87	4·1°	4·0°	3·7°	—	2·44	2·54
He^{2+}	Cu	20	3·60	5·1°	5·0°	4·9°	—	1·52	1·58
He^{2+}	Au	20	3·60	5·9°	5·5°	5·4°	—	2·72	2·97

3.3.2 Rigid Lattices

In Table 3.1 we illustrate the results for $\langle 100 \rangle$ axial channeling in a simple two-dimensional copper lattice[7]. The angles ψ_1 and ψ_2 are the characteristic angles (see equations (2.34) and (2.35)) proposed by Lindhard[8] and the theoretical values of ρ_{crit} derived from the equation:

$$U_1''(\rho_{\text{crit}}) = 5E/d^2 \tag{3.15}$$

and ψ_{crit} from equation (3.14) (d is the spacing between the atoms in a string).

In the low energy regime the correlation between ψ_2 and ψ_{crit} is good—a one-to-one agreement is not necessary since ψ_2 is only a characteristic angle and ψ_{crit} is dependent to a certain extent on the row geometry. The agreement with ψ_1 at higher energies is not as good, principally because ψ_1 has been calculated assuming a value of $\rho_{\text{crit}} = 0.7a$ (where a is the Thomas–Fermi screening radius), which is limited by thermal vibrations rather than intrinsic effects. It is found that equation (3.15) is of general applicability and may be used for any choice of interatomic or continuum potential.

The validity of equation (3.15) for three-dimensional axes has been established by considering $\langle 001 \rangle$ rows of copper.[7] The results are shown in Figure 3.1 for a variety of ions as well as the theoretical curves given by the approximate solution to equation (3.15) cited in equation (2.29). We see that for small ρ_{crit} the agreement is good, theoretical values being slight over-estimates. Hence equation (2.29) applies at high energies, as a rule of thumb for $E \gg Z_1 Z_2 e^2/a$ (e is the electronic charge).

When ρ_{crit} becomes comparable to the channel radius (~ 1.34 Å for $\langle 100 \rangle$ copper) the channeled ions interact with several rows simultaneously so that the single row approach is no longer valid. Thus at low values of $(Ea/Z_1 Z_2 e^2)$ and high values of ρ_{crit} the theoretical curve is a definite under-estimate. The simulations have a definite experimental relevance since they determine the fraction of a perfectly aligned external beam which undergoes channeling and which does not. The latter is given by $Nd\pi\rho_{\text{crit}}^2$ (where N is the atomic density). This fraction is closely associated with the minimum yield $\chi_{\min}$, in a channeling experiment and will be discussed in (3.5).

Varelas and Biersack[36] have conducted similar calculations for protons and He$^+$ ions in Al, Au, Cu, MgO and LiF. The variation of the critical approach distance as a function of energy is illustrated in Figure 3.1(b). Their results indicate consistently higher values than the analytical estimates of Lindhard[8] and Ibel and Sizmann.[37] However, the analytical expression

$$\frac{4E}{d^2} = \frac{\{U_1''(\rho_{\text{crit}})\}^2 U_1(\rho_{\text{crit}})}{\{U_1'(\rho_{\text{crit}})\}^2}$$

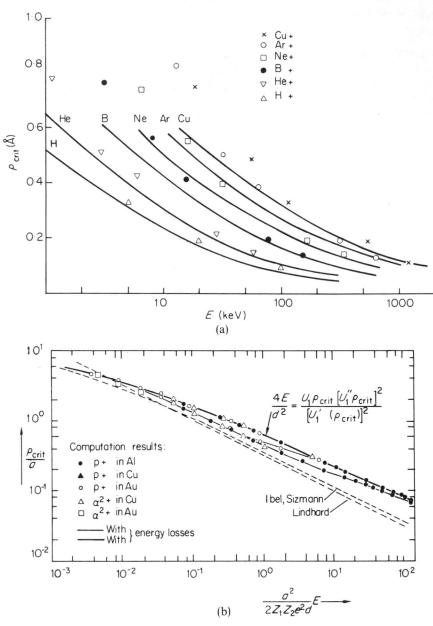

Figure 3.1. (a) The critical approach distance to $\langle 001 \rangle$ rows in copper for a variety of ions as a function of energy. The solid line represents the theoretical values given by equation (2.29). (b) The critical approach distance ρ_{crit} plotted as a function of energy. The computational results are compared with analytical estimates (after Varelas and Biersack

derived by Varelas and Biersack[36] shows excellent agreement with the calculated curves.

3.3.3 Vibrating Strings

Any displacement of atoms from their equilibrium site by thermal vibrations is bound to alter the transverse energy of the channeled ion. Under these conditions, for a stable orbit, the ions must remain further away from the centre of a string than they would for the case of a rigid lattice. Thus ρ_{crit} increases with increasing temperature and since the continuum potential $U_1(\rho_{crit})$ decreases with increasing ρ_{crit}, then from equation (3.14) this leads to a decrease in ψ_{crit}.

It is also obvious that any evaluation of ρ_{crit} must depend on the depth considered. Thus a given rate of change of $E_\perp$ may be tolerable over 200 Å, but the cumulative effects after, say, 1000 Å may render a large fraction of the initially channeled ions unstable (i.e. they become dechanneled). The simulated trajectories indicate that the transition from stable to unstable trajectories (i.e. dechanneling) occurs in two stages: dechanneling over the first few orbits and thereafter a slower dechanneling over a much longer range. By distinguishing between stable and unstable trajectories over a distance approximately equal to $5\rho_{ch}/\psi_{crit}$ (ρ_{ch} is the channel radius) one obtains values of ρ_{crit} which have a significance over larger penetration depths and certain consistencies begin to appear in the results.

A plot of ψ_{crit} against x_{rms} is shown in Figure 3.2 for 5, 20 and 100 keV protons in $\langle 001 \rangle$ copper (two-dimensional lattice). All three curves may be described by an exponential

$$\psi_{crit}(x_{rms}) = \psi_{crit}(0) \exp\left(-\frac{c_1 x_{rms}}{a}\right) \qquad (3.16)$$

A far more fundamental relationship becomes evident when we plot ρ_{crit} against x_{rms} (Figure 3.3). In all cases a linear relationship is obtained:

$$\rho_{crit}(x_{rms}) \simeq \rho_{crit}(0) + c_2 x_{rms} \qquad (3.17)$$

Values of the parameters c_1 and c_2 are listed in Table 3.2. We should note at this point that the values of ρ_{crit} calculated from equation (3.15) agreed to within ($\pm 10\%$) of the values obtained directly from the inspection of the 'critical' trajectories.

These results demonstrate that the change in the critical angle results primarily from a change in the critical approach distance instead of a change in the continuum row potential itself as was proposed by Erginsoy.[9]

Figure 3.4 shows the results obtained for a full three-dimensional lattice for 5, 20 and 100 keV protons in $\langle 001 \rangle$ copper. In this case one has the approximate fit given by

$$\rho_{crit}^2(x_{rms}) = \rho_{crit}^2(0) + (c_3 x_{rms})^2 \qquad (3.18)$$

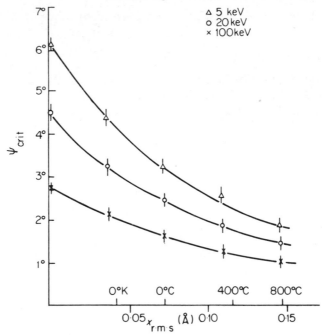

Figure 3.2. The critical angle as a function of x_{rms} for protons between two parallel $\langle 001 \rangle$ rows in a two-dimensional copper lattice. The solid line is the best fit

Table 3.2. Effect of thermal vibrations on two-dimensional rows

Z_1	Z_2	E (keV)	E/A	d (Å)	d/a	c_1	c_2
H	Cu	5	1·74	3·60	24·8	2·4	2·9
		20	6·96	3·60	24·8	2·3	2·7
		100	34·8	3·60	24·8	2·2	2·4
H^+	Au	20	1·89	3·60	34·0	1·5	3·3
		20	1·89	4·07	38·4	1·6	2·9
		20	1·89	4·87	45·9	1·5	2·0
He^{2+}	Cu	20	3·38	3·60	25·4	2·5	3·1

Table 3.3. Critical parameters for planar channeling. Protons in copper; $\{100\}$ planes

E (keV)	R/a	$\bar{d}$ (Å)	Theoretical		Computed		
			y_{crit}/a	ϕ_{crit}	y_{crit}/a	ϕ_{crit}	ϕ_r
5	4·65	4·85	2·32	3·40°	2·55	3·15°	6·19°
10	4·04	5·55	1·86	2·80°	1·82	2·80°	4·37°
20	3·40	6·60	1·49	2·20°	1·43	2·25°	3·10°
100	1·96	11·4	0·92	1·15°	0·98	1·15°	1·38°
300	0·97	23·2	0·78	0·71°	0·72	0·72°	0·80°

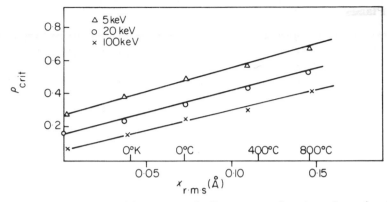

Figure 3.3. The critical approach distance as a function of x_{rms} for the same lattice as in Figure 3.2

where $c_3 \simeq 2$. This curve is also given in Figure 3.4 for comparison. Note that while in the two-dimensional case ρ_{crit} varies linearly with x_{rms}, in the full crystal the relationship is quadratic. The constant c_3 increases with penetration depth considered, or at fixed depth it should decrease slowly with energy.

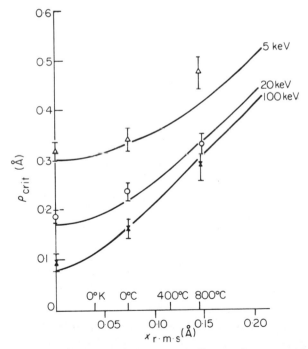

Figure 3.4. The critical approach distance for protons in $\langle 001 \rangle$ copper as a function of x_{rms}. The solid line is the empirical relationship (3.18) with $c_3 = 2$

3.3.4 Planes

The geometry of a planar channel is in many ways quite similar to that of two parallel rows in a two-dimensional lattice but in this case we denote the critical angle by ϕ_{crit} and the critical approach distance by y_{crit}. Values of y_{crit} evaluated for protons in (100) copper[7] are plotted in Figure 3.5 as a

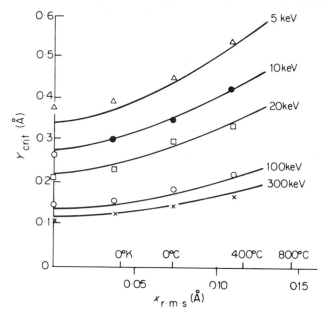

Figure 3.5. The critical approach distance y_{crit} computed for protons in (100) copper as a function of x_{rms}. The solid line is the theoretical curve (3.19)

function of both energy and rms thermal vibrational amplitude. The values are also listed in Table 3.3. The fit between the computed and theoretical values of y_{crit} is seen to be quite close.

For vibrating planes one finds as a purely empirical result of computer simulation that the ratio of $y_{crit}(x_{rms})$ to $y_{crit}(0)$ depends only on the value of x_{rms} and is independent of energy. This yields the simple formula for calculating y_{crit}:

$$y_{crit}(x_{rms}) = y_{crit}(0)\{1\cdot0 + 0\cdot86(x_{rms}/a)^{\frac{3}{2}}\} \qquad (3.19)$$

where the functional dependence on (x_{rms}/a) is the best numerical fit and has not necessarily been given a definite theoretical interpretation. For protons in (100) copper equation (3.19) is valid only for $400\,eV < E < 330$ keV. Preliminary results[10] at 500 keV do in fact indicate a disagreement with equation (3.19). Furthermore, measurements of the planar critical angles for ions of several MeV in tungsten[11] and silicon[12] indicate a better fit with

$$y_{crit}(x_{rms}) \simeq 2x_{rms}.$$

The examples discussed above illustrate one of the chief advantages of computer simulation mentioned in the introduction. Computer simulation may be used to test the validity of analytical models and also to provide the necessary numerical parameters. Furthermore, the analytical model may be extended by the formulation of empirical equations, such as (3.17), (3.18) and (3.19) which extend the applicability of the analytical models.

3.4 BLOCKING AND CHANNELING HALF ANGLES

The parameters discussed in the preceding section are fundamental to the theory of channeling. They are, however, essentially theoretical parameters which cannot be evaluated directly by experiment. We now consider the evaluation of parameters which can be directly measured by experiment and as such we indicate the second application of channeling—named the direct measurement of experimental parameters. Undoubtedly, the most extensive simulation of experiments has been carried out by Barrett.[13,14] In this work the primary objective has been to use realistic parameters (i.e. energy loss, interatomic potential beam divergence, oxide surface films etc.) to evaluate $\psi_{\frac{1}{2}}$ (and χ_{min}) and to compare these results directly with experimental values.

3.4.1 Half Angles ($\psi_{\frac{1}{2}}$)

This, the most frequently measured experimental parameter, arises as follows: if we measure, say, the yield of some close interaction process (i.e. Rutherford scattering or nuclear reaction) as the angle between the incident beam and the crystal axis is varied, then we get the familiar channeling dip. The half angle is the dip angle between the two symmetrical points corresponding to a yield of 50% normal. According to the rule of reversibility (Section (2.6)) such a curve is equivalent to the blocking profile across an axis or plane. Barrett has selected the following set of conditions for an extensive series of calculations of $\psi_{\frac{1}{2}}$: (i) protons in tungsten, (ii) energies between 0·4 and 10 MeV, (iii) beam divergence and mosaic spread equal to zero, (iv) a depth range of 0–5500 Å or less, (v) a tilt plane 12° from (11$\bar{2}$) for the [111] axial dip or a tilt plane 0·75° from [111] for the (01$\bar{1}$) planar dip. Figure 3.6 shows the temperature and energy effects for axial and planar channeling. The best fit curve (full line) is

$$\psi_{\frac{1}{2}} = k\{U_1(mx_{rms})/E\}^{\frac{1}{2}} \tag{3.20}$$

where U_1 is the static string potential (equation 2.6) and k and m are adjustable fitting parameters. For protons, the overall best fit for all energies is

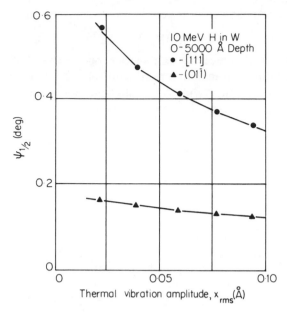

Figure 3.6. Half angles as a function of thermal vibration amplitude for axial and planar channeling. The points are the results of computer calculations and the solid lines from equations (3.21) and (3.22) (after Barrett[13])

$k = 0.80$ and $m = 1·2$. These must be regarded as adjustable parameters which show slight variations with, for example, depth, row geometry, beam divergence, etc. However, the close agreement found for several different materials implies that the variation is minor. Equation (3.20) may be rewritten[14]

$$\psi_{\frac{1}{2}} = k\psi_1\{\zeta(mx_{rms}/a)\}^{\frac{1}{2}} \qquad (3.21)$$

where $\zeta(mx_{rms}/a)$ is a function of the interatomic potential chosen. This equation relates the half angle to the Lindhard angle ψ_1, a correction factor k and a temperature factor $\{\zeta(mx_{rms}/a)\}^{\frac{1}{2}}$.

In an analogous way the results for planar channeling may be expressed as:

$$\phi_{\frac{1}{2}} = k[\{Y_1(mx_{rms}) + Y_1(d_p - mx_{rms}) - 2Y_1(0·5\,d_p)\}/E]^{\frac{1}{2}} \qquad (3.22)$$

Y_1 is the static plane potential (equation 2.37) and d_p the interplanar spacing. In this case, the best fit from Barrett's calculated calculations gives $k = 0·72$ and $m = 1·6$ for 10 MeV protons in (01$\bar{1}$) tungsten.

In addition to a temperature dependence, equations (3.19) and (3.20) predict an energy dependence of the form $E^{-0·5}$. The results for axial

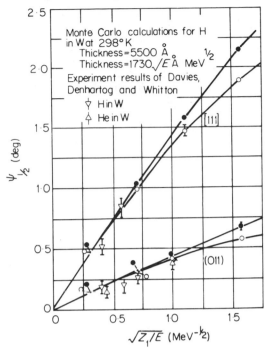

Figure 3.7. Dependence of axial and planar half angles in tungsten on charge and energy of the incident ions (after Barrett[13])

channels (Figure 3.7) set for a constant number of wavelengths (i.e. not constant thickness) predicts a relation proportional $E^{-0.47}$. The difference between this result and $E^{-0.5}$ is small compared with the experimental uncertainties. If $\psi_{\frac{1}{2}}$ is calculated for constant thickness rather than number of wavelengths no single power law can be fitted to the computed results.

Comparison of these results with experiments is summarized in Tables 3.4 and 3.5. In the axial case the calculated values for tungsten tend to be slightly less than the measured ones, particularly for heavy ions. The agreement between computed and measured values is good for other target materials, in particular for the semiconductors Si, Ge and C (diamond). The silicon and germanium agreement is quite close while the computed values for tungsten are consistently too high. Barrett suggests that this is due to the fact that the silicon and germanium results are extrapolated back to zero depth whereas tungsten is not. The problem is that $\psi_{\frac{1}{2}}$ decreases with increasing depth due to multiple scattering.

In the planar results (Table 3.5) the agreement for germanium and silicon is good, whereas the computed values for tungsten are slightly higher than the experimental values.

Table 3.4a. Comparison of calculated and experimental values of $\psi_{\frac{1}{2}}$ for axial channeling. The calculated values were obtained from equation (3.21) using $k = 0.8$ and $m = 1.2$ (after Barrett[13])

Target	Direction	Ion	Energy (MeV)	$\psi_{\frac{1}{2}}$(deg) Calculated	Measured	Ref.
Al	⟨011⟩	H	1.4	0.45	0.42	30
Al	⟨011⟩	H	0.4	0.84	0.90 ± 0.10	31
W	⟨111⟩	He	5.49	0.87	0.85 ± 0.05	32
			6.00	0.83	0.80 ± 0.05	
			7.68	0.74	0.75 ± 0.10	
Au	⟨011⟩	I	60.0	0.80	1.00 ± 0.15	29
W	⟨002⟩	H	2.0	0.95	1.00 ± 0.07	11
			3.0	0.78	0.79 ± 0.07	
			6.0	0.55	0.55 ± 0.07	
		He	2.0	1.34	1.39 ± 0.07	
			10.0	0.60	0.67 ± 0.07	
		C	10.0	0.98	1.10 ± 0.07	
			30.0	0.57	0.64 ± 0.07	
		O	10.0	1.13	1.23 ± 0.07	
			30.0	0.65	0.70 ± 0.07	
		Cl	10.0	1.60	1.82 ± 0.07	
			30.0	0.93	1.00 ± 0.07	
	⟨111⟩	H	3.0	0.83	0.85 ± 0.07	
			6.0	0.59	0.52 ± 0.07	
	⟨024⟩	H	3.0	0.52	0.51 ± 0.07	
		He	10.0	0.40	0.42 ± 0.07	
		C	30.0	0.38	0.36 ± 0.07	
		Cl	30.0	0.62	0.70 ± 0.07	
Si	⟨011⟩	H	3.0	0.30	0.26 ± 0.07	11
Au	⟨011⟩	Cl	20.0	0.84	1.10 ± 0.07	11
Si	⟨011⟩	H	0.25	1.03	1.02 ± 0.06	28
			0.5	0.73	0.68 ± 0.06	
			1.0	0.51	0.53 ± 0.06	
			2.0	0.36	0.36 ± 0.06	
		He	0.5	1.03	1.10 ± 0.06	
			1.0	0.73	0.75 ± 0.06	
			2.0	0.51	0.55 ± 0.06	
Ge	⟨011⟩	He	1.0	0.93	0.95 ± 0.06	28
C(dia)	⟨011⟩	H	1.0	0.53	0.54 ± 0.06	28
		He	1.0	0.75	0.75 ± 0.06	

Andersen and Feldman[17a] have used a single-string binary collision model to compare calculated $\psi_{\frac{1}{2}}$ values with both the continuum model and the halfway plane models. In general, they find that the three models are in good agreement. Feldman[17b] using the same model has compared the computer values with experimental values and again finds agreement between the two to within 20%.

Table 3.4b. Comparison of calculated and experimental values of $\phi_{\frac{1}{2}}$ for planar channeling. The calculated values were obtained from equation (3.22) using $k = 0.72$ and $m = 1.6$ (after Barrett[13])

Target	Direction	Ion	Energy (MeV)	$\phi_{\frac{1}{2}}$(deg) Calculated	Measured	Ref.
W	(002)	H	2·0	0·25	0·22	11
			3·0	0·20	0·17	
			6·0	0·14	0·12	
	(011)	H	2·0	0·32	0·26	
			3·0	0·26	0·22	
			6·0	0·18	0·18	
	(002)	·He	2·0	0·35	0·27	
			10·0	0·16	0·14	
	(011)	He	2·0	0·45	0·38	
			10·0	0·20	0·15	
	(002)	C	10·0	0·25	0·20	
			30·0	0·14	0·12	
	(011)	C	10·0	0·32	0·28	
			30·0	0·18	0·16	
	(002)	O	10·0	0·29	0·25	
			30·0	0·17	0·15	
	(011)	O	10·0	0·36	0·33	
			30·0	0·21	0·17	
	(002)	Cl	10·0	0·40	0·30	
			30·0	0·23	0·22	
	(011)	Cl	10·0	0·50	0·42	
			30·0	0·29	0·25	
Si	(004)	H	3·0	0·07	0·07	11
	(022)	H	3·0	0·10	0·09	
Au	(002)	Cl	20·0	0·27	0·31	11
	(022)	Cl	20·0	0·21	0·24	
	(111)	Cl	20·0	0·30	0·32	
Ge	(022)	He	0·5	0·44	0·40	28
			1·0	0·31	0·30	
			1·9	0·23	0·23	
	(004)	He	1·0	0·23	0·18	
	(224)	He	1·0	0·18	0·20	

3.5 DECHANNELING

Under conditions of near alignment, an incident ion beam may be divided into two components, a channeled and nonchanneled fraction. At the incident surface an abrupt division takes place corresponding to those particles injected into regions around the atomic strings—the random component (i.e. at distances closer than ρ_{crit} to vibrating string), and those injected at the remaining area of the channel corresponding to the channeled fraction. As we saw in Chapter 2, there will be some particles lying in between these two extremes (quasi-channeled ions) which are stable for possibly one or two oscillations but are intrinsically unstable—we will regard such particles as nonchanneled. The remaining channeled ions are slowly de-channeled due to the nonperfect nature of crystals, the dechanneling taking place over relatively long distances in the crystal.

3.5.1 Minimum Yield χ_{min}

The minimum yield corresponds to the yield of some close interaction process when the incident beam is aligned exactly with the axes or plane. A theoretical background to the contributions to χ_{min} was discussed in Chapter 2. Two factors contribute to the minimum yield; firstly the distribution of ions among the transverse energy states, and secondly, each transverse energy state can be assigned an effective factor which reflects the fraction of time spent by a particular trajectory in the region where a close interaction may occur. From our previous discussion this corresponds to regions where $\rho < \rho_{crit}$ for a string and $y < y_{crit}$ for a plane. Both these parameters are readily calculated by computer simulation.[14]

Consider a beam of ions incident onto a crystal surface. The close interaction yield from the surface monolayer is unaffected by the alignment, since these atoms are not screened. Similarly, when the surface is covered by an amorphous oxide, the yield from this region is equal to the random yield. Once past the surface the random and quasi-channeled components contribute to χ_{min}. Generally, the random component increases with depth due to dechanneling. For this reason, the measurement of χ_{min} with depth has been used to calculate dechanneling rates (see Chapter 7). For the moment we will ignore a depth dependence arising from the oscillatory nature of the channeled ions.

In Figure 3.8 we illustrate Barrett's[14] computed values of minimum yield as a function of temperature for 3 MeV protons in tungsten. The calculated minimum yields are averaged between 340 and 1370 Å (thus avoiding the surface peak) and the results can be fitted to the curve

$$\chi_{min} = Nd\pi\{C(\Delta)\rho_{rms}^2 + C'(\Delta)a^2\} \tag{3.23}$$

where Δ is the variance of the gaussian distribution of the beam directions used in the simulations and ρ_{rms} is the root mean square thermal displacement in a plane. A least-squares fit to the calculations for several ion-target combinations, for $\Delta = 0°$ yields values of $C(0) = 3·0 \pm 0·2$ and $C'(0) = 0·2 \pm 0·1$. Thus

$$\chi_{min} \simeq Nd\pi C(\Delta)\rho_{rms}^2 \tag{3.24}$$

The energy dependence of χ_{min} is illustrated in Figure 3.9 for two different crystal temperatures. The depth interval sampled is chosen to include the

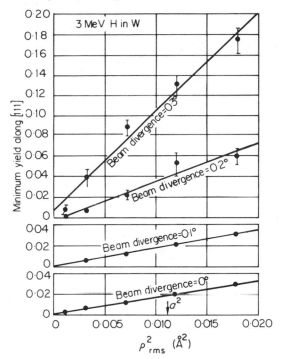

Figure 3.8. Minimum yield for axial channeling in tungsten as a function of thermal vibrational amplitude (after Barrett[13])

same number of oscillations of the channeled ions at each energy. In the high energy limit $\chi_{\min}$ approaches the constant value given by equation (3.24) but increases with decreasing proton energy below 1 MeV.

Closely associated with the minimum yield, is the effective number of surface layers, L, which contribute to a normal yield. Barrett[13] defines L as the area under the surface peak, divided by the row spacing d. Figure 3.9(a) shows the energy dependence of L and Figure 3.9(b) its temperature dependence. At high energies L approaches the limit:

$$L \simeq \zeta = \kappa x_{\mathrm{rms}}/\psi_{\frac{1}{2}}d \tag{3.25}$$

where the proportionality constant κ is determined empirically to be 2·2. In the limit of high energies $L \propto E^{\frac{1}{4}}$ and, after inserting the temperature dependence of $\psi_{\frac{1}{2}}$, $L \propto x_{\mathrm{rms}}^{1\cdot2}$ for large vibrational amplitudes.

At very low energies only one collision is necessary to deflect a non-channeled ion clear of the first row so that L must approach the limit 1 at low energies. The simplest empirical function which describes L at both high and low energies is:

$$L = (1 + \zeta^2)^{\frac{1}{2}} \tag{3.26}$$

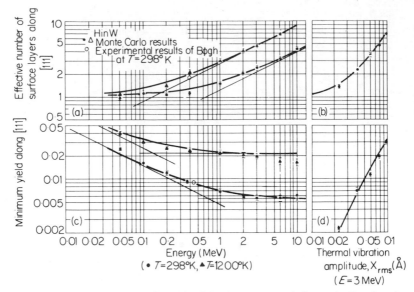

Figure 3.9. Dependence of χ_{min} and L on energy and thermal vibrational amplitude (after Barrett[13])

The continuous curves plotted in Figures 3.9(a) and (b) are derived from equation (3.26).

By assuming a relationship of the form

$$\chi_{min} \propto L\psi_{\frac{1}{2}} \propto L/\zeta \tag{3.27}$$

equation (3.24) can be generalized to:

$$\chi_{min} = C(\Delta)Nd\pi\rho_{rms}^2(1 + \zeta^{-2})^{\frac{1}{2}} \tag{3.28}$$

This equation is plotted as the continuous curves in Figure 3.9(c) and (d) and is seen to give a good overall fit with the computed results.

Comparisons between equation (3.28) and experimental results are somewhat hazardous.[13] A close agreement is found with what are felt to be the two most accurate determinations of χ_{min} in tungsten: Bøgh's[18] value of $0.9 \pm 0.1\%$ for 0.4 MeV protons along $\langle 111 \rangle$, and Appleton and Feldman's[19] value of $1.0 \pm 0.1\%$ for 1 MeV He$^+$ along $\langle 011 \rangle$. The agreement, however, may be rather fortuitous, hingeing on the choice of ρ_{rms} which is not accurately known. Moreover, the values of L given by computer simulation are less than the experimental values in both cases.

3.5.2 Depth Dependence of χ_{min}

We conclude this discussion of minimum yield by considering, in detail, its depth dependence. This example illustrates one important advantage of computer simulation over the continuum model: computer simulation considers the details of the ion trajectories and not an average effect as deduced from the continuum model.

The minimum yield as measured at a fixed depth behind the crystal surface shows strong oscillations which arise from the oscillatory nature of the quasi-channeled ions. Thus Figure 3.10[13] plots the close collision probability P as a function of depth for 0·4 MeV protons along the $\{1\bar{1}1\}$

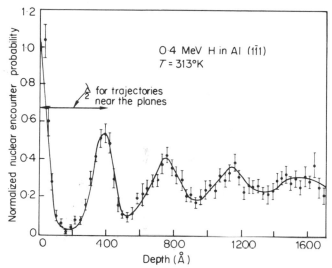

Figure 3.10. Close collision probability as a function of distance from the surface for planar channeling (after Barrett[13])

planes in aluminium. At the surface the normalized yield must approach unity as the ions strike the surface atoms with random impact parameters and a few collisions are required to establish any correlations. This region constitutes the 'surface peak'. Just behind the surface the ions sort themselves out into channeled trajectories, which being rather unlikely to undergo close collisions need not concern us for the moment, and the more relevant nonchanneled (mostly quasi-channeled) trajectories which do. The latter land in the vicinity of a plane from which they are immediately repelled, producing a minimum yield just behind the surface as they cross the empty interplanar channel and a second maximum as they all strike the next plane in phase. If these ions may be characterized by a wavelength λ then successive maxima should appear at depths corresponding to $\lambda/2$, λ, $3\lambda/2$, etc., and minima, at $\lambda/4$, $3\lambda/4$, etc. The oscillation tends to die away with increasing depth as the trajectories grow out of phase due to both the strong multiple scattering experienced by nonchanneled ions and the basic variations in the wavelength due to the wide range of transverse energies associated with the different ions.

Van Vliet[3] has derived a simple analytical expression for the wavelength

$$\frac{\lambda}{2} = \left(\frac{\pi}{2}\right)\frac{d_p}{\phi_r} \tag{3.29}$$

where ϕ_r is given by

$$E\phi_r^2 = 2\pi a(Nd_p)Z_1Z_2e^2$$

For the system in Figure 3.10 this yields a value of $\lambda/2 \simeq 400$ Å in fair agreement with the computed value.

The axial situation is more complex, since neither channeled nor quasi-channeled trajectories have a simple oscillatory motion. The computed depth dependence of χ_{min} is illustrated in Figure 3.11. The yield of unity at the surface and the first deep minimum are retained but the subsequent peaks are irregular. The first maximum corresponds to quasi-channeled ions striking a nearest-neighbour row; the second, to second nearest-neighbour rows; and so forth. The damping is considerably faster in the axial case.

A simple estimate for the distance of the first peak below the surface is:[3]

$$\lambda/2 \simeq d_1/\psi_1 \tag{3.30}$$

where d_1 is the distance between neighbouring rows. For the system illustrated in Figure 3.11 equation (3.30) yields $\lambda/2 \simeq 325$ Å, in qualitative agreement with Barrett's calculations.

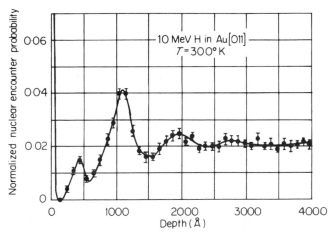

Figure 3.11. Close collision probability as a function of distance from the surface for axial channeling (after Barrett[13])

It is clear from this discussion that the equilibrium minimum yield, χ_{min}, may either be measured at a sufficiently deep depth or else be averaged over some interval which is large compared to the wavelength.

3.5.3 Dechanneling at Defects

The dechanneling considered so far has been confined to a crystal whose only defect has been thermal vibrations and surface oxide films. We will now consider the contribution of lattice defects to dechanneling. Morgan and Van Vliet[20a] define the dechanneling probability as:

$$P = \frac{(n_{ch} - n'_{ch})}{n_0} \tag{3.31}$$

and a dechanneling cross section as

$$\sigma = PA_0 \tag{3.32}$$

where n_{ch} and n'_{ch} are the channeled fluxes in a perfect and imperfect crystal respectively, n_0 is the total flux and A_0 the channel cross-sectional area. Clearly, from the discussion of the flux distribution across a channel (Chapter 2) P and σ are sensitive functions of the position of a defect in the channel. A detailed discussion of the dechanneling owing to stacking faults, interstitials and dislocations is given in references (20a and b). We will confine our discussion to one example—dechanneling at dislocations.

Dechanneling near a dislocation differs from that at, say, a stacking fault or an interstitial atom in that the primary cause is the curvature of entire rows (or planes) rather than scattering from a single displaced atom.

Consider a dislocation with a burgers vector $\mathbf{b} = a_0/2 \langle 110 \rangle$ lying along $\langle 100 \rangle$ in a fcc lattice (Figure 3.12). This dislocation could never exist in real copper but was chosen because of its relative simplicity to provide some insight into elementary analytic ideas regarding dislocation dechanneling. For the same reasons, the pure edge dislocation of burgers vector $\mathbf{b} = a_0/2 \langle 010 \rangle$ Figure 3.12(b) was generated.

By confining the ions to a (100) plane normal to the dislocation the effect of the dislocations in a two-dimensional lattice was studied. An analysis of the trajectories of the ions in the two-dimensional crystal suggest two qualitative conditions for dechanneling:

   (i) The degree of distortion in the channel walls must exceed some critical value.

   (ii) The ion must interact closely with a wall at some point where the minimum distortion is exceeded.

In Figure 3.13 the dechanneling probability P in a rigid lattice is shown as a function of r_0, the distance of the mid-point of the channel from the dislocation for 5, 20, 100 and 500 keV protons in copper. Note firstly that

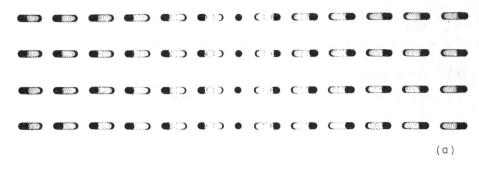

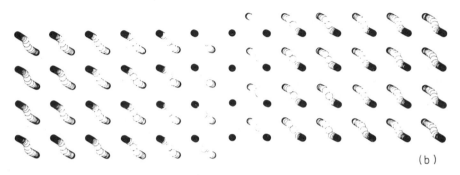

Figure 3.12. (a) A dislocation of Burgers vector $\mathbf{b} = (a_0/2)\langle 110 \rangle$ lying along $\langle 100 \rangle$ and viewed down $\langle 001 \rangle$. The rows in the central (010) plane terminate at the dislocation, all others twist by it. The radius of each atom is 0·18Å. (b) A pure edge dislocation of Burgers vector $\mathbf{b} = (a_0/2) \langle 010 \rangle$ lying along $\langle 100 \rangle$ and viewed down $\langle 001 \rangle$

as one moves towards more gradual distortion (large r_0) P decreases. Secondly, the width clearly increases with energy. At no point is dechanneling 100% complete, the maximum in all cases being approximately 50%. At high energies, for example, we see that 500 keV protons can pass between two atoms in the very distorted rows near the core without suffering a large angle deflection—hence the decrease in P for small r_0. In Figure 3.13 the results are also compared with an analytic theory proposed by Quéré.[21] Here the dislocation is surrounded by a cylinder of radius $(\lambda/2)$ within which dechanneling is inevitable where

$$\frac{\lambda}{2} = \left(\frac{2Eb}{3\cdot 6 U'_1(\rho_{\text{crit}})} \right)^{\frac{1}{2}} \tag{3.33}$$

The solutions of $\lambda/2$ at 5, 20, 100 and 500 keV taking $\rho_{\text{crit}}(\chi_{\text{rms}}) \simeq \rho_{\text{crit}}(0)$ are indicated. It is seen that $\lambda/2$ is typical of the range of dechanneling, illustrating

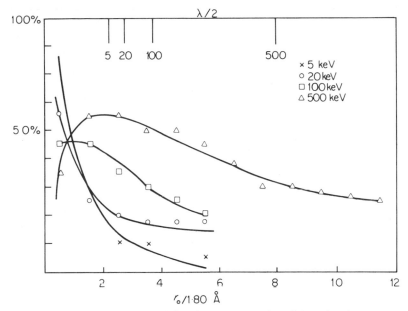

Figure 3.13. The dechanneled fraction near an edge dislocation in a two-dimensional lattice as a function of r_0, the separation between the channel and the dislocation. The theoretical widths, $\lambda/2$, given by Quéré (reference 21) are indicated

that Quéré's criterion produces a good estimate of the minimum distortion required. However, the dechanneling cylinders about a dislocation are neither sharply defined nor 100% opaque; any theory of dislocation dechanneling must also consider the probability of a close interaction (i.e. condition (ii)).

Dechanneling out of $\langle 100 \rangle$ axial channels in three dimensions is similar to that in two. The important difference from two dimensions (Figure 3.14) is that the probability per channel decreases with energy and is consistently lower—a few per cent. at 500 keV. The reason lies in the second condition given above. Now provided that the wavelength of the channeled ion is considerably greater than the range of the distortion parallel to the axis the probability of the two intersecting, P', is the ratio of the distorted row projected onto the transverse plane relative to the cross-sectional area per row (Nd). Thus

$$P' \simeq \rho_{\mathrm{crit}} Nd |\mathbf{b} \times \mathbf{n}| \tag{3.34}$$

where $\mathbf{n}$ is a unit vector in the channel direction. P' is plotted as a solid line in Figure 3.14 (using $\rho_{\mathrm{crit}} = \rho_{\mathrm{crit}}(0)$). At $E > 100$ keV the correlation between P' and P is good but at low energies P' is definitely an underestimate.

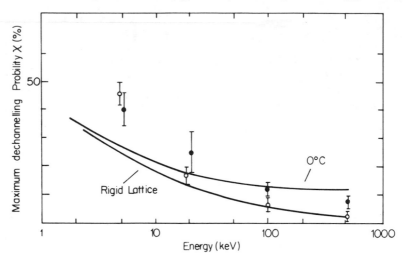

Figure 3.14. The maximum dechanneled fraction from $\langle 001 \rangle$ channels near a dislocation Burgers vector $\mathbf{b} = a_0/2\langle 110 \rangle$, for a rigid lattice. The solid line is equation (3.34)

This example illustrates clearly how computer simulation can provide an 'ideal' experiment which can provide a test for the validity of analytic estimates (i.e. such as equations (3.33) and (3.34)). Furthermore, computer simulation could provide a similar function in understanding much of the present experimental and theoretical work on dechanneling (Chapter 7). A region in which computer simulation would be advantageous is binary crystals, where theoretical studies are difficult due to the existence of two different atomic species.

3.6 FLUX ENHANCEMENT

We can understand the origin of flux peaking in the following way. An ion entering the crystal at a point $\mathbf{r}_{in}$ with a transverse angle ψ_{in} has a transverse energy:

$$E_{\perp} = U(\mathbf{r}_{in}) + E\psi_{in}^2 \qquad (3.35)$$

and, neglecting multiple scattering for the moment its subsequent motion is restricted to those portions of the transverse plane where $U(\mathbf{r}) < E_{\perp}$.

For perfect alignment ($\psi_{in} = 0$) it is evident that the centre of the channel where $U(\mathbf{r}_0) = 0$ is accessible to all ions and consequently samples a higher-than-normal flux. Similarly, positions of higher $U(\mathbf{r})$ near the atomic strings are only accessible to those ions of high transverse energy (those which had initially landed closer to the atomic strings in fact) and therefore they sample a much reduced flux of channeled ions. This, in qualitative terms, is the origin of the considerable variation observed in Figure 3.15.

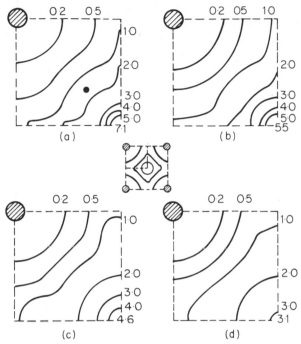

Figure 3.15. Computed contour plots of the flux of 1 MeV α-particles in the ⟨001⟩ copper transverse plane for perfect alignment and averaged between 0 and 1080 Å. (a) With no multiple scattering; (b) lattice vibrations corresponding to 0 °C; (c) 0 °C plus inelastic multiple scattering with $(\partial E/\partial z)_{inel} = 42$ eV Å^{-1} and $c = 1$; (d) 0 °C plus a beam collimation of $\pm 0.23°$

These arguments can be made more precise since we know from the work of Lindhard[8] that once statistical equilibrium has been reached ions of a given $E_\perp$ have an equal probability of being found anywhere within their accessible area. Knowing the distribution of $E_\perp$ (e.g. from equation 3.35) one may calculate the integrated probability of finding an ion at any position in the transverse plane. Such quantitative calculations were described in detail by Van Vliet[23] and verified by comparable computed profiles, not only in the ideal case of perfect alignment but also for finite angles of alignment and multiple scattering.

Using the continuum model[22,23] the average flux outside an equipotential contour passing through the point **r**, for the case of perfect alignment is

$$F(\mathbf{r}) = [1 + \ln\{A_0/A'(\mathbf{r})\}] \qquad (3.36)$$

where A_0 is the cross-sectional area of the channel and $A'(\mathbf{r})$ the cross-sectional area inside the equipotential curve passing through $\mathbf{r}$.

The maximum value occurs in the centre of the channel where $U(\mathbf{r}_0)$ is a minimum. In these circumstances:

$$F(\mathbf{r}_0) = \{1 + \ln(A_0/\Delta A)\} \tag{3.37}$$

ΔA does not go to zero since flux must be sampled over a finite area. For example, if one were to measure the backscattering yield from an interstitial whose equilibrium site coincided with $U(\mathbf{r}_0) = 0$, thermal vibrations would cause it to sample the mean flux over an area $\Delta A \simeq \pi\rho_{rms}^2$. Alternatively, multiple scattering would cause the best channeled ions to be spread over the flat region in the centre of the channel. Van Vliet[3] related the area ΔA to the continuum potential and the multiple scattering by $\Delta A = (\pi\delta E'_\perp/k)$ where $\delta E'_\perp$ is the fluctuation in transverse energy near the potential minimum and k is defined by equation (3.38). The maximum flux $F(\mathbf{r}_0)$ is therefore solely determined by three factors: the size of the channel, the curvature of the continuum potential at the minimum and the degree of multiple scattering. There are several basic causes of multiple scattering: (i) beam collimation, (ii) random deflections in amorphous surface layers (see Section 2.5), (iii) thermal vibrations, (iv) inelastic multiple scattering and (v) imperfections such as radiation damage and impurities. The continuum model assumes that statistical equilibrium has been reached and hence ignores the possibility of variations with depth. For comparison with the analytical equations the computed profiles are averaged over a depth interval between the surface and some distance (~ 1080 Å) into the solid.

3.6.1 Comparison of Analytical and Computer Calculations

We will now compare the theoretical expressions based on the continuum model with the results of computer simulation.

Figure 3.15 shows the four different profiles plotted as intensity contours for 1 MeV α-particles in the $\langle 001 \rangle$ copper channel. Only one quarter of the channel is shown but the rest of the profile is symmetric as illustrated by the inset. Figure 3.15(a) represents a rigid lattice with no thermal vibrations, no inelastic scattering and perfect collimation; Figure 3.15(b), (c) and (d) illustrate the effects of introducing thermal vibrations, inelastic scattering and finite beam collimation respectively. Due to the statistical nature of the results the contours are somewhat qualitative but the large variation of the flux across the channel is quite evident.

The relative flux is greatest in the centre of the channel and is very sharply peaked; the figure in the lower right hand corner gives the maximum flux, $F(\mathbf{r}_0)$ recorded in each case. Note also that the flux is minimized near the row positions. The primary effect of introducing dispersive forces is to reduce $F(\mathbf{r}_0)$ and to spread the central peak.

The similarity between the potential contours of Figure 2.3 and the flux contours of Figure 3.15 is indicative of the close relationship which exists between the two.

Van Vliet[23] has conducted a detailed comparison between the computer calculations and the continuum model and concludes that there is no reason to doubt the validity of expressions (3.36) and (3.37). Hence, for symmetric channels, the flux at a given position $\mathbf{r}$ is uniquely determined by the transverse potential at the point $U(\mathbf{r})$.

3.6.2 Depth Oscillations

So far we have ignored any possible variation of the flux profile with depth, as we have considered only the equilibrium flux as averaged over long depth intervals. It is, however, obvious that the flux must be uniform at the incident surface and that the flux variation requires a certain depth to develop. We saw earlier that striking depth effects are known in planar channels where the channeled trajectories have an oscillatory motion with a well defined wavelength (also see Chapter 6). An ion which enters a crystal parallel to the plane, crosses the halfway plane at depths of $\lambda/4$, $3\lambda/4$, etc., and returns to its entry point at $\lambda/2$, λ, etc. This 'in phase' motion causes the planar flux profile to vary between a uniform distribution and one which is strongly peaked at the centre of the channel, the separation being $\lambda/4$.

A comparable behaviour is not normally expected for axial channels where the ions normally execute complex two-dimensional oscillations so that the simple concept of a fixed wavelength is not generally applicable. However, best channeled ions which are confined near the centre of a single channel can execute a simple oscillatory motion. Van Vliet[23] shows that in those channels where $U(\mathbf{r})$ is symmetrical about the midchannel axis (e.g. $\langle 001 \rangle$ in fcc crystals but not $\langle 101 \rangle$) the continuum potential near the axis is given to a good accuracy by the 'harmonic bowl approximation':

$$U(\mathbf{r}) = k(\mathbf{r} - \mathbf{r}_0)^2 \qquad (3.38)$$

For the Lindhard standard potential one may show that:

$$k \simeq \frac{2Z_1 Z_2 e^2}{d} \left(\frac{3C^2 a^2}{\rho_{ch}^4} \right) \qquad (3.39)$$

where ρ_{ch} is the channel radius and $C^2 \simeq 3$. In this limit each trajectory has a strict wavelength given by

$$\lambda = 2\pi (E/k)^{\frac{1}{2}} \qquad (3.40)$$

and oscillates back and forth through the channel axis, giving rise to similar flux oscillations to the oscillations in planar channels. Substituting equation

(3.39) into (3.40) gives:

$$\frac{\lambda}{4} \simeq \frac{\rho_{ch}}{\psi_1}\left(\frac{\pi^2 \rho_{ch}^2}{3nC^2 a^2}\right)^{\frac{1}{2}} \tag{3.41}$$

where n is the number of strings surrounding the channel.

The computed flux profiles confirm the existence of strong depth oscillations in the midchannel flux along axial channels. This is illustrated in Figure 3.16 for 1 MeV α-particles along $\langle 001 \rangle$ copper. With no multiple

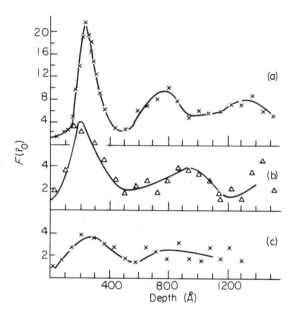

Figure 3.16. The variation of $F(\mathbf{r}_0)$ with depth for 1 MeV α-particles in $\langle 001 \rangle$ copper. (a) No multiple scattering; (b) 0 °C, inelastic multiple scattering and beam collimation of $\pm 0.06°$; (c) as (b) but with a beam collimation of $\pm 0.23°$

scattering, curve (a), an initial peak of more than 20 times normal is found while, although the subsequent oscillations are partially damped out, at least two more peaks are evident. Multiple scattering, curves (b) and (c), was found to reduce the magnitude of the initial oscillation but not to remove it entirely. The subsequent oscillations were, however, damped out and the midchannel flux tended toward its equilibrium value.

To investigate the validity of equation (3.41) Morgan and Van Vliet[24] calculated the depth of the first maximum below the surface, i.e. $\lambda/4$, as a function of energy for α-particles in $\langle 001 \rangle$ copper and is plotted in Figure

3.17. The curve shows the predicted $E^{\frac{1}{2}}$ dependence: the upper curve, which corresponds to no multiple scattering, yields slightly higher values of $\lambda/4$ (see below). The dashed line represents the theoretical result (3.41) and is seen to be in good agreement with the computed results.

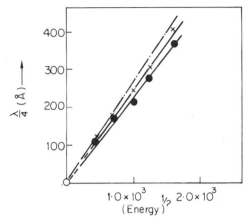

Figure 3.17. The computed value of $\lambda/4$ plotted as a function of $E^{\frac{1}{2}}$, the dot-dash curve is equation (3.41). The upper curve $(+)$ is for $c = 0$ and the lower curve ($\bullet$) for $c = 0.5$

The damping effect of multiple scattering is shown in greater detail in Figure 3.18. The inelastic multiple scattering is varied by varying the parameter c. With $c = 0$ (no inelastic deflections) the flux at the first peak $\simeq 18$; for $c = 0.25$, $F(\mathbf{r}_0) \simeq 9.5$; and for $c = 0.5$, $F(\mathbf{r}_0) \simeq 7.5$. Further increases in c to 0.75 or 1.0 causes a negligible reduction in the first peak. Hence, the initial introduction of multiple scattering tends to disperse the best channeled ions over the region of low $U(\mathbf{r})$ so that the trajectories do not necessarily pass through $\mathbf{r}_0$. Further increases cause relatively smaller changes due to the progressively steeper slope of $U(\mathbf{r})$. It should also be noted that inelastic multiple scattering tends to shift the first peak towards the surface (i.e. decrease $\lambda/4$) as the increase in the multiple scattering with depth progressively undermines the wavelike nature of the best channeled trajectories.

It must be stressed that the depth oscillation observed at $\mathbf{r}_0$ is quite distinct from the surface peak and subsequent minima generally found in the close collision yield from lattice atoms discussed earlier. This is associated with those ions of the high transverse energy $E_\perp$ which landing near the atomic rows are immediately deflected toward the empty channel centre. The depth of the minimum behind the surface is approximately (ρ_{ch}/ψ_1) which is generally smaller by a factor of two or three than the spacing of the

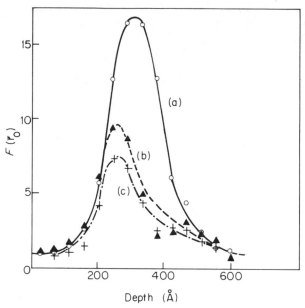

Figure 3.18. Showing in more detail the first peak of $F(r_0)$ for 1·5 MeV α-particles in $\langle 001 \rangle$ copper. Curve (a) $c = 0$, (b) $c = 0·25$, (c) $c = 0·5$

best channeled oscillations. It should also be appreciated that these two cases are both extremes and represent long and short wavelength limits for channeled trajectories.

3.7 RANGE CALCULATIONS

The first application of computer simulation to channeling was the calculations of Robinson and Oen[2] of ion ranges in solids, indeed, the discovery of the channeling phenomenon has been attributed to these calculations. In this respect we see the full advantage of computer calculations. Not only can the final range distribution be determined, but by monitoring the details of each trajectory the factors influencing the final profile may be related directly to the profile; such parameters as dechanneling, multiple scattering, energy loss, etc.

In all the work published so far, the only imperfections introduced into the calculations have been thermal vibrations. The range of an ion in a solid depends on the energy loss of the ions. Thus the calculations depend on the values of the interatomic potentials and electronic energy loss inserted into the computer calculations; references (2, 33–35) contain the detailed results of many calculations of ion ranges for a variety of incident ions in crystal lattices.

An example of such profile shown in Figure 3.19 as taken from the early work of Robinson and Oen.[2] In all cases the inelastic energy loss is a function of the length of an ion's trajectory and does not take into account variations in the nature of the trajectory (i.e. whether it is confined to the centre of a channel or whether it is moving in a random direction). A more realistic simulation would utilize an inelastic energy loss which was calculated at each binary collision and where the inelastic energy loss is a function of the impact parameter. Clearly, this modification only becomes important when electronic stopping plays an important role in the range profile.

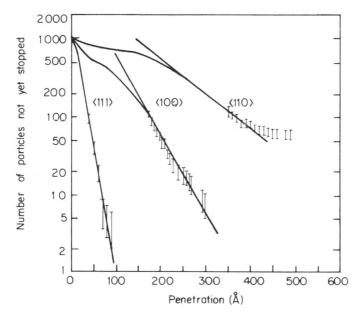

Figure 3.19. Range profiles for 5 keV copper atoms in a static copper lattice. (Born–Mayer potential) (after Robinson and Oen[2])

Bierman and Van Vliet[25] have considered the effect of including an inelastic energy loss as a function of the impact parameter. The range profiles were calculated for 19 keV Ne$^+$ ions in copper for a number of different orientations using experimental values of the stopping power function determined from gas phase collisions. Typical range profiles are shown in Figure 3.20. The random profile is also shown. In all these curves perfect beam collimations are assumed and a lattice temperature corresponding to 0°C. The long aligned penetration depths were attributed to the almost complete absence of nuclear stopping and to a lesser extent to the reduction in electronic stopping. Owing to this relative insensitivity to the electronic

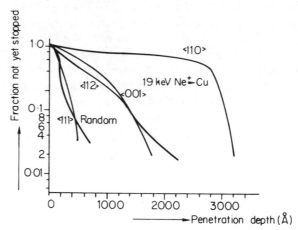

Figure 3.20. Range profiles of 19 keV Ne⁺ ions in copper
for different orientations. The lattice temperature is 0 °C.
Perfect alignment is assumed (after Bierman and Van
Vliet[25])

stopping, there is therefore no startling effect due to the introduction of
inelastic losses which are sensitive to the impact parameter.

One very interesting result arising from these calculations is illustrated
in Figure 3.20. A well defined range is found in the $\langle 101 \rangle$ channels but not in
either the $\langle 100 \rangle$ or the $\langle 112 \rangle$ channels. This has been attributed to some
process which tends to focus or funnel a large fraction of the ions towards the
centre of the channel. Some insight into this effect is gained from Figure 3.21

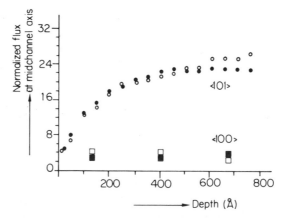

Figure 3.21. $F(\mathbf{r}_0)$ for 19 keV Ne⁺ ions in $\langle 100 \rangle$ and $\langle 101 \rangle$ copper. In the $\langle 101 \rangle$
channel $F(\mathbf{r}_0)$ increases with depth due to the 'funnelling' effect (after Bierman
and Van Vliet[25])

where the maximum flux near the midchannel axis is shown as a function of depth in the crystal. As we saw earlier, multiple scattering tends to decrease the flux with depth,* this is found to be correct in the symmetric $\langle 100 \rangle$ channel. For the $\langle 101 \rangle$ channel however the maximum flux increases with depth as 'funnelling' accentuates the normal flux until it reaches a value of almost 25 times normal. A similar result has been observed[26] for 2·0 MeV α-particles, illustrating that the effect, although less pronounced, still occurs for high energy particles. The origin of the effect has not been established. It is not, however, the 'damping' effect mentioned by Lindhard[8] since it is independent of the inelastic losses. The importance of the funnelling effect, if confirmed, on the resulting range profiles, is obvious. This example illustrates clearly one crucial advantage of computer simulation, not only can the range profile be predicted, but the individual process contributing to that profile may be extracted from the iterative process.

In this chapter we have selected a number of examples which illustrate the main features of computer simulation. The technique, although not as elegant as analytical models, has played an important role in the development of our understanding of the channeling process, particularly in situations where the continuum model is inapplicable.

References

1. Gibson, J. B., Goland, A. N., Milgram, M. and Vineyard, G. H., *Phys. Rev.*, **120**, 1229 (1960).
2. Robinson, M. T. and Oen, O. S., *Phys. Rev.*, **132**, 2385 (1963).
3. Van Vliet, D., *Computer studies of channeling in crystals*, to be published in *Radiation Effects in Solids*, Gordon and Breach, New York.
4. Torrens, I. M., *Interatomic Potentials* (to be published by Academic Press).
5. Morgan, D. V. and Van Vliet, D., *Harwell Report*, R6283.
6. Morgan, D. V. and Van Vliet, D., *Can. J. Phys.*, **46**, 503 (1968).
7. Morgan, D. V. and Van Vliet, D., *Radiation Effects*, **8**, 51 (1971).
8. Lindhard, J., *Dansk. Vid. Selsk., Mat. Fys. Medd.*, **34**, 14, (1965).
9. Erginsoy, C., *Phys. Rev. Lett.*, **15**, 360 (1965).
10. Van Vliet, D., Private Communication.
11. Davies, J. A., Denhartog, J. and Whitton, J. L., *Phys. Rev.*, **165**, 345 (1968).
12. Appleton, B. R., Erginsoy, C. and Gibson, W. M., *Phys. Rev.*, **161**, 330 (1967).
13. Barrett, J. H., *Phys. Rev.*, **166**, 219 (1968).
14. Barrett, J. H., *Phys. Rev.*, **3**, 1527 (1971).
15. Morgan, D. V. and Van Vliet, D., *Radiation Effects*, **5**, 157 (1970).
16. Andersen, J. U., *Dansk. Vid. Selsk. Mat. Fys. Medd.*, **36**, 7 (1967).
17. (a) Andersen, J. U. and Feldman, L. C., *Phys. Rev. B*, **1**, 2063 (1970).
    (b) Feldman, L. C., *Ph.D. Thesis*, Rotgers University (1966).
18. Bøgh, E., *Phys. Rev. Lett.*, **19**, 61 (1967).

---

* The depth interval over which this flux is average is relatively large and the oscillation described earlier is masked. We are thus observing the change in the average flux.

19. Appleton, B. R. and Feldman, L. C., *Atomic Collision Phemomena in Solids*, Amsterdam, North-Holland, p. 417 (1970).
20. (a) Morgan, D. V. and Van Vliet, D., see ref. 19, p. 476.
    (b) Van Vliet, D., *Phys. stat. sol.* (a), 521 (1970).
21. Quéré, Y., *Phys. stat. sol.* **30**, 173 (1968).
22. Andersen, J. U., Andreasen, O., Davies, J. A. and Uggenhoj, E., *Radiation Effects*, **7**, 1 (1970).
23. Van Vliet, D., *Radiation Effects*, **10**, 137 (1971).
24. Morgan, D. V. and Van Vliet, D., *Radiation Effects*, **12**, 203 (1972).
25. Bierman, D. J. and Van Vliet, D. *Physica*, **57**, 221 (1972).
26. Alexander, R. B., *Ph. D. Thesis*, University of Oxford.
27. E. Bøgh, Private Communication.
28. Picraux, S. T., Davies, J. A., Eriksson, L., Johansson, N. G. E. and Mayer, J. W., *Phys. Rev.*, **180**, 873 (1969).
29. Noggle, T. S. and Barrett, J. H., *Phys. stat. solidi*, **36**, 761 (1969).
30. Andersen, J. U., Davies, J. A., Nielsen, K. O. and Andersen, S. L., *Nucl. Instrum. Methods*, **38**, 210 (1965).
31. Bøgh, E. and Uggerhoj, E., *Nucl. Instrum. Methods*, **38**, 216 (1965).
32. Domeij, B. and Bjorkquist, K., *Phys. Lett.*, **14**, 127 (1965).
33. Oen, O. S. and Robinson, M. T., *J. appl. Phys.*, **35**, No. 8, 2515 (1964).
34. Harrison, D. E., Leeds, R. W. and Gray, W. L., *J. appl. Phys.*, **36**, No. 10, 3154 (1965).
35. Harrison, D. E. and Greiling, D. S., *J. appl. Phys.*, **38**, No. 8, 3200 (1967).
36. Varelas, C. and Biersack, J., *Nucl. Instrum. and Meth.*, **79**, 213 (1970).
37. Ibel, K. and Sizmann, R., *Phys. stat. sol.*, **29**, 403 (1968).

CHAPTER IV

Energy Loss of Channeled Ions: Low Velocity. Ions

J. M. POATE

4.1 INTRODUCTION

An ion traversing a solid loses energy by virtue of the Coulomb interactions along its path. For high energy particles such as protons or α-particles, it is relatively easy to characterize the interactions between the incident particle and electrons of the solid; the energy loss resulting in the ionization or excitation of the atoms of the solid. At MeV energies the probability of energy loss to nuclear recoils is so small that only the electronic processes need be considered. The theory of energy loss in this region was first established by Bohr[1] who classified the problem in terms of the impact parameter of the collisions. Phenomenologically Bohr was able to divide the electronic processes into close, or single-particle and distant, or resonant collisions. The resonant collisions can be identified with the excitation of plasmons in the solid. The problem was placed on a more accurate quantum mechanical basis by Bethe and Bloch[2] giving rise to the Bethe–Bloch formulation which is that commonly used in stopping predictions at high energy. The agreement between theory and experiment is very good being of the order of a few per cent.

Stopping power is inversely proportional to the square of the incident velocity at the high energies, however, with decreasing energies the stopping power passes through a maximum as the traversing ions capture electrons and become neutralized. According to Bohr the maximum should occur at velocities of approximately $v_0 Z_1^{2/3}$, where v_0 is the Bohr velocity of $2\cdot2 \times 10^8$ cm s^{-1}, corresponding to 25 keV/nucleon. At lower energies the whole problem of the inelastic electronic interactions not only becomes more complex but elastic or nuclear collisions begin to play a strong role in the energy loss process. These elastic recoils can completely dominate the stopping process at energies of a few keV. The theoretical treatment in this low velocity region is largely due to the Aarhus school of Lindhard and his collaborators.[3] It is construed that the electronic and nuclear contributions to the energy loss are independent (obviously for very violent collisions these

115

effects are completely correlated but within the framework of the theory, and its emphasis on averaging procedures, the two effects can be considered independent). The elastic scattering cross sections are calculated from screened Coulomb potentials and the electronic contributions are calculated from the concept of the ion moving through an electron gas, with electron densities derived from the Thomas–Fermi model. Lindhard and Scharff[4] were able to show the electronic stopping powers in this low energy region to be proportional to the ion velocity and to depend monotonically upon the atomic numbers of the projectile and the stopping material, Z_1 and Z_2. The unified approach to the low energy stopping by Lindhard, Scharff and Schiøtt[3]— now generally referred to as LSS theory—is successful in predicting stopping to within 20%.

Energy loss theory, before the discovery of channeling, was only concerned with descriptions of particles traversing random arrays of target atoms. However, the recognition that energetic ions could be steered between atomic rows by a series of gentle correlated collisions gave considerable impetus to the study of energy losses. Channeled ions move, by definition, through regions of low electron density where electronic energy losses are expected to be smaller than that for an equivalent thickness of amorphous material. Moreover the fact that channeled ions are restrained from close encounters with the target atoms means that the contribution of nuclear recoils to the energy loss is drastically reduced. This is important in the low energy region as it allows one to measure experimentally electronic energy loss without interference from nuclear recoils. The first experiments on low energy ions in polycrystalline aluminium[5] and high energy ions in single crystal silicon,[6] did indeed show a reduction in energy loss. The results for the high energy channeled losses can be understood semiquantitatively from Bohr's model of close and distant collisions (see Chapter 5). From low energy channeling measurements two distinct features emerge: not only are the energy losses reduced but they depend strongly upon Z_1, the Z of the incoming ion. The main purpose of this review is to present the current status of theory and experiment in this low energy region. Planar channeling experiments of energetic ions through very thin crystals are at the stage now where trajectories can be parameterized and energy losses associated with discrete trajectories. From such measurements it is possible to deduce stopping powers as a function of channel impact parameter; this is an important development and will be covered by Datz *et al.* in Chapter 6. Here we will be mainly concerned with the particular case of well channeled ions which are assumed to be constrained to trajectories near the centre of the channel; almost all the low energy experiments and theory fall within this category. In the next sections we will firstly outline the relevant experiments and then discuss the theory.

In addition to the intrinsic physical interest in this field it is worthwhile indicating some of the technological implications which have given impetus to such studies. Energy loss theory and experiment have traditionally been concerned with MeV protons and α-particles, the region of interest in nuclear physics. However, in the early 1960s much effort was devoted to elucidating the mechanisms responsible for radiation damage in nuclear reactor cladding and structural materials. Damage can arise from neutrons causing recoils of the individual atoms of the host material with energies of, typically, 10 keV. Experiments were therefore performed to simulate the damage with heavy ions of the requisite energy from heavy ion accelerators. The realization of the possibilities of ion implantation as a technological tool also underlined the need to know the energy loss characteristics of heavy ions entering solids and has stimulated heavy ion measurements in single crystal semiconductors.

4.2 TRANSMISSION STUDIES

Experimentally there is an obvious delineation between differential energy loss in transmission measurements and range determinations. As transmission experiments tend to be more illustrative of the stopping process we will consider them first. A good example of the techniques and the physical processes involved is given by the recent experiment of Eisen *et al.*[7] Stopping power measurements were made in silicon crystals for α-particle energies between 0·1 and 18 MeV. As channeled energy losses were required over the whole energy range fairly stringent experimental conditions were required. To cover the whole energy range two single-ended and one tandem Van de Graaff accelerators were used with the helium beams collimated to an angular divergence of 0·03°. A beam of known energy was then passed through suitably thin crystals and the emergent energy measured by an electrostatic analyser for the energy region below 400 keV and by silicon surface-barrier detectors for higher energies. The emergent beam is viewed with a sufficiently small solid angle so that only well channeled or randomly transmitted particles which have not suffered large angle elastic collisions are detected. Figure 4.1 shows the stopping power measurements for a random and ⟨110⟩ direction. The energy losses here are determined from the peaks of the transmitted spectra. The experiment clearly demonstrates two features of the electronic stopping process: the turnover in the stopping power curve in the region of the Bohr velocity and the reduction in energy loss for channeled particles.

It is a very real problem obtaining thin single crystals for transmission studies; and to date low energy measurements have only been performed in silicon and gold. For this reason relevant transmission experiments through amorphous materials are discussed here. A comprehensive review by Northcliffe and Schilling[8] lists the data and constructs universal stopping power and range tables.

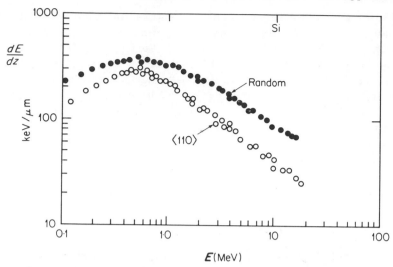

Figure 4.1. The stopping power measurements of Eisen *et al.*[7] for He^+ ions in random and $\langle 110 \rangle$ channeling directions in thin Si crystals; after Eisen *et al.* 1972. E is the incident ion energy corrected for finite crystal thickness. Each plotted point represents the average of several experimental points. The maximum in the stopping power curve occurs at an energy of about 500 keV and the corresponding Bohr prediction, $v_0 Z_1^{2/3}$, is 225 keV

The first indication that stopping powers in the low energy region were not simple monotonic functions of the Z_1 of the projectile came from the work of Teplova *et al.*[9] in 1962 in which the ranges of various ions in air were measured. The work was put on a quantitative footing by the Canadian group of Ormrod, MacDonald and Duckworth in a series of experiments.[10–13] Stopping powers, as a function of Z_1, were measured in thin foils of boron, carbon and aluminium and in gaseous nitrogen and argon. The data were taken for constant velocity, in the region of $0.5\,v_0$, where one expects a simple Z_1, Z_2 monotonic dependence. The initial measurements were made in carbon and a marked oscillatory dependence of the stopping as a function of Z_1 was observed. The data for boron and aluminium revealed a similar oscillatory dependence with the maxima and minima in the same positions as that observed for carbon. Although the gaseous stopping data of nitrogen and argon give the same oscillatory behaviour, the magnitudes are some 40% lower than corresponding foil stopping data.

The Aarhus group of Fastrup, Hvelplund and Sautter[14,15] extended the measurements in carbon foils up to $Z_1 = 39$ and at velocities of 0.63 and $0.91\,v_0$. Figure 4.2 shows the data for $v = 0.63\,v_0$; the smooth curve is the LSS prediction for the energy loss.

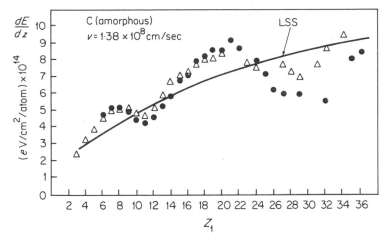

Figure 4.2. Random stopping cross section measurements of Hvelplund and Fastrup[15] for various incident ions, Z_1, of velocity $0.63 v_0$ in thin carbon foils; after Hvelplund and Fastrup 1968 and Cheshire *et al.* 1969. The dots are experimental points and the triangles are theoretical predictions[29] using the modified Firsov model for singly stripped incident ions.
The smooth curve is the LSS prediction for electronic stopping

More recently Hvelplund[16] has measured stopping powers of various ions in helium, air and neon. A plot of stopping powers against Z_2 constructed from Hvelplund's results and those mentioned previously is shown in Figure 4.3. Obviously any attempt to arrive at a measurement of Z_2 dependence will encounter severe normalization problems. In a Z_1 measurement the target material is fixed, but in a Z_2 measurement the incident ion is fixed and the stopping material varied. The accuracy of the Z_2 measurement therefore depends upon the accuracy with which target thicknesses can be measured. Moreover comparison has to be made between gaseous and solid targets; and, as mentioned previously, the gaseous targets appear to have anomalously low stopping cross sections. But as Hvelplund points out, the difference between adjacent gaseous and amorphous stopping materials falls with increasing energy. Allowing for these uncertainties, the results of Hvelplund clearly show a Z_2 dependence very similar in nature to the dependence on Z_1.

The first systematic transmission studies for low energy channeled ions were carried out by Eisen.[17] His stopping cross section measurements for ions of velocity $0.68 v_0$ in the $\langle 110 \rangle$ channels of $0.55\ \mu m$ silicon crystals are shown in Figure 4.4. The oscillations in stopping are now seen to become very pronounced but still retaining the same relative position as the amorphous studies. The experimental points for $Z_1 > 26$ have been taken from unpublished data of Eisen for ions of velocity $0.45 v_0$ and scaled up

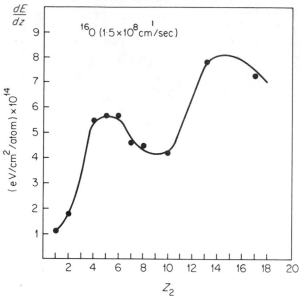

Figure 4.3. Hvelplund's plot[16] of the stopping cross sections for 200 keV ^{16}O ions moving through a variety of stopping materials, Z_2; after Hvelplund 1971. The smooth curve has been drawn to guide the eye

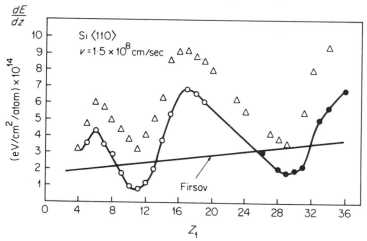

Figure 4.4. Eisen's stopping cross section measurement[17] for a variety of ions, Z_1, channeled along the $\langle 110 \rangle$ axis of thin silicon crystals; after Eisen 1968 and Cheshire *et al.* 1968. The open circles refer to data taken at velocities of $0.68 \, v_0$; the filled-in circles refer to data taken at velocities of $0.45 \, v_0$ which have been scaled up ($\times \frac{3}{2}$) for comparison with the higher velocity data. The triangles are theoretical predictions[29] using the modified Firsov model for singly stripped incident ions. The line is the average Firsov prediction with $R = 2$ Å

$(\times \frac{3}{2})$ in the ratio of the velocities. The transmitted energy spectra were measured with an electrostatic analyser, and the leading edges of the spectra were assumed to give the energy loss of the best channeled particles. This is a satisfactory assumption as it can be shown experimentally that the positions of the leading edges of the transmitted spectra depend sensitively on the trajectory of the ion in the channel. Eisen also measured energy loss as a function of energy and fitted the stopping power P to an expression of the form $dE/dz = -kE^P$. Instead of P having the theoretically assumed value of 0·5, proportional to velocity, the values were seen to oscillate around 0·5, being approximately out of phase with the stopping power dependence. These values of the exponent are shown in Figure 4.5. Also plotted are the

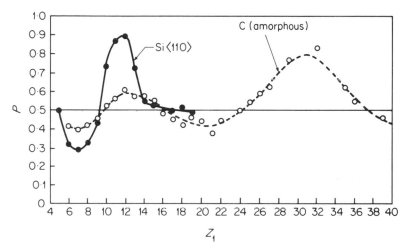

Figure 4.5. A plot of the exponent P, from the expression $dE/dz = -kE^P$, as a function of Z_1 for two sets of data; the silicon channeling data of Eisen[7] and the amorphous carbon data of Hvelplund and Fastrup[15]

values of Hvelplund and Fastrup[15] for amorphous carbon. The values of P certainly depend upon the range over which they are obtained and are not quantitative enough to draw any strong conclusions on correlations with Z_1. Nevertheless there is definitely an oscillatory dependence of the stopping exponent about 0·5 which seems to be apparently out of phase with the stopping power dependence. If stopping power measurements are made over a wide enough range it is probable[15] that the value of P will not be far from 0·5.

In a more recent experiment Bøttiger and Bason[18] have channeled a variety of ions along the $\langle 110 \rangle$ axis of gold crystals approximately 1000 Å in thickness. The ions were analysed using a double focusing magnetic spectrometer, and the leading edges of the transmitted energy spectra were used to evaluate the stopping powers. Figure 4.6 shows their results for ions of $7 \leqslant Z_1 \leqslant 54$ of velocity $1.5 \times 10^8 \, \mathrm{cm \, s^{-1}}$. The relative magnitude and position of the oscillations agree well with the data for silicon.

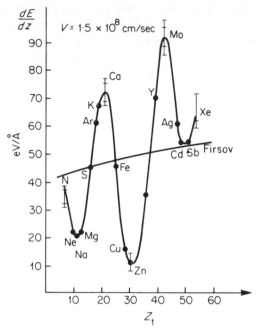

Figure 4.6. The stopping power measurements of Bøttiger and Bason[18] for the channeling of ions with velocity $0.68 \, v_0$ in $\langle 110 \rangle$ gold crystals; after Bøttiger and Bason 1969. The smooth curve has been drawn to guide the eye. The line is the average Firsov prediction for channeled energy loss

4.3 RANGE STUDIES

The measurements of the ranges of heavy ions implanted into single crystal, polycrystalline and amorphous materials have contributed significantly in the past ten years to our knowledge and understanding of energy loss processes. Such measurements usually involve the implanting of radioactive ions and the subsequent sectioning of the target to count either the residual activity or that removed in the sectioning. The techniques and results are covered by J. L. Whitton in Chapter 8 on range profiles.

Range studies have the great advantage that a wide range of single crystal materials are amenable to the stripping or sectioning techniques. However, once,the ranges have been determined the problem remains of extracting the stopping powers (of course the knowledge of the detailed shape of the range profile is of importance in its own right, for example in connection with ion implantation in semiconductors). The ranges of the channeled ions can depend critically upon such factors as the dechanneling probability. The detection of dechanneled particles is not usually a problem in transmission experiments owing to the stringent collimation imposed upon the transmitted particles. Nevertheless stopping power data can be obtained from ranges under conditions which would have been inaccessible to transmission experiments. Indeed J. L. Whitton is carrying out range experiments on a variety of fcc and bcc crystals to extract the Z_2 dependence of the stopping.

Chapter 8 covers comprehensively the many factors governing the shape of range profiles and the way in which the electronic stopping cross sections may be extracted. An excellent example and precursor of many experiments was that of Eriksson et al.,[19] who determined the range profiles of many ions, as a function of energy, implanted in tungsten along the $\langle 100 \rangle$ and $\langle 110 \rangle$ directions. The experiment demonstrates very nicely features of the stopping and channeling process: the suppression of nuclear stopping in a channeling direction, thus allowing electronic stopping to dominate; the $E^{1/2}$ dependence from measurements over a wide energy range; and the strong oscillatory Z_1 dependence of the electronic stopping (Figure 8.14).

It will be worthwhile here to summarize briefly the results for the Z dependence of the electronic stopping from both transmission and range measurements. Amorphous or random stopping measurements have been carried out in thin foils of boron, carbon and aluminium and channeling measurements have been carried out in silicon, tungsten, and gold. If stopping powers are plotted for fixed velocity, Z_1 oscillatory behaviour is observed. The position of the oscillations are relatively insensitive to the choice of target material. However, the amplitudes of the oscillations are significantly enhanced if the ions are channeled. The position of the minima occur at approximately $Z_1 = 3, 11, 30$ and 50. Although corresponding studies of the Z_2 dependence of the stopping are still in their infancy, oscillatory behaviour is observed which is similar to the Z_1 oscillations.

4.4 THEORIES

The first calculation of the slowing down of slow particles was performed by Fermi and Teller[20] in 1947. They were concerned with the capture of negative muons by atoms and wanted to show that the time for a muon to be stopped and captured to form a mumesic atom was shorter than the natural decay time. The muon was considered to be moving in a degenerate electron gas with velocity v much smaller than the maximum velocity v_F of the electrons. In a muon–electron collision the change of electron velocity will be of the order v. As the electrons belong to a degenerate gas, all collisions for which the final velocity of the electron lies inside the occupied zone of the velocity space will be forbidden by the Pauli principle. Therefore, only electrons with velocities within v of v_F will be capable of colliding. With this restriction they arrived at an expression for the stopping power of muons in a degenerate electron gas:

$$-\frac{dE}{dz} = \frac{2}{3\pi} \frac{m^2 e^4}{\hbar^3} v \ln \frac{137 v_F}{c}. \tag{4.1}$$

This was the first derivation to show the proportionality of energy loss and velocity.

Lindhard,[21] somewhat later, considered the properties of an electron gas in more detail and arrived at essentially the same derivation. As he pointed out the velocity dependence of the stopping is quite understandable if one compares stopping with electrical resistance, an analogous physical process, where according to Ohm's law the time rate of momentum loss is proportional to the electrical current or average electron velocity. Indeed there is another analogue from frictional phenomena where the viscous force generally exhibits a linearity between its magnitude and the relative velocity, for example Stokes' law.

Using a Thomas–Fermi treatment to estimate the electronic density distribution and a dielectric formulation of energy loss, Lindhard and Scharff[4] found the electronic stopping to be

$$S_e = \xi_e \times 8\pi e^2 a_0 \frac{Z_1 Z_2}{Z} \frac{v}{v_0} \tag{4.2}$$

where

$$Z = (Z_1^{\frac{2}{3}} + Z_2^{\frac{2}{3}})^{\frac{1}{2}} \tag{4.3}$$

and ξ_e is of order 1–2 ($\xi_e \simeq Z_1^{\frac{1}{6}}$). It was the intent of the authors[22] to develop a general and simple framework for the stopping theory which held to an accuracy of 10 or 20%, and indeed the theory has been successful in predicting random stopping powers to that accuracy. Not unexpectedly, however, their theory is inappropriate for channeled energy losses: absolute magnitudes can be in error by factors of four or five and no provision is made to explain the Z_1 and Z_2 oscillations.

The central theme of the Lindhard theory is that the electronic density distribution of the stopping medium can be considered in terms of an electron gas. This approach has many attractions; nevertheless, there is an alternative way of viewing the stopping process, namely that of a series of binary encounters. Firsov[23] has formulated such an approach. He was concerned with calculating mean electron excitation energies in slow binary atomic collisions. As a quantum mechanical treatment of the problem is prohibitively difficult, he decided to treat the process on a semiclassical basis. It is assumed that electrons of the incident ion's electron cloud penetrate within the effective range of the target atom potential and lose the excess energy associated with their translational motion. Similarly electrons of the target atom acquire energy when they enter the range of the incident atom potential. In this way a fraction of the kinetic energy of the incident atom is transferred to the electrons and expended on their detachment from the colliding atoms. As Firsov points out, it is difficult to justify this approach theoretically; rather any justification should come from the quantitative agreement between the calculation and experiment.

The energy transfer, E, is calculated from the electron momentum flux through a plane midway between the colliding atoms

$$E = m \int \mathbf{R} \cdot d\mathbf{R} \int \tfrac{1}{4} n v_e \, dS \qquad (4.4)$$

where $\mathbf{R}$ is the internuclear separation. The surface integral represents the flux across the plane; n is the number of electrons at this surface with an average velocity v_e (the factor $\tfrac{1}{4}$ is an effusion coefficient).

The energy loss as a function of impact parameter can then be derived

$$E(R) = \frac{(Z_1 + Z_2)^{\frac{5}{3}} 4 \cdot 3 \times 10^{-8} v}{\{1 + 3 \cdot 1 (Z_1 + Z_2)^{\frac{1}{3}} 10^7 R\}^5} \quad \text{(eV)} \qquad (4.5)$$

where v is the incident atom velocity in cm s^{-1}. Teplova *et al.*[9] used this expression to calculate stopping in gases by integrating over impact parameter, for example,

$$-\frac{dE}{dz} = \int_0^\infty E(R) 2\pi R \, dR. \qquad (4.6)$$

A similar integration by Hvelplund and Fastrup[15] yielded the result

$$\frac{dE}{dz} = 5 \cdot 15 \times 10^{-15} (Z_1 + Z_2)(v/v_0) \quad \text{eV cm}^2/\text{atom} \qquad (4.7)$$

using the definition for Z given in equation (4.3). This expression predicts stopping values very similar to those of Lindhard and Scharff.

The significance of $E(R)$ became evident on realization that it predicted, quite accurately, average channeled stopping powers if a value of R equal to the channel radius was used. Several authors, Cheshire *et al.*,[24] Bhalla and Bradford,[25] Winterbon[26] and El-Hoshy and Gibbons[27] realized that the Z_1 oscillations could be correlated with the electronic shells of the colliding atoms and modified the Firsov formula to embody such effects. Electron densities were calculated from Hartree–Fock wavefunctions instead of

Thomas–Fermi densities as used in the original calculations. D. E. Harrison's plot[28] of electronic radial distribution clearly shows the variation in the electronic densities which can be correlated with the Z_1 oscillations. The calculations of Cheshire *et al.*[29,30] will be outlined. The surface integral was modified to be

$$m \int \frac{1}{4} n v_e \, dS = \frac{1}{4} m \int \left(\sum_\alpha v_\alpha |\psi_\alpha|^2 + \sum_\beta v_\beta |\psi_\beta|^2 \right) dS \qquad (4.8)$$

where the $\psi_{\alpha,\beta}$ are the Slater orbitals of the projectile and target,

$$\psi_\alpha = r^{n_\alpha - 1} \exp(-\xi_\alpha r) Y_{l_\alpha m_\alpha}(\theta, \varphi)$$

and $v_{\alpha,\beta}$ are the corresponding velocities. To obtain an approximate v_α we equate $\frac{1}{2} m v_\alpha^2$ to the expectation value of the kinetic energy for each orbital. The dividing surface was modified to intersect the line joining the projectile and target atom at the point where the electron density is a minimum.

Figure 4.7 shows a comparison between the Firsov energy loss function $E(R)$ and the modified function for collision between B^+ and Si. Silicon is assumed to be in its neutral ground state, and the corresponding Slater orbitals were obtained using the formulae of Clementi and Raimondi.[31] The Slater orbitals for singly stripped B were generated using the previous technique from Clementi's tabulation[32] of wavefunctions. The modified Firsov function shows structure owing to the overlap of electronic shells when the colliding atoms have a separation of 1–2 Å which corresponds to typical radii of major channels. At small impact parameters the two expressions converge.

The following examples illustrate the success of the theory in explaining the Z_1 oscillations. Figure 4.4 shows a comparison between the energy loss data of Eisen[17] for ions channeled in silicon and theory. The target atoms were assumed to be in their ground state with the projectile atoms singly ionized as this was considered to be the most likely charge state traversing the crystal. The calculations were carried out for a single impact parameter of 2 Å corresponding to the mean radius of the axial channel. This use of a single impact parameter can be justified on the grounds that the experiment involves measurement of the best channeled particles which, presumably, spend most of their time near the centre of the channel. The positions of the oscillations are predicted well, but the magnitude of the stopping cross section is overestimated (use of neutral atoms for the projectiles raises the magnitude of the calculations by some 20%). A further check on the theory was made to explain the amorphous carbon energy loss measurements of Hvelplund and Fastrup.[15] In the experiment only ions with an angular spread of 0·17° about the beam were detected. Calculations were therefore carried out integrating over all impact parameters greater than those which would produce an angular deflection of less than 0·17°. The fit is surprisingly good as shown in

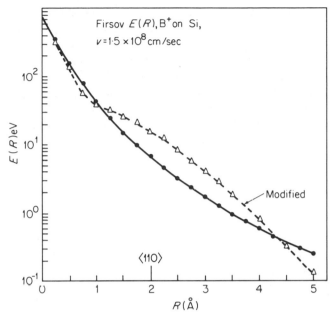

Figure 4.7. A comparison between the Firsov energy loss function, $E(R)$, and the modified Firsov function incorporating Hartree–Fock predictions for the electron densities. The colliding atoms were taken to be B^+ and silicon with a relative velocity of $0.68\,v_0$. The structure owing to overlap of electronic shells at typical distances encountered in channeled energy loss measurements is apparent. The mean $\langle 110 \rangle$ channel radius in silicon is indicated. The triangles and dots refer to values of the impact parameter, R, at which the Firsov functions were calculated and the lines represent the continuous functions

Figure 4.2. Further checks have been made on the theory[30] using unpublished silicon channeling data of Eisen and the gold channeling experiment of Bøttiger and Bason.[18]

Although the original Firsov model was constructed on an *ad hoc* basis, it predicts amorphous and average channel stopping cross sections well. Moreover the modified Firsov model predicts quite accurately the position of the Z_1 oscillations (as the theory is symmetric, identical Z_2 oscillations are also expected) and the impact parameter dependence of the stopping. The oscillatory behaviour is caused by variations in the electronic densities of the colliding atoms at radial distances which correspond to large impact parameter processes. The effect is therefore due to variations in the size of the colliding atoms. Here the nomenclature 'size' can be misleading if compared with the more usual solid state sense which has a different definition.

Finneman and Lindhard[33] have proposed an interesting model to explain the Z_1 oscillations. In an analogy with the Ramsauer–Townsend effect (see, for example, Mott and Massey[34]) where resonance effects in low energy electron scattering are observed as a function of incident velocity and target atom, they have constructed transport equations for the scattering of electrons from a Lenz–Jensen potential. The transport equation gives the electron momentum change in the scattering and therefore the energy loss. By choosing electron densities corresponding to densities at the channel radii, their quantum mechanical evaluation of the transport cross section shows oscillatory behaviour very similar to that observed experimentally. Full details of these calculations are not yet available.

4.5 PROJECTIONS

It would now appear that experiment and theory of channeled energy loss in the low velocity region have reached some maturity. Nevertheless it is not too difficult to catalogue areas which need attention. Perhaps the most obvious case is that of the energy dependence of the stopping. All theories predict the stopping to be proportional to velocity, whereas the experimental data indicate definite deviations from this dependence. Unfortunately there does not appear to be enough data to stimulate a detailed theoretical inspection. Similarly the problem of the Z_2 oscillations needs more experimental attention.

The recent experiment of Eisen *et al.*[7] has indicated that the ratio of channeled to random stopping for α-particles in silicon has an interesting energy dependence; in fact it peaks at the same position as the random stopping. The reason for this is not clear, but it raises interesting possibilities. For example channeled energy loss in the high energy region is expected (Chapter 5) to be affected strongly by plasmon excitation, whereas in the low energy region the interactions should only depend upon local electron densities and not plasmon excitation. Measurements of the energy loss ratios for heavy ions may therefore give information on the energy loss mechanisms and the local electron density in channels.

The subject of energy loss in channeling has, owing to the constraints of theory and experiment, been divided into low and high energy sections with no real discussion of the intermediate region. For this reason the electron gas approach to energy loss for both random and channeled stopping of Lindhard and his collaborators is most powerful in that it can lead to a unified theory of stopping. Nevertheless the success of the Firsov model in calculating energy loss lends credence to the technique of isolating by theory and experiment, the energy loss as a function of impact parameter for individual collisions and in this way constructing channeled energy loss.

References

1. Bohr, N., *Phil. Mag.*, **25**, 10 (1913); Bohr, N., *Dan. Vid. Selsk., Mat. Fys. Medd.*, **18**, No. 8 (1948).
2. Fano, U., *Ann. Rev. Nucl. Sci.*, **13**, 1 (1963).
3. Lindhard, J., Scharff, M. and Schiøtt, H. E., *Dan. Vid. Selsk. Mat. Fys. Medd.*, **33**, No. 14 (1963).
4. Lindhard, J. and Scharff, M., *Phys. Rev.*, **124**, 128 (1961).
5. Davies, J. A., Brown, F. and McCargo, M., *Can. J. Phys.*, **41**, 829 (1963).
6. Dearnaley, G., *IEEE Trans. Nucl. Sci.* **NS–11**, 249 (1964).
7. Eisen, F. H., Clark, G. J., Bøttiger, J. and Poate, J. M., Proceedings of the Gausdal Conference, *Radiation Effects*, **13**, 93 (1972).
8. Northcliffe, L. C. and Schilling, R. F., *Nuclear Data Tables*, **A7**, 233 (1970).
9. Teplova, Ya. A., Nikolaev, V. S., Dmitriv, I. S. and Fateeva, N. L., *Soviet Phys.—JETP*, **15**, 31 (1962).
10. Ormrod, J. H. and Duckworth, H. E., *Can. J. Phys.*, **41**, 1424 (1963).
11. Ormrod, J. H., MacDonald, J. R. and Duckworth, H. E., *Can. J. Phys.*, **43**, 275 (1965).
12. MacDonald, J. R., Ormrod, J. H. and Duckworth, H. E., *Z. Naturforsch*, **21a**, 130 (1966).
13. Ormrod, J. H., *Can. J. Phys.*, **46**, 497 (1968).
14. Fastrup, D., Hvelplund, P. and Sautter, C A , *Dan. Vid. Selsk., Mat. Fys. Medd.*, **35**, No. 10 (1966).
15. Hvelplund, P. and Fastrup, B., *Phys. Rev.*, **165**, 408 (1968).
16. Hvelplund, P., *Dan. Vid. Selsk., Mat. Fys. Medd.*, **38**, No. 4 (1971).
17. Eisen, F. H., *Can. J. Phys.*, **46**, 561 (1968).
18. Bøttiger, J. and Bason, F., *Radiation Effects*, **2**, 105 (1969).
19. Eriksson, L., Davies, J. A. and Jespersgaard, P., *Phys. Rev.*, **161**, 219 (1967).
20. Fermi, E. and Teller, E., *Phys. Rev.*, **72**, 399 (1947).
21. Lindhard, J., *Dan. Vid. Selsk., Mat. Fys. Medd.*, **28**, No. 8 (1954); Lindhard, J. and Scharff, M., *ibid*, **27**, No. 15 (1953).
22. Lindhard, J., *Proc. R. Soc.*, **A311**, 11 (1969).
23. Firsov, O. B., *Soviet Phys. JETP*, **9**, 1076 (1959).
24. Cheshire, I. M., Dearnaley, G. and Poate, J. M., *Phys. Lett.*, **27A**, 304 (1968).
25. Bhalla, C. P. and Bradford, J. N., *Phys. Lett.*, **27A**, 318 (1968).
26. Winterbon, K. B., *Can. J. Phys.*, **46**, 2429 (1968).
27. El-Hoshy, A. H. and Gibbons, J. F., *Phys. Rev.*, **173**, 454 (1968).
28. Harrison, D. E., *Applied Phys. Lett.*, **13**, 277 (1968).
29. Cheshire, I. M., Dearnaley, G. and Poate, J. M., *Proc. R. Soc.*, **A311**, 47 (1969).
30. Cheshire, I. M. and Poate, J. M., *Atomic Collision Phenomena in Solids*, Ed. by D. W. Palmer, M. W. Thompson and P. D. Townsend, Amsterdam, North Holland, 388 (1970).
31. Clementi, E. and Raimondi, D. L., *J. Chem. Phys.*, **38**, 2686 (1963).
32. Clementi, E., *Tables of Atomic Functions*, California IBM San Jose (1965).
33. Finneman, J. and Lindhard, J., to be published.
34. Mott, N. F. and Massey, H. S. W., *The Theory of Atomic Collisions*, O.U.P., Chapter XVIII (1965).

CHAPTER V

Energy Loss of Channeled Ions: High Velocity Ions

G. Dearnaley

5.1 INTRODUCTION

In this chapter we shall be concerned with experimental observations and theoretical explanations of the energy loss of channeled ions in the so called Bethe–Bloch region of atomic stopping. Ion velocities are then well in excess of the maximum orbital electron velocities $Z_1 e^2/\hbar$ and the ion under such circumstances may be considered to remain a bare nucleus. The limiting energy for this situation to hold rises as $M_1 Z_1^2$, and corresponds to about 100 keV for protons and 1·6 MeV for helium ions: it is not surprising therefore that experimental evidence is confined to species with $Z_1 \leqslant 2$.

Since the ion can be considered as an unscreened point charge interacting with a simple Coulomb potential the problem is a classically simple one, yet despite the fact that some of the earliest and most quantitative measurements in the study of channeling related to the energy loss of such fast moving ions, there is still no fully acceptable theory of the behaviour observed. This is all the more surprising in view of the fact that the Bethe–Bloch theory accounts so well for the energy loss in amorphous media. Nor has the problem escaped attention: it has been the subject of at least twelve theoretical papers and four distinct methods of approach have so far been put forward. It is timely to review what progress has been made over this past decade, and to assess how close we may be to resolving the matter. Although a good deal of experimental work has already been carried out, we shall see that there is a need for further data and several desirable experiments will be outlined.

From the practical point of view, a better understanding of the energy loss of fast channeled ions is useful in the study of crystals by means of a channeled probing beam (see Chapter 13). It is often important to know, from the energy spectrum of backscattered ions, the depth at which a collision occurred, and this is possible only if the energy loss of the channeled ions can be calculated.

5.2 THE EXPERIMENTAL DATA

The first experiments to measure the effect of channeling upon the energy loss of ions in the Bethe–Bloch region were carried out by Dearnaley[1] in 1963. This work was stimulated by the earlier transmission experiments at lower energies carried out by Nelson and Thompson,[2] and using the same geometrical arrangement protons of 2·1 MeV were collimated on to a thin crystal of silicon cut normal to the $\langle 111 \rangle$ direction. By rotation about this axis, which was suitably inclined to the proton beam, variations in the transmitted energy spectrum could be measured with a silicon surface-barrier detector placed behind the absorber crystal. Reductions of up to 50 % in energy loss were observed when the protons travelled along the $\langle 110 \rangle$ axial channels (Figure 5.1). It was proposed that the ions were steered by an effective potential of a nearly square-well form, and so caused to spend a greater time in regions of low electron density. It was deduced that the acceptance angle

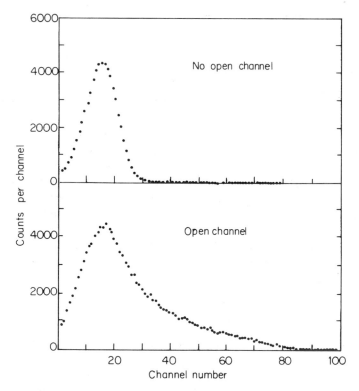

Figure 5.1. (a) Pulse height distribution for protons of incident energy 2·1 MeV transmitted through a thin silicon crystal (above) under random conditions and (below) under channeling conditions. The minimum proton energy loss is about half the random loss (after Dearnaley[1])

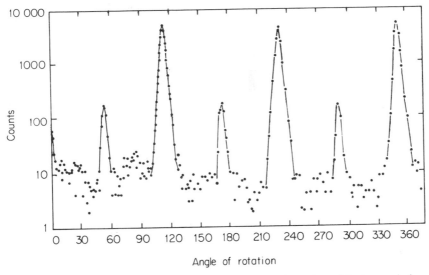

Figure 5.1. (b) The distribution with regard to crystal orientation of the transmission of protons with an energy exceeding a fixed threshold (approximately corresponding to channel 30 in the upper figure) (after Dearnaley[1])

over which channeling would be feasible should decrease with the square root of energy, but the value of this critical angle and the influence of thermal lattice vibration upon it were not derived.

Only axial channeling was described in this paper, although it was conceived at the time that planar channels in a lattice would also lead to a reduced energy loss, and this was demonstrated soon afterwards by Erginsoy et al.[3] Experiments with an increased angular definition of the incident protons (Gibson et al.,[4] Sattler and Dearnaley,[5] Farmery et al.[6]) showed that well channeled particles could be resolved as a distinct group in the transmitted spectrum, and so a precise measurement of their energy loss became feasible (Figure 5.2). About this time a number of laboratories equipped with electrostatic accelerators turned their attention to ion channeling, and so the available data grew rapidly. There were, however, many new phenomena to be observed, and much of the experimental work was concerned with the angular dependence of channeling phenomena and the spatial distribution of particles emerging from a crystal.[7,8] Early work[5] did reveal, however, that the energy loss of well channeled protons in germanium could be as much as a factor of three less than the random value and, as we shall see, this had important theoretical consequences.

As a result of experiments with moderately low energy (375 keV) protons, Eisen[9] observed that the minimum energy loss along the $\langle 110 \rangle$ axis in silicon was a little less than that measured along the $\{111\}$ and $\{110\}$ planes

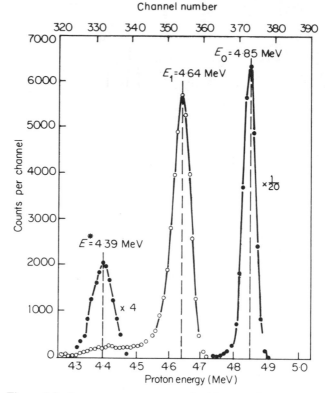

Figure 5.2. Energy spectra measured with a small diameter detector for protons of energy 4·85 MeV transmitted through a 50 μm thick silicon crystal at a random orientation (left-hand peak) and under channeling conditions along the {111} plane (central peak). The right-hand peak shows the spectrum of incident protons. All three spectra correspond to the same number of incident protons (from Gibson *et al.*[4])

which intersect in that axis. This result disagreed with the findings of Gibson *et al.*[4] at higher energies, which had suggested that the width of the broadest planar channel determines the energy loss along an axis. This point is still debated and the energy differences under discussion are very slight. It would appear[10,11] that the likelihood of measuring any difference depends upon such factors as the ion energy, the crystal thickness and its perfection, as these influence the probability of an ion trajectory remaining within a single axial channel (i.e. undergoing 'proper channeling'). The question as to whether the interaction of ions with a crystal lattice can be fully understood in terms of their interaction with a single atomic row or string[12] has been considered by Barrett,[13] with the aid of computer simulation of the ion motion. Differences in critical angles are indeed found, indicating that the ion interacting with a three-dimensional array of rows behaves differently from a set of equivalent ion–string collisions. This being so, the energy loss experienced along axial and planar channels may also be slightly different.

In measurements of the energy loss of the transmitted ions, two types of detector arrangement are feasible and have been employed. In one case a wide acceptance angle detector, placed close to the absorber, measures the emergent spectrum of all the ions while in the other case a finely collimated detector placed at a large distance from the absorber measures the energy of selected components of the emergent beam. Appleton *et al.*[7] have compared the results for the case of 3 to 11 MeV protons channeled through silicon. Figure 5.3 shows the total transmission spectra for two planes and an axis, and

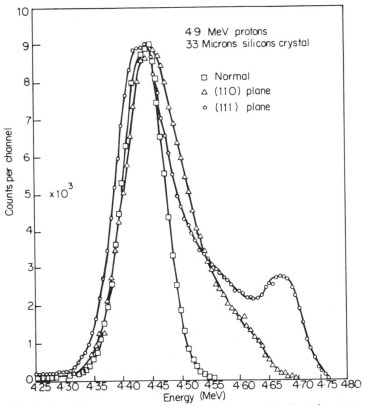

Figure 5.3. (a) Transmitted energy spectra, normalized to the same peak height, for 4·9 MeV protons incident parallel to the {111} and {110} planes and along a random direction, measured in a large area detector immediately behind the silicon absorber, the thickness of which was 33 μm (from Appleton *et al.*[7])

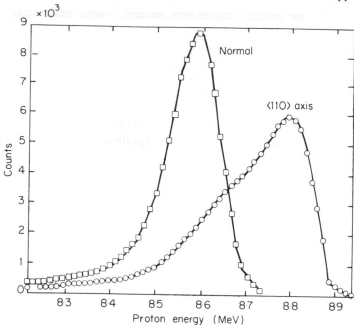

Figure 5.3. (b) Transmitted energy spectra, corresponding to the same number of incident protons of 9·0 MeV directed parallel to the ⟨110⟩ axis and a random direction in a 48 μm thick silicon crystal (from Appleton *et al.*[7])

it is obvious that in the former case the probability of an ion remaining channeled throughout the 33 μm thick crystal is quite small: in axial channeling the probability is much larger and very few of the protons suffer an energy loss as large as the normal value. Another feature is apparent in Figure 5.3(a), and this is the presence of some particles in the channeled spectrum with an energy loss greater than normal. This so called 'high loss component' has been studied in detail by Appleton *et al.*[7] who showed that the fractional probability of the high and low loss components, derived by subtraction of the normal from the channeled spectrum, varies with angle of incidence relative to the crystal plane (Figure 5.4). It is now believed that these high energy-loss particles are 'quasi-channeled', that is, that they enter the crystal at such an angle that they make a number of very close collisions with the atomic rows before suffering a dechanneling encounter, after which their energy loss reverts to normal. By travelling for some way in a trajectory which lies in a region of consistently high electron density, these ions lose energy more rapidly than in random collisions.

A finely collimated detector, on the other hand, can be used to measure the energy loss of those particles which emerge close to the crystal axis or plane, and have suffered no net deflection. Such very well channeled ions exhibit a

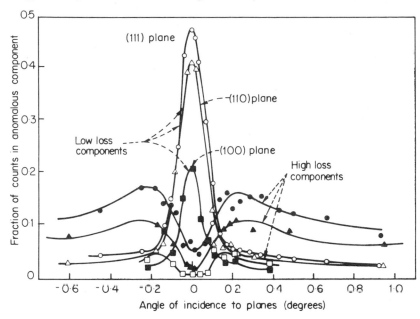

Figure 5.4. The fraction of high-loss and low-loss particles in the transmitted energy distributions for 2·8 MeV protons as a function of the angle of incidence relative to the {111}, {110} and {100} planes of a 33 μm thick silicon crystal (from Appleton *et al.*[7])

well defined energy loss, as shown in Figure 5.5, again taken from the paper by Appleton *et al.*[7] Table 5.1 shows their results for different incident proton energies and here the ratio of the channeled to the random energy losses $(E_1 - E_c)/(E_1 - E_N)$ shows no significant variation with proton energy. The width of the channeled distribution is noticeably less than that of the random spectrum, and this is to be expected since the statistical fluctuation in energy loss from random collisions will exceed that in a well defined channeled trajectory with a restricted range of impact parameters. In fact, the standard deviation in the distribution, after correction for the instrumental resolution, is approximately proportional to the energy loss. We shall return to this point later, but here remark that the true straggling, or statistical fluctuation in energy loss, might be expected to be lower than this measured value.

Several comparisons have been made of ion channeling in the Bethe–Bloch region using different crystals. The idea here was to investigate whether differing degrees of ionic bonding would influence the minimum energy loss by modifying the electron orbitals of the absorber atoms. Since the lattice structure and atomic number of the material will also affect the energy loss, it is necessary to choose similar crystal structures and lattice dimensions.

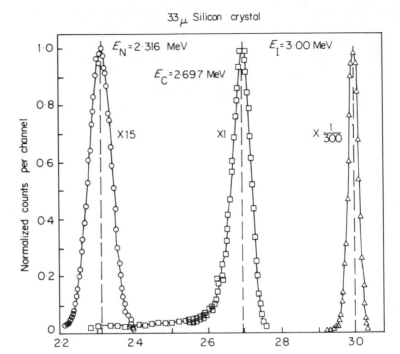

Figure 5.5. Energy spectra recorded in a small acceptance detector for 3·0 MeV protons transmitted in a random direction (left-hand spectrum) and parallel to the {111} planes (central spectrum) of a 33 μm thick silicon crystal. The right-hand spectrum shows the incident energy distribution (from Appleton *et al.*[7])

Sattler and Dearnaley[14] and Sattler and Vook[15] compared the diamond lattice semiconductors Si, Ge and GaAs and the zinc blende structures InAs, GaSb, AlSb and InSb. The conclusion arrived at was that the dominant factors which determine the energy loss are the channel dimensions and the atomic number of the absorber. The dependence upon the atomic number Z_2 of the absorber was found to be the same in the cases of well channeled ions and randomly transmitted ions. The nature of the bonding between the atoms has no significant effect. In the data published it appears that the ratio of the most probable channeled energy loss to the random energy loss, in several materials, increases slowly with ion energy. However, since the channeled energy peak becomes narrower with increasing energy in an absorber of given thickness, it follows that the minimum observed energy loss ratio, i.e. corresponding to the upper limit of transmitted energies, may vary more slowly or not at all with ion energy. Clark *et al.*[16] extended the survey of

Table 5.1. Energy loss of channeled and normal protons in silicon crystals

Crystal	E_I (MeV)	$E_C\{111\}$ (MeV)	$E_C\{110\}$ (MeV)	E_N (MeV)	$(E_I-E_C)/(E_I-E_N)$ $\{111\}$	$(E_I-E_C)/(E_I-E_N)$ $\{110\}$
33 μm Si	2·81	2·49	—	2·08	0·44	—
	2·97	2·68	—	2·28	0·41	—
	3·00	2·70	2·60	2·32	0·44	0·59
	4·43	4·21	4·11	3·94	0·45	0·65
	6·53	6·36	6·29	6·16	0·46	0·65
	7·00	6·84	6·79	6·64	0·43	0·58
	8·58	8·44	8·39	8·27	0·47	0·61
	9·00	8·88	8·82	8·71	0·43	0·62
	11·00	10·90	10·85	10·75	0·42	0·59
48 μm Si	7·00	6·78	—	6·48	0·43	—
	9·03	8·82	—	8·59	0·49	—

Energy loss of channeled and normal protons in germanium

Crystal	E_I (MeV)	$E_C\{111\}$ (MeV)	$E_C\{110\}$ (MeV)	E_N (MeV)	$(E_I-E_C)/(E_I-E_N)$ $\{111\}$	$(E_I-E_C)/(E_I-E_N)$ $\{110\}$
25 μm Si	3·00	2·67	2·52	1·95	0·31	0·45
	5·00	4·78	4·68	4·33	0·33	0·48
	7·00	6·83	6·76	6·47	0·33	0·45
	9·00	8·84	8·79	8·54	0·34	0·46
	11·00	10·86	10·81	10·61	0·33	0·47

Most probable energy loss values for protons transmitted along various planar channels in silicon and germanium (after Appleton *et al.*[7]). E_I is the incident proton energy, E_C the channeled energy and E_N the energy measured along a random crystal direction.

different absorbers to the body-centred cubic metals Fe, Mo and W and to the ionic insulators NaCl, CsI and MgO. These authors faced a question which had been largely overlooked in earlier work, namely whether the more significant measurement is the energy corresponding to the peak of the channeled energy spectrum or the leading edge of that peak, i.e. the minimum measurable energy loss. Clark *et al.*[16] inferred, for several reasons, that the width of the peak is dominated by differences in ion trajectory, but tabulated both the minimum and the most probable channeled energy loss. As in the work of Sattler *et al.*[14,15] there was no significant dependence of the behaviour upon material properties and the dominant factor was shown to be the size of the channel. In the light of this result, Clark *et al.*[16] plotted the minimum channeled energy loss for 4 MeV protons as a function of the width of a series of axial channels, for Si and Ge (Figure 5.6). These materials have closely

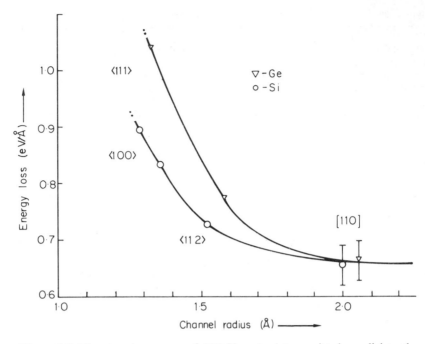

Figure 5.6. The stopping power of 4 MeV protons transmitted parallel to the major crystal axes in Si and Ge as a function of the channel radius corresponding to each axis (after Clark *et al.*[16])

similar lattice parameters and the same crystal structure. The result showed that for large channels the energy losses converge upon what may be an asymptotic limit, called the 'isotropic component.' The difference between the stopping powers of Si and Ge in the noncrystalline case was attributed to the different numbers of core electrons, while the 'isotropic' channeled component was associated with the effect of the valence electrons distributed throughout the channel.

Other energy loss measurements of light ions in metals were reported by Gibson *et al.*[17] who investigated the transmission of 400 keV protons and 800 keV helium ions through thin gold crystals. The helium ion energy is somewhat below the Bethe–Bloch region, though a similar behaviour was reported for both species of ion. The minimum energy loss observed along an axial direction was found to be the same as that for incidence along the most open plane which intersects that axis, a result in disagreement with that of Eisen[9] in silicon. The minimum channeled energy loss measured for 800 keV helium ions appeared to be about 60 % of the random energy loss. Machlin *et al.*[18] channeled protons of about 100 keV energy through very thin gold crystals of high quality. Owing to the breadth of the transmitted energy spectra, it is difficult to draw conclusions regarding the minimum energy loss, but it does seem clear that the ratio of the channeled loss to the random loss is at least 0·7. We shall see later that this ratio is in agreement with the results of work in silicon.

An investigation of the effect of temperature on the energy loss of 1·5 MeV protons channeled through silicon was made by Fujimoto *et al.*[19] and it was shown that the intensity of the channeled component diminishes with temperature by about a factor of two between 300 and 560 K but that the minimum and most probable energy losses do not vary.

It is only recently that an extended and systematic study of the energy loss of helium ions channeled in silicon has been performed by Eisen *et al.*,[10] covering the energy region between 100 keV and 18 MeV with the use of four different accelerators. Several interesting features emerged from these experiments. From measurements with crystals of different thicknesses it was shown that the width of the channeled energy spectrum is not determined by straggling (which increases as the square root of thickness) but by variations in ion trajectory (leading to a linear dependence on thickness). As shown in Figure 5.7, the $\langle 110 \rangle$ axial channeled spectrum is very broad and as much as

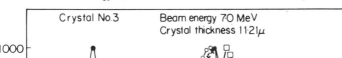

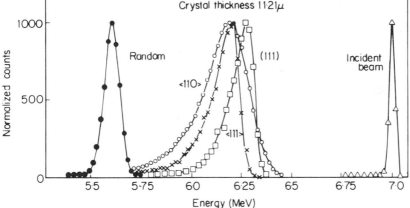

Figure 5.7. Energy spectra recorded in a small acceptance angle detector for 7·0 MeV ^{4}He ions transmitted through an 11·2 μm thick silicon crystal along the directions indicated (after Eisen *et al.*[10])

two to three times as wide as the random spectrum. A small but significant difference was observed between the minimum energy losses along the $\langle 110 \rangle$ axis and the $\{111\}$ planes which intersect it. The most probable energy losses also occur at different energies, a result which contrasts with those of

Appleton *et al.*[7] and Della Mea *et al.*,[20] but these authors were for the most part concerned with protons. Eisen *et al.*[10] point out that the ⟨110⟩ axis in silicon is broader than the width of the {111} planar channel and that when this is no longer the case, as in the ⟨111⟩ and ⟨100⟩ axes and their corresponding planes, the minimum energy losses for axial and planar channels are equal. This disagreement, while probably not of any great significance experimentally or theoretically, deserves to be resolved. The stopping power of He$^+$ ions in silicon[10] is shown as a function of energy in Figure 4.1, and the ratio α_{hkl} of the various channeled stopping powers to the most probable random value is shown in Figure 5.8. It is particularly interesting to see that in each case α passes through a maximum at the same energy as that corresponding with the maximum stopping power.

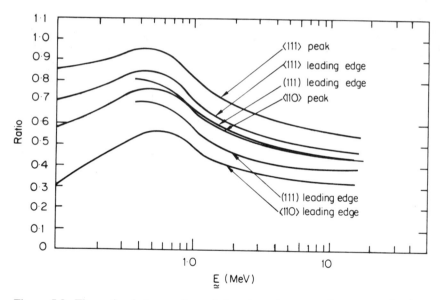

Figure 5.8. The ratios between channeled and random stopping powers for ^{4}He ions transmitted through silicon crystals, as a function of ion energy (after Eisen *et al.*[10])

Della Mea *et al.*[20] have also measured the energy loss of H$^+$, D$^+$ and He$^+$ ions channeled through silicon. They found that although the leading edges of the ⟨110⟩ and {111} transmitted spectra appear rather different the most probable (or peak) energies were the same and the spectra could be fitted with Gaussian distributions centred on the same value of energy. Della Mea *et al.*[20] therefore concluded that the minimum energy loss is effectively the same for the axis and the most open intersecting plane, and chose to present the most probable energy losses. Figure 5.9 shows their values together

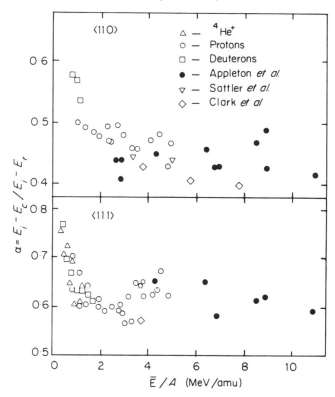

Figure 5.9. The ratio between channeled and random stopping power for various ions in silicon for the ⟨110⟩ and ⟨111⟩ directions, as a function of the mean energy per mass number (after Della Mea et al.[20])

with other experimental results for the ratio of channeled to random energy losses for protons, deutrons and helium ions as a function of the energy per nucleon, E/A. Once again there is a marked energy dependence of this ratio, and in further unpublished work Eisen[21] has shown that for protons also the ratio rises towards a maximum at about the same energy as that at which the random energy loss is a maximum.

This observation, which may well turn out to be a general one for all ions and absorbers, helps to explain the variation in channeled energy losses reported earlier.[17,18] When the energy lies near to that at which the stopping power is a maximum the channeled value may attain as much as 90% of the random value (Figure 5.8). Since helium ions of about 1 MeV energy are widely used as probes for the investigation of damage distributions and impurity location in crystals this fact is of practical importance besides stimulating a theoretical explanation.

5.3 THEORIES

The earliest suggestion of a procedure by which the electronic stopping cross section of fast channeled ions might be calculated was put forward by Nelson and Thompson.[2] A maximum effective impact parameter b_0 was defined in terms of the impulse approximation

$$b_0 = \frac{Z_1^2 e^4 M_1}{m E_1 I} \tag{5.1}$$

corresponding to the transfer of an energy I, the mean ionization energy. By inserting appropriate values of the electron density near to the ion trajectory into an integral with respect to impact parameter from zero to b_0 the energy loss dE/dz would be derived. The calculation was not carried out, but if it had been, the result for well channeled ions would almost certainly have been too low because of the fact that excitation of electrons, involving impact parameters much greater than b_0, can be expected to make a contribution to the energy loss.

Lindhard[12] considered the application of the Bethe–Bloch formula to channeled ions: by this we have

$$\frac{dE}{dz} = \frac{4\pi Z_1^2 e^4}{m v^2} Z_2 N \ln\left(\frac{2 m v^2}{I}\right) \tag{5.2}$$

in which N is the atomic density and I the average excitation potential. Lindhard followed Bohr[22] in dividing the collisions into the two classes consisting of close encounters, with large momentum transfers to individual electrons, and distant resonance collisions in which the impact parameter exceeds the electron orbital radius. Then, by the 'equipartition rule' of Lindhard and Winther,[23] it is argued that at asymptotically high ion velocities the close and distant collisions make an equal contribution to the energy loss. Therefore in the case of well channeled protons of MeV energies one would expect the energy loss to fall by up to a factor of two, as the ion trajectory can lie beyond the electron orbitals of all the atoms of the absorber.

In fact, as we have seen in Section 5.2, experiments showed[5] that the energy loss could be reduced by as much as a factor of three in germanium when protons travel along the $\langle 110 \rangle$ axial channel. This result destroys the usefulness of the simple concept of 'close' and 'distant' collisions and the elegant application of the equipartition rule. The earlier experiment, by Dearnaley,[1] in which it was shown that the efficiency of ionization per unit energy loss remains unaltered under channeling conditions suggested also that a distinction between the physical processes of interaction involved in close and distant encounters with the electrons of the absorber was a false one. Otherwise it is difficult to see why the balance between ionization and other dissipative processes such as phonon production should be independent of the ion trajectory.

Lindhard[12] had treated the electrons of the absorber as constituting an electron gas, with momentum and energy distributions derived from the Thomas–Fermi model. The next attempt to formulate the solution to the problem of energy loss was put forward by Erginsoy[24] along rather different lines. Erginsoy distinguished between the core electrons, whose spatial distribution follows that of the atomic nuclei, and the conduction electrons, which are more uniformly distributed. Erginsoy adopted an impact parameter formulation in which the average energy loss per electron is given by

$$\overline{\Delta E} = \Delta z N \int \bar{\varepsilon}(b)v(b)\,\mathrm{d}b \qquad (5.3)$$

where $\bar{\varepsilon}(b)$ is the average energy transferred to an electron in a collision with impact parameter b and $v(b)$ is the probability distribution of impact parameters. Bloch had shown that in first-order perturbation theory the Coulomb interaction between a heavy particle and a light electron yields

$$\bar{\varepsilon}(b) = \frac{2}{m}\left(\frac{Z_1 e^2 \omega}{v^2}\right)\{K_0^2(\omega b/v) + K_1^2(\omega b/v)\} \qquad (5.4)$$

in which K_0 and K_1 are zero and first-order modified-Bessel functions, and ω equals the change in electron excitation energy divided by $\hbar$. In uncorrelated collisions we have simply

$$v(b)\,\mathrm{d}b = 2\pi b\,\mathrm{d}b \qquad (5.5)$$

and

$$-\frac{\mathrm{d}E}{\mathrm{d}z} = 2\pi N \int_{b_{\min}}^{b_{\max}} \bar{\varepsilon}(b)b\,\mathrm{d}b. \qquad (5.6)$$

At sufficiently high velocities, for which $\omega b_{\max}/v \ll 1$, this reduces to the well known form

$$-\frac{\mathrm{d}E}{\mathrm{d}z} = \frac{4\pi Z_1^2 e^4 N}{mv^2}\ln\left(\frac{b_{\max}}{b_{\min}}\right). \qquad (5.7)$$

The value of $b_{\min}$ is chosen to correspond to the maximum momentum transfer $2mv$ to the electron, so that $b_{\min} = \hbar/2mv$. The maximum impact parameter $b_{\max}$ is determined by an adiabaticity criterion

$$b_{\max} = \frac{\hbar v}{E_f - E_i}. \qquad (5.8)$$

where E_i is the initial and E_f the final electron energy. This will clearly vary with the electron shell involved, and Erginsoy[24] calculated $b_{\max}^{K} \simeq 0.1$ Å for 5 MeV protons in silicon while $b_{\max}^{L} \simeq 2.0$ Å. On this basis K-shell excitation should be completely suppressed by channeling, while up to 85 % of the L-shell excitation could be eliminated in the case of well channeled 5 MeV protons along the $\langle 110 \rangle$ axis.

Next, Erginsoy considered energy losses to the weakly-bound valence electrons, which he assumed to be completely unaffected by channeling in the case of metals and only weakly influenced by it in semiconductors. He treated the valence electrons as a uniform gas with a complex dielectric constant, as had Lindhard[25] earlier. The energy loss is derived from an integration of the imaginary part of the reciprocal of the dielectric constant with respect to the energy transfer value. Numerical integration along the lines proposed by Lindhard and Winther[23] led Erginsoy to conclude that the asymptotic limit of equipartition between single-particle and damped-plasmon excitation is valid at high velocities where $v > v_F$, the Fermi velocity of the electron gas.

Thus, by neglecting core excitation by 3 MeV protons channeled along the $\langle 111 \rangle$ silicon axis, Erginsoy[24,26] divided the energy losses to valence electrons into two contributions

$$-\frac{dE}{dz} = \frac{4\pi Z_1^2 e^4 N}{m v^2} L \tag{5.9}$$

where

$$L = Z_{val} \ln \frac{v}{v_F} + Z_{loc} \ln(2 m v v_F / \hbar \omega_p).$$

Here the first term is due to collective or plasmon excitation of the valence electrons ($Z_{val} = 4$ in silicon) while the second term is due to single particle excitation of valence electrons distributed uniformly throughout the channel, and $\hbar \omega_p$ is the energy of the plasmon (or elementary collective excitation of the electron gas) which has a value close to 16 eV in silicon. From experimental measurement[7] of the minimum energy loss Erginsoy arrived at $Z_{loc} \simeq 4$ for the $\{111\}$ planar channel, and this figure was compared with the result derived by assuming that the spread in energy of the transmitted proton spectrum is due entirely to straggling, or statistical fluctuation in the mean energy loss, and a formula of Bohr's then yields $Z_{loc} \simeq 4.1$. This excellent agreement was considered to confirm the validity of the model.

However, as we have seen in Section 5.2, Eisen[9,10] and other experimental physicists[18] have adopted a different interpretation of the width of the channeled energy spectrum. Rather than attributing the width to straggling, they argue that it comes about because slightly different proton trajectories through the crystal correspond to different energy losses. In their view the straggling in energy for a given, well channeled ion trajectory is small in comparison. This argument would lead to much smaller values of Z_{loc} and spoil the agreement that Erginsoy deduced with the value derived from the channeled energy loss. Moreover, the results of the calculation appear to have been less satisfactory in the case of germanium than in the possibly fortuitous case of silicon.

Luntz and Bartram[27] treated the problem of core excitation by channeled ions a little differently, within the framework of Erginsoy's model. A different adiabatic criterion led to a calculated result for the energy loss of α-particles channeled in CsI crystals which fell below the experimental value measured subsequently by Clark *et al.*[28] It is clear that almost any required result could be derived from a suitably chosen starting criterion.

Brice[29] adopted a very different approach in which the transmission of a particle through the lattice is treated as a series of diffractions of a plane wave at each independent scattering event. The atoms are considered to behave as a core consisting of the nucleus and tightly bound electrons able to move independently of a negatively charged shell of loosely bound electrons. Application of phonon theory in a simple harmonic approximation then yields four bands of phonon-like excitations, two of which correspond to acoustical and optical phonons and the other two correspond to the outer electrons and the core moving out of phase. These latter modes introduce a time-dependent polarization of the lattice and it is the lower energy set of these polarization modes which are identified with the plasmon excitations. Brice assumed that the energy loss of a well channeled ion takes place predominantly to the collective excitations of the solid: he did not set out to calculate the random stopping power. The number of loosely bound outer electrons was treated as an adjustable parameter, to be derived by comparison with experiment. The energy loss was derived in terms of the transition probability $W(\mathbf{K}, \Delta\omega)$ that the incident ion undergoes a change of momentum $\hbar\mathbf{K}$ with a simultaneous change in energy $\hbar\Delta\omega$, the first-order time-dependent perturbation theory. Brice made a number of approximations including the assumption that the displacement of the cores is negligible compared with the corresponding displacement (in a polarization mode) of the outer electrons. He set the frequency spectrum of these modes equal to the average frequency ω_p, with $\hbar\omega_p = 20\,\text{eV}$ (for germanium). The stopping power is then given by

$$-\frac{dE}{dz} = \frac{1}{v}\sum_{\mathbf{K}} ne\ W(\mathbf{K}, n)$$

in which n is the number of plasmons excited in a collision:

$$\hbar\Delta\omega = n\hbar\omega_p.$$

Brice[29] was able to calculate the contours of equal stopping power as a function of the angle of incidence of a proton near to the $\langle 110 \rangle$ axis in germanium, with the result shown in Figure 5.10. The values are nearly constant along the planes but fall sharply along the axis. However, when account is taken of the energy dependence of the transition probabilities the

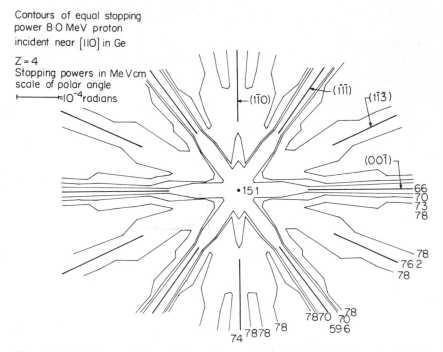

Figure 5.10. Polar projection of the contours of equal stopping power for 8 MeV protons incident near the ⟨110⟩ direction in germanium. The pole of the projection is the ⟨110⟩ axis. These contours were determined using the 'unbroadened' probability function (after Brice[29])

difference becomes less marked, and Brice obtains a value along the ⟨110⟩ axis which is 4% lower than that along the {111} plane for the case of 7 MeV protons in germanium. It is interesting that Eisen *et al.*[10] observed a 5% difference for helium ions of this energy channeled along these directions in silicon, although Della Mea *et al.*[20] working at lower energies did not detect any difference. Brice was able to fit the experimental data of Sattler and Dearnaley[14] with values of the number N of loosely-bound electrons as follows:

$$N\langle 110 \rangle = 5 \cdot 6 \pm 0 \cdot 3$$

$$N\langle 112 \rangle = 7 \cdot 2 \pm 0 \cdot 3$$

$$N\langle 111 \rangle = 8 \cdot 6 \pm 0 \cdot 4.$$

These values, while larger than the number (4) of valence electrons, show a degree of anisotropy which is consistent with the idea that in narrower channels fewer electrons are screened from the polarizing influence of the channeled ion. Brice went on to calculate the straggling in energy of the transmitted channeled particles, arriving at values which were appreciably smaller than those which had been observed[14] at the time with rather poor angular collimation but not at all inconsistent with recent measurements in silicon[10,20] if one considers the width of the Gaussian fitted to the leading edge of a planar channeled spectrum.

The method adopted by Brice treats all channeled collisions as giving rise to collective excitations. This does not resolve the question as to why the ionization efficiency remains the same under channeling conditions[1] unless, following Klein,[30] one assumes that even under random conditions the dominant mode of electron excitation in silicon is collective. With such an assumption it is easier to understand the small values of Fano factor which are observable in semiconductor charged particle detectors.[31]

A different approach to the problem of energy loss has been published in a preliminary form by Bonsignori and Desalvo.[32,33] Their method treats the energy absorption in terms of a complex dielectric constant, as had Lindhard[25] and Erginsoy[24] previously, but takes account of the fact that the electron gas is not spatially uniform but has the lattice periodicity. A general expression for the spatially periodic dielectric constant had been derived earlier by Falk.[34] The contributions of single particle collisions and collective plasmon excitation towards the overall energy loss were calculated both for channeling and random conditions. Bonsignori and Desalvo found a significant departure from equipartition, the ratio of stopping powers arising from close and distant collisions being about 1·5 for 100 keV protons in gold. Of course, this energy is well below those considered by Erginsoy[24] and it is doubtful whether the equipartition rule of Lindhard and Winther[23] should apply even in a spatially uniform electron gas under these conditions. The effect of channeling along the ⟨100⟩ direction was found to be very small, and approximately equal (1·3 %) for both single particle and collective excitation. The theoretical value for the energy loss in a random direction was about a factor 4·5 times lower than that measured[18] experimentally. The reduction in the ⟨100⟩ channeled energy loss of 100 keV protons in gold compared with the random value is about 10 % (for the most probable energy loss) or 30–40 % (for the minimum channeled energy loss) according to the data of Machlin *et al.*[18] Bonsignori and Desalvo[33] attribute the disagreement largely to the fact that the theory in its present form takes account only of s-electrons, and propose that an extension of the model to the tight-binding case, so as to handle d-electrons, would lessen the discrepancy. It would also be interesting to apply their approach to the higher energy data now available in silicon since here the channeled energy loss is far better defined experimentally and, moreover, the results could be compared with those of Erginsoy.[24,27] An interesting aspect of the Bonsignori and Desalvo treatment is that channeling appears to reduce the probability of both single particle and collective excitations equally. This would explain the constancy of the ionization probability per unit energy loss measured by Dearnaley.[1]

The most recent attack upon the problem of the energy loss of channeled ions is that which has been embarked upon by Dettmann and Robinson.[35] Their approach takes us back to first principles and offers the hope of an entirely unified treatment which does not oblige us to divide the collisions into near and distant events. Moreover, it can be applied equally to the case of well channeled ions and the random energy loss, so that the experimentally measured ratios between these values may be compared with theory. The energy transfer $\varepsilon(v, b)$ to an electron in a given shell is calculated as a function of the ion velocity, $\mathbf{v}$, and the impact parameter b (assumed constant throughout the collision) in first-order Born approximation. It is assumed therefore that the ion velocity $\mathbf{v}$ is much greater than the electron orbital velocities. The one-electron Hamiltonian contains the sum of two Coulomb potentials, owing to the target nucleus and the moving ion, and in first-order Born approximation $\varepsilon(v, b)$ can be represented by an average over the ground state of the electron. The treatment outlined so far corresponds to that adopted by Bloch, but differs subsequently in not being restricted to distant collisions.

Bloch introduced a cut-off radius b_{min}, because his large impact parameter calculation of the ground state average diverges for small b. The cut-off was chosen to correspond to the maximum classical energy transfer to an electron. In contrast to Bloch, Dettman and Robinson avoided the large impact parameter approximation and found for $b \leqslant a_0$ much smaller values for $\varepsilon(v, b)$, extending to $b = 0$ in a flattish maximum. For large b the energy transfer approaches Bloch's result falling as $1/b^2$ up to $b_c = a_0(v/v_0)$, where v_0 is the Bohr velocity. The energy loss dE/dz is obtained by summing $\varepsilon(v, b)$ over all electron shells, but there is no need to distinguish core and valence electrons. The approach is essentially analytic, since only an integration for intermediate b values is carried out numerically. It will be most interesting to witness how well this approach explains the experimental data on channeled energy losses, and furthermore what new experimental tests will become apparent.

5.4 FURTHER WORK

We are now in a better position to judge what further experimental and theoretical work is necessary in the study of the energy loss of fast channeled ions. As the theory develops, of course, other experiments are likely to suggest themselves, but we shall pick out a few which appear significant from a present day standpoint.

The theories differ with regard to their predictions of the behaviour of very fast particles: in some the ratio α_{hkl} of the minimum channeled energy loss to the random value is expected to reach an asymptotic value close to those already measured, while in others a steady, though slow, decrease is expected. A measurement of α in, say, silicon or germanium at proton energies in the range 100 to 500 MeV would therefore be desirable. The precision with which the crystals must be oriented is not likely to be too severe a problem, but beam collimation will be more difficult. Whitehead[36] has suggested the use of active collimators, i.e. small scintillator crystals or other fast detectors arranged in coincidence on either side of the absorber. A coincidence signal then gates the transmitted energy detector.

Further experiments are needed to verify whether α_{hkl} always reaches a maximum at the energy corresponding to the maximum stopping cross section. If so, this gives a straightforward result for theoretical consideration. A comparison, for example, might be carried out with ^{3}He and ^{4}He ions in the same crystal.

It should be settled whether the minimum energy loss in an axial channel can differ from those in the intersecting major planar channels, but the result is not likely to be of great significance either way. It is more that the present discrepancies are untidy.

The more interesting experiments seem to be those in which some change is made to the electronic system in a crystal and a measurement is made of the minimum channeled energy loss, which may be most sensitive to the state of the loosely bound outer electrons. Energy might be pumped optically into a crystal of suitable material, of the type used in lasers, and a study made to find whether a moving ion can release this energy and so suffer a reduced energy loss. The amount of stored energy within a given channel seems easily detectable in terms of the energy of a single proton.

Another experiment would be to investigate whether the transition to the superconducting state brings about a change in the energy loss of a fast particle transmitted through a thin single crystal. If the long range interaction is with pairs of electrons rather than single particles, there may be a change in the minimum channeled energy loss. Without further theoretical consideration this difficult experiment would probably be too speculative.

References

1. Dearnaley, G., *IEEE Trans. on Nucl. Sci.* **NS–11**, no. 3, 249 (1964).
2. Nelson, R. S. and Thompson, M. W., *Phil. Mag.*, **8**, 1677 (1963).
3. Erginsoy, C., Wegner, H. E. and Gibson, W. M., *Phys. Rev. Lett.*, **13**, 530 (1964).
4. Gibson, W. M., Erginsoy, C., Wegner, H. E. and Appleton, B. R., *Phys. Rev. Lett.*, **15**, 357 (1965).
5. Sattler, A. R. and Dearnaley, G., *Phys. Rev. Lett.*, **15**, 59 (1965).
6. Farmery, B. W., Nelson, R. S., Sizmann, R. and Thompson, M. W., *Nucl. Instrum. & Meth.*, **38**, 231 (1965).
7. Appleton, B. R., Erginsoy, C. and Gibson, W. M., *Phys. Rev.*, **161**, 330 (1967).
8. Dearnaley, G., Mitchell, I. V., Nelson, R. S., Farmery, B. W. and Thompson, M. W., *Phil. Mag.*, **18**, 985 (1968).

9. Eisen, F. H., *Phys. Lett.*, **23**, 401 (1966).
10. Eisen, F. H., Clark, G. J., Bøttiger, J. and Poate, J. M., *Radiation Effects* (to be published).
11. von Jan, R., *Phys. Rev. Lett.*, **18**, 303 (1967).
12. Lindhard, J., *Dansk Vid. Selsk.*, *Mat. Fys. Medd.*, **34**, No. 14 (1965).
13. Barrett, J. H., *Phys. Rev. B*, **3**, 1527, (1971).
14. Sattler, A. R. and Dearnaley, G., *Phys. Rev.*, **161**, 244 (1967).
15. Sattler, A. R. and Vook, F. L., *Phys. Rev.*, **175**, 526 (1968).
16. Clark, G. J., Morgan, D. V. and Poate, J. M., *Atomic Collision Phenomena in Solids*, Ed. D. W. Palmer *et al.*, Amsterdam, North-Holland, p. 388 (1970).
17. Gibson, W. M., Rasmussen, J. B., Ambrosius-Olesen, P. and Andreen, C. J., *Can. J. Phys.*, **46**, 551 (1968).
18. Machlin, E. S., Petralia, S., Desalvo, A., Rosa, R. and Zignani, F., *Phil. Mag.*, **22**, 101 (1970).
19. Fujimoto, F., Komaki, K., Ozawa, K., Mannami, M. and Sakurai, T., *Phys. Lett.*, **29A**, 332 (1969).
20. Della Mea, G., Drigo, A. V., Lo Russo, S., Mazzoldi, P. and Bentini, G. G., to be published (1972).
21. Eisen, F. H. (to be published).
22. Bohr, N., *Dansk. Vid. Selsk.*, *Mat. Fys. Medd.*, **18**, No. 8 (1948).
23. Lindhard, J. and Winther, A., *Dansk. Vid. Selsk.*, *Mat. Fys. Medd.*, **34**, No. 4 (1964).
24. Erginsoy, C., Proc. Conf. on Solid State Physics with Accelerators, Brookhaven 1967 (Brookhaven National Laboratory Report BNL 50083 (C-52)).
25. Lindhard, J., *Dansk. Vid. Selsk.*, *Mat. Fys. Medd.*, **28**, No. 8 (1954).
26. Erginsoy, C., *Interaction of Radiation with Solids*, Ed. A. Bishay, N.Y., Plenum Press, p. 341 (1967).
27. Luntz, M. and Bartram, R. H., *Phys. Rev.*, **175**, 468 (1968).
28. Clark, G. J., Dearnaley, G., Morgan, D. V. and Poate, J. M., *Phys. Lett.*, **30A**, 11 (1969).
29. Brice, D. K., *Phys. Rev.*, **165**, 475 (1968).
30. Klein, C. A., *J. Phys. Soc. Japan, Suppl.*, **21**, 307 (1966).
31. Dearnaley, G., *Proc. Conf. on Semiconductor Nuclear Particle Detectors and Circuits, Gatlinburg 1968* (National Academy of Sciences publication 1593).
32. Bonsignori, F. and Desalvo, A., *Lett. al Nuovo Cimento*, **1**, 589 (1969).
33. Bonsignori, F. and Desalvo, A., *J. Phys. Chem. Solid*, **31**, 2191 (1970).
34. Falk, D. S., *Phys. Rev.*, **118**, 105 (1960).
35. Dettmann, K. and Robinson, M. T., private communication (1972).
36. Whitehead, C., private communication (1970).

CHAPTER VI
Detailed Studies of Channeled Ion Trajectories and Associated Channeling Potentials and Stopping Powers

S. DATZ, B. R. APPLETON and C. D. MOAK

6.1 INTRODUCTION

The trajectories of ions channeled in crystalline solids are governed by their interaction with the interatomic potentials of the atoms making up the planes or axial 'strings'. In the case of planar channeling, the particles oscillate in a well-defined two-dimensional potential and the trajectories depend only upon the initial transverse energy. For axial channeling the motion is in general less defined. If the particle has insufficient transverse energy to penetrate the potential maxima between adjacent strings, it will be confined to motion between a single set of strings ('hyperchanneling'). Even less constrained is the motion of axially channeled particles which have sufficient transverse energy to penetrate the potential maxima between strings. These particles may wander between axial channels, their only constraint being the avoidance of small impact parameters with a single string potential. A crude approximation to the potentials is all that is necessary for the determination of Rutherford scattering dips and many other observables in channeling. But, aside from the intrinsic interest in associating a channel potential with fundamental atomic properties, a detailed potential aids in the understanding of delicate features such as 'flux peaking'.

In any case, if the detailed trajectory of the particle could be determined, the potentials could be obtained directly. In all the studies carried out thus far the label identifying a given trajectory has been the energy loss experienced by the ion along its path. Ions moving close to atomic planes and axes go through regions of higher electron density and hence lose more energy. Since planar channeling is the simplest to treat and has been most studied to date, this will be treated first in Section 6.2. Recent studies on 'hyperchanneling' potentials will be treated in 6.3.

6.2 PLANAR CHANNELING

To date, planar channeling potentials and stopping powers have been studied by several groups using somewhat different methods.

All the experiments carried out involve the use of very thin crystals and energy sensitive detectors as schematically indicated in Figure 6.1. An ion

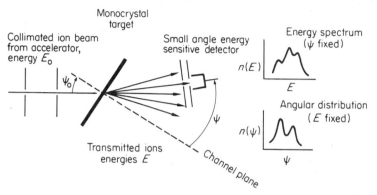

Figure 6.1. Outline of the experimental arrangements for studying the energy losses of ions in thin single crystal targets

from some suitable accelerator is passed through a thin single crystal which can be adjusted to a given tilt angle, and rotated in its mounting plane. An energy-sensitive detector subtending a small solid angle is placed in line with the emergent beam or is rotated about the crystal position.

All the methods are based upon the same fundamental model. For this reason, we will first describe the general model and discuss the various contributions in chronological order.

6.2.1 First-Order Model for Planar Channeling[1,2]

Consider a planar channel consisting of two parallel closely packed lattice planes (Figure 6.2). In this channel a charged particle of energy E experiences a repulsive force from the atoms in each wall. We assume a smooth 'planar potential', $V(x)$, where x is the distance from the midplane. The particle motion reduces to a two-dimensional problem in which the direction of incidence into the channel lies in the xz plane. The walls are assumed to be rigid. Since for channeled particles the maximum angle between particle trajectory and the plane is approximately $1°$, the z component of the velocity is assumed to be constant. Thus,

$$\frac{d^2x}{dz^2} = -\frac{1}{2E}\frac{\partial V(x)}{\partial x}$$

(6.1)

where E is the particle energy. We can express the potential, in general, in

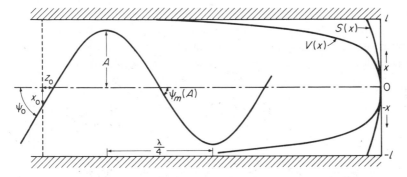

Figure 6.2. Trajectory of an ion entering a planar potential at x_0 with and angle ψ_0. The potential $V(x)$ and stopping power $S(x)$ functions are indicated on the right

the form of a polynomial expansion

$$V(x) = V(0) + a_1 x^2 + b_1 x^4 \ldots \qquad (6.2)$$

in which case the equation of motion is that of an undamped anharmonic oscillator. The consequences can be seen in Figure 6.3, which shows trajectories for three different impact parameters with respect to the channel

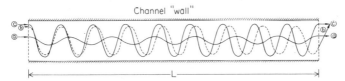

Figure 6.3. Trajectories for three entrance points. Particles a and c will reach a detector in line with the beam

midplane. Ions entering close to the midplane are deflected only weakly (path a); the wavelength λ of their oscillatory motion is comparatively long and their amplitude A is comparatively small. Those entering closer to the edge of the channel (paths b and c) are repelled more strongly and have paths of larger amplitude and shorter wavelength. Assuming sinusoidal motion, we can approximate the solution by the periodic form

$$x = A_1 \sin\{(z - z_0)/\hat{\lambda}\} + A_3 \sin\{3(z - z_0)/\hat{\lambda}\} \qquad (6.3)$$

where $A = A_1 + A_3$ is the oscillation amplitude and $\hat{\lambda} = \lambda/2\pi$. Neglecting higher-order terms in equation (6.2), we insert (6.3) and (6.2) into (6.1). Setting the coefficients of $\sin\{(z - z_0)/\hat{\lambda}\}$ and $\sin\{3(z - z_0)/\hat{\lambda}\}$ to zero yields

$$\hat{\lambda}^{-2} = a + \tfrac{3}{4}bA_1(A_1 + A_3) \simeq a + \tfrac{3}{4}bA^2 \qquad (6.4)$$

where $a = a_1/E$ and $b = 2b_1/E$. For any realistic potential the anharmonicity decreases with decreasing amplitude so that in the region of minimum amplitude, i.e. for parallel entrance close to the centre of the channel, the wavelength should be independent of amplitude. We designate this as the A_0 region.

The energy loss suffered by particles in this energy range is primarily caused by inelastic collisions with electrons and, since the electron density close to the atomic planes is higher than at the channel centre, the stopping power $s(x)$ increases with x. The energy loss of particle b (Figure 6.3) should be larger than that of a because it has penetrated the atomic planes more deeply and more often. We can therefore associate (for a given crystal thickness) an energy loss with a trajectory and a given entrance point and hence with a given wavelength and amplitude.

As the entrance point is moved from the midplane toward the channel wall, the energy loss increases and if the energy spectrum of all the emerging particles were measured it would be structureless. However, if the detector subtends a small angular aperture, only those trajectories which leave the crystal within that angle would be detected. This introduces a boundary condition for the wavelength of the detected particles

$$\lambda_a = (L \pm z_0 \mp z_e)/n \qquad (6.5a)$$

$$\lambda_b = (L \pm z_0 \pm z_e)/(n + \tfrac{1}{2}) \qquad (6.5b)$$

where n is an integer, L is the length of the crystal, and z_0 and z_e are the lengths associated with the phase shifts of incidence angle ψ_0 and exit angle ψ_e. (For sinusoidal motion $z = \lambda \cos^{-1}(\lambda\psi/A)$.) An early example of a real spectrum is shown in Figure 6.4.

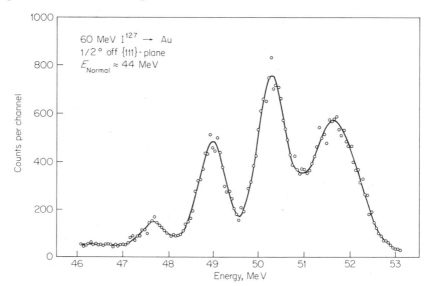

Figure 6.4. The energy spectrum observed in 60 MeV ^{127}I ions transmitted through {111} planar channels in a 0·7 μm thick Au crystal. The beam was incident 0·5° from the plane. The detector was collinear with the incident beam

6.2.2 'Perfect Crystal' Experiment

Let us first consider the information obtainable from an ideal experiment in which we employ a flawless crystal of perfectly uniform thickness. If the experiment were arranged with the detector and beam in line and the planar channel aligned with the detector beam axis, i.e. $\psi_0 = \psi_e = 0$, then $\lambda_a = L/n$, $\lambda_b = L/(n + \frac{1}{2})$. If the path length is changed, the wavelengths (and associated energy loss) of each detected group shifts until at $\Delta L = L_2 - L_1 = \lambda$; $(dE/dz)_{n,L} = (dE/dz)_{(n+1),L}$. A measurement of the ΔL required to return to a given wavelength group is a direct measure of the wavelength, and the dependence of λ upon stopping power is established. Tilting the crystal plane with respect to the beam axis (i.e. $\psi_0 = \psi_e \neq 0$) splits the λ_b solutions but the λ_a solutions are unchanged as long as the detector is kept on the beam axis.

The population of a given group depends on the length of the region δx_0 which leads to exit angles included within the detector aperture $\delta \psi_e$. With $\psi_0 = \psi_e = 0$ and a fixed small aperture detector, two effects combine to suppress the intensity of high amplitude groups. First, the variation in exit angle with entrance point $\delta \psi_e/\delta x_0$ increases with both decreasing wavelength and increasing amplitude; and, second, a steeper potential gradient at larger amplitude causes a more rapid decrease in wavelength within a given x_0 zone.

The amplitude of an oscillation is a function of the transverse energy,

$$V(A) = V(x_0) + E \sin^2 \psi \tag{6.6}$$

so that for each amplitude there is a critical angle $\psi_c(A)$ corresponding to $x_0 = 0$ beyond which the intensity at A vanishes. When the crystal is tilted, the population of a given sidegroup increases until it drops abruptly to zero at $\psi_c(A)$. Here again two effects are responsible for the increase. As ψ is increased from 0 to $\psi_c(A)$ the phase shift passes from $\frac{1}{4}\lambda$ to 0 and $d\psi/dz$ goes from maximum to minimum. The second effect again is related to the steepness of the potential. As ψ is increased the entrance point x_0 for a given A is moved to smaller values and since $\partial V(x_0)/\partial x_0$ decreases with x, a larger range of x_0 values will be included in $\delta V(A)$. The width of the energy-loss group is related to the gradient of the stopping power function within the δA region contributing to the peak.

If the functional form of the trajectory is known, a measurement of the cut-off angle $\psi_c(A)$ for a given wavelength group would give the amplitude and the exact path is determined. For example, if the wave were sinusoidal, we should have

$$A = \lambda \psi_c(A). \tag{6.7}$$

Even if the functional form is not known, it could be measured by an experiment in which the detection angle (ψ_e) is changed to compensate for a thickness change which deflects the emergence angle for a given dE/dz

group; e.g. start with $L = L_1$ and $\psi_0 = \psi_e = 0$ and focus attention on one dE/dz group. Now if the pathlength is changed slightly, the group emergence angle also changes. By rotating the detector to find the emergence angle, it could be followed up to $\psi_c(A)$. This procedure determines the detailed trajectory shape over a quarter wavelength.

In practice, the spectra are strongly affected by even small amounts of mosaic spread in the specimens used. The principal effect is to eliminate structure attributable to the half integral groups (equation 6.5b).[2] For example, when the detector is in line with the beam direction, $\psi_0 = \psi_e$ and $z_0 = z_e$. The phase factors for the integral solutions cancel and the structure is unaffected by a spread in ψ; but for the half integral solutions the phase factors add, and slight variations in ψ cause drastic changes in the transmitted wavelength. Mosaic spread (variation in ψ) is equivalent to variation in entrance and exit points. For the integral solutions a shift in entrance angle is exactly compensated by a shift in exit angle. On the other hand, with the half integral solutions, a decrease in entrance angle causes an increase in exit angle, and since $\psi_c(A) < 0.5°$, a distribution in ψ of greater than $0.1°$ can completely eliminate the observation of structure attributable to half integral solutions. The effects of length uncertainties and radiation damage are discussed in references 2 and 3.

(1) *Wavelength or Frequency Determinations (Variable Channel Length Method)*

Of all the possible variables which can be determined, the one which has proved most amenable to quantitative measurement is that of wavelength or frequency. Several methods have been used. The pathlength through a given planar channel can be varied within a single specimen by simply varying the tilt angle of the crystal with respect to the beam axis. This method, developed by the Oak Ridge group,[1,2] is exemplified in Figure 6.5 where we have plotted $\Delta E/\Delta z$ spectra for ^{127}I ions channeled in a $\{111\}$ plane for several different thicknesses. (These spectra were taken at various rotation angles with respect to the plane and no significance should be attached to the relative group populations.) The position marked A_0 corresponds to the minimum-energy-loss group. The particles in this group have been confined to trajectories near the channel midplane where the potential is almost harmonic. The definition of the A_0 group depends on the detector aperture $\delta\psi_e$; e.g. for $\psi_0 = \psi_e = 0$ all amplitudes having $\psi_c(A) \leqslant \delta\psi_e$ will be detected without regard to integral wavelength conditions. Thus, in the $1\ \mu m$ wavelength region for $\psi_e \simeq 0.001$ rad, all particles with $A < 0.15$ Å will be detected (see equation (6.7)). The peak position of the A_0 group should not vary, and the upper right-hand leading edge of this group represents the energy loss for particle paths very close to the midplane. The peaks of the intermediate groups move with changing pathlength as predicted by equation (6.5a).

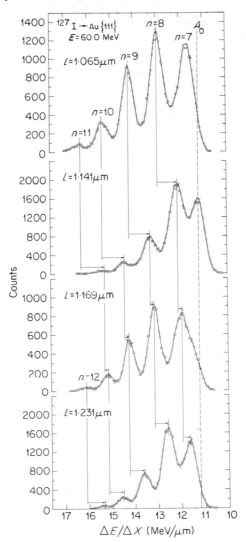

Figure 6.5. Energy spectra observed in ^{127}I ions transmitted through $\{111\}$ channels of a Au crystal as a function of the pathlength

Plots of the peak positions against $\Delta E/\Delta z$ for 60 MeV ^{127}I in the $\{100\}$ and $\{111\}$ planar channels of Au are shown in Figure 6.6. The vertical distance between lines represents the energy spacing between groups at a fixed thickness. To first order, the horizontal interval between the n and $(n + 1)$ groups is the wavelength of the particles in the n group at the corresponding

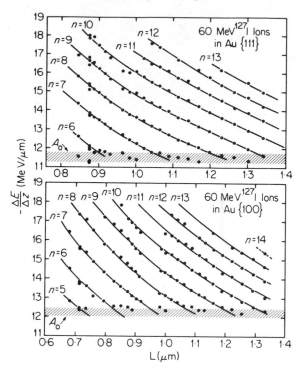

Figure 6.6. Energy loss at the group peak positions for 60 MeV ^{127}I ions channeled in $\{111\}$ and $\{100\}$ planes of Au as a function of pathlength. The symbol n designates the number of particle oscillations along the path

value of $\Delta E/\Delta z$ on the abscissa. Approximate values of n (within one unit) can be obtained from the measured spacing at a given thickness. Actually, since the particle loses a significant fraction of its energy, its trajectory is somewhat altered so that, strictly speaking, the condition for observation is an integral number of periods. To the degree that the parallel and transverse energies are separable, there are two effects of energy loss on the path. First, since the particle is oscillating in a transverse potential with a fixed frequency, a decrease in longitudinal velocity will decrease the wavelength (equation (6.9)). On the other hand, a loss in transverse energy will decrease the amplitude and increase the wavelength. For reasons which are discussed in detail by Robinson,[3] the first effect predominates and results in a slight shrinkage in horizontal spacing with increasing thickness. To correct for energy loss in the crystal and obtain the initial stopping power,[3] we assume some form of the energy dependence of the stopping power E^p then

$$(-dE/dz)_{E=E_0} = \frac{E_0^p}{(1-p)Z}(E_0^{1-p} - E^{1-p}) \quad p \neq 1$$

$$= \frac{E_0}{Z} \ln \frac{E_0}{E} \qquad p = 1 \qquad (6.8)$$

Moreover, since the transverse energy is only slightly affected, frequency is defined by

$$\omega = \frac{n(1-2p)(E_0^{1-p} - E^{1-p})}{(1-p)(E_0^{\frac{1}{2}-p} - E^{\frac{1}{2}-p})} \qquad p \neq 1 \qquad (6.9)$$

where n is the integral number of oscillations which the ion makes in passing through the channel. Note that ω is not a true frequency, since a factor $(2m)^{\frac{1}{2}}$ (m, the mass of the ion) has been omitted from the definition.

When properly corrected, all values of n should yield the same relationship of initial stopping power to oscillation frequency. An example of this is shown for 60 McV I ions in Au in Figure 6.7. The linearity of the curves obtained was at first surprising (and quite useful in easing data treatment). Although we have used the case of 60 MeV I in Au, the linearity of the ω

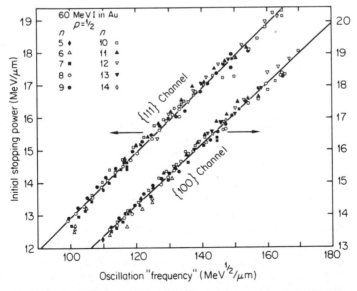

Figure 6.7. The dependence of the stopping power of 60 MeV ^{127}I ions on their transverse oscillation frequencies in two planar channels of Au

against (dE/dz) curves is by no means unique to this case and has been found for all systems studied[2,4,5,6,7] (see Table 6.1). Hence, we have the empirical relationship $(-dE/dz)_{E=E_0} = \alpha + \beta\omega$.

Table 6.1. Curvature parameters, screening lengths and ionic charges deduced for various ions in Au, Ag and Si

Target	Ion (ref.)	Energy (MeV)	Curvature parameter (eV Å^{-3})			Screening length (l/b)Å	Derived ion charge Z_2
			y_{111}	y_{100}	y_{111}/y_{100}		
Au	^{4}He (2)	3	64	108	0·59		2·0
	^{16}O (5)	10	154	253	0·61		4·8
	^{127}I (2, 5)	15	837			0·31	26·0
		21·6	915	1600	0·57		30·0
		60	1288	2077	0·62		40·0
Ag	^{127}I (6)	21·6	833	1420	0·59	0·31	33·0
		60	1042	1696	0·62		41·0
			y_{111}	y_{110}			
Si	^{1}H (7)	0·4	{14·5 10·9	30·1		0·45	0·9

(2) *Frequency Determination (Energy Variations Method)*

Since the oscillation 'frequency'

$$\omega = n(2m)^{\frac{1}{2}}v/L, \qquad (6.10)$$

it is quite clear that variation in initial stopping power with ω can equally well be studied using a fixed pathlength L and a variable velocity v. However, this approach requires more accurate knowledge of the energy dependence of the stopping power and an energy-independent potential (which may not be the case when heavy ions which have charge states that vary with energy are used as projectiles.) It was in fact to investigate the effect of variable potential that the energy-dependent structure studies reported by Datz *et al.*[4] were undertaken. The same experimental set-up is used. The expectations of the model are indicated in Figure 6.8. In the upper portion of the figure, we have pictured a trajectory with $n = 1$ for some high velocity. A trajectory with lower amplitude at that same velocity will have $n < 1$ and will not be collected by the detector. As we decrease the velocity, the higher amplitude particle has $n > 1$ and is diverted. The lower amplitude particle has $n = 1$ and is focused. If for the sake of simplicity we assume that dE/dz for a given amplitude is simply proportional to velocity, we would expect that a plot of $(dE/dz)/E^{\frac{1}{2}}$ against $E^{\frac{1}{2}}$ would resemble that in the lower half of the figure. The random stopping power would be a horizontal line as would the so-called

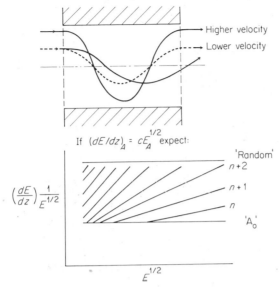

Figure 6.8. Simplified model for expected energy dependent behaviour of channeled ion energy loss spectra

A_0 group (minimum stopping power in the channel). As the velocity decreases, the amplitude, and hence the stopping power for a given n value, also decreases. At lower velocities, we also expect to find a greater number of groups contained between the A_0 and the random limits. In an experiment using beams of ^{127}I at 15, 21·6, 29·4, 38·4, 48·6 and 60·0 MeV (equal velocity increments) we measured the energy loss spectra for a fixed pathlength (0·5 μm) in the {111} and the {100} planes of Au. The data for the {111} and {100} planes are plotted in Figure 6.9. The fact that the random energy loss line is not horizontal in the plot may be caused by the nonzero intercept of the linear portion of the velocity against stopping power curve.[8] The ratio of the least energy loss to the random energy loss varied from 0·40 to 0·51 for {111} and 0·46 to 0·56 for {100} for energies ranging between 15 and 60 MeV respectively. The qualitative behaviour expected from Figure 6.8 is in fact observed. However, at lower velocities, we were not able to observe as many high amplitude groups as anticipated. At 15 MeV structure in the {100} spectrum was barely discernable. This effect may be attributable to the increased probability for dechanneling of high amplitude groups at lower longitudinal velocities. The data given in Figure 6.9 can also be cast into the form of a (dE/dz) against ω plot. Because of the variable E_0 in these experiments, it is more convenient to plot (dE/dz)/E_0^p against ω. In this case we are fixing L and varying v. The parameters of these curves are strongly dependent on the choice of p.

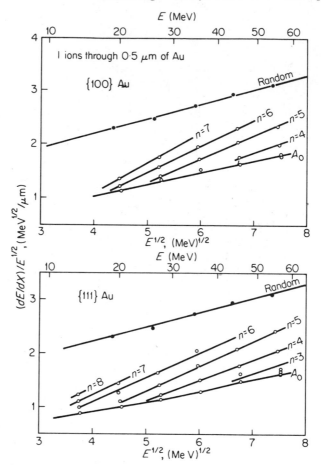

Figure 6.9. Energy loss at group peak positions for 15–60 MeV I ions channeled in {111} and {100} planes of a 0·5 μm thick Au crystal as a function of velocity

For the case of ions with variable charge and uncertain p, this is clearly not the method of choice for *initial* determination of potential parameters. However, for light ions especially and for cases where L can not be conveniently varied (as is the case for hyperchanneling studies), the method has its merits.

(3) *Frequency Determination (Equipopulation Method)*

A third method was developed by Eisen and analysed by Eisen and Robinson.[7] This was used to analyse energy loss spectra of 0·4 MeV protons transmitted through silicon. These spectra did not contain resolvable energy peaks. The alternative procedure was to measure the angular distribution for portions of the emergent energy spectra as a function of pathlength. The basis of the technique may be understood by reference to Figure 6.10. There are always two trajectories with the same amplitude, distinguished only by their having entered the crystal on opposite sides of the channel midplane, but in general the two exit angles ψ_1 and ψ_2 will bear no particular relation to each other. However, in the special case illustrated in Figure 6.10a, when $z = (n \pm \frac{1}{4})$,

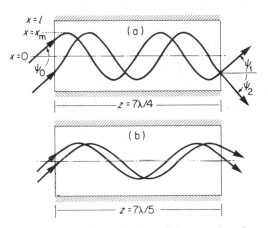

Figure 6.10. Determination of the wavelengths (or periods) of oscillation of planar channeled ions by measurement of the angular distributions of monoenergetic transmitted ions

n an integer, then $\psi_1 = \psi_2$. In this case, if a small-acceptance-angle detector, arranged to respond only to ions in a selected narrow energy interval corresponding to this pair of trajectories, is rotated about the crystal axis (Figure 6.1), it will map out an emergent angular distribution of two equally intense peaks symmetrically disposed about the channel plane. The second spectrum of Figure 6.11 corresponds to this condition. The change in pathlength required to bring the same (dE/dz) portion of the energy spectrum

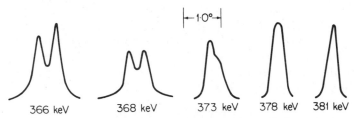

Figure 6.11. Some angular distributions reported for protons transmitted through the {111} channels of thin Si crystals. The initial energy of the ions was 400 keV

back to this condition corresponds to the wavelength of the particles. The same procedure (equations (6.8) and (6.9)) may be used to correct the data back to the initial stopping power of the ion, and a plot of $(dE/dz)_{E=E_0}$ against ω may be obtained. Here again the stopping power was found to be accurately proportional to the channel oscillation frequency.

6.2.3 Potentials and Stopping Power Functions from Frequency against Stopping Power Data

Robinson[3,9,10] developed a method for the extraction of atomic and planar potentials from the data obtained in the above experiments. We start with the definition of a smoothed planar potential

$$V_1(\mathbf{x}) = 4\pi\kappa\rho l \int_{\bar{x}}^{\infty} rV(r)\,dr \qquad (6.11)$$

where $V(r)$ is the interaction potential between an ion and a lattice atom separated by a distance r, ρ is the atomic density of the target, l is the half-width of the planar channel, $\mathbf{x}$ is the length of the normal from the ion to the plane of lattice atoms, and κ is a factor allowing for the possibility that there may be parallel channels of different widths in some crystal structures (e.g. Si). The channeled ions move between a pair of planes so that the planar channel potential (see Figure 6.2)

$$V_2(x) = V_1(l + x) + V_1(l - x), \qquad -l \leqslant x \leqslant l \qquad (6.12)$$

The ion oscillates with a frequency

$$\omega^{-1} = 2 \int_0^{x_m} \{V_2(x_m) - V_2(x)\}^{-\frac{1}{2}}\,dx, \qquad (6.13)$$

where x_m is the oscillation amplitude. The model stopping power is

$$S(x, E) = s_0 + s_1\{\sigma(x) - 1\} \qquad (6.14)$$

where s_0 and s_1 depend on the ion energy. The spatial dependence of the stopping power is described by the function $\sigma(x)$, normalized so that $\sigma(0) = 1$. The value of s_0, the stopping power at the channel centre, may be taken directly from the experiments by measuring the minimum energy loss (leading edge) of the spectrum. The stopping power of the channeled ions is found by averaging equation (6.14) over the oscillatory motion, weighting each portion of the channel cross section by the time which the ion spends there.

Empirically, it has been found that the data are well represented by the straight line (see listing in Table 4.1)

$$(-dE/dz)_{E=E_0} = \alpha + \beta\omega \qquad (6.15)$$

as illustrated by Figure 6.7. Because of the apparent generality of this observation, it has been adopted as an empirical first principle, the consequences of which are to be explored. When equation (6.14) is averaged over the motion of the ion and the result is compared with equation (6.15), it is found that as long as the number of half oscillations executed by the ion is integral, the intercept and slope of the straight line are given by*

$$\alpha = s_0 - s_1 \tag{6.16}$$

$$\beta = 2s_1 \int_0^{x_m} \sigma(x)\{V_2(x_m) - V_2(x)\}^{-\frac{1}{2}}\, dx. \tag{6.17}$$

From equation (6.17) it is easily found that,[9] as long as β is constant,

$$\sigma(x) = \frac{d}{dx}\left[\frac{2}{V_2''(0)}\{V_2(x) - V_2(0)\}\right]^{\frac{1}{2}} \tag{6.18}$$

where the primes represent differentiation with respect to x.

The experimental data may be used to evaluate the curvature parameter,

$$
\begin{aligned}
y &\equiv 2\pi^2(s_0 - \alpha)^2/\beta^2 l \\
&= V_2''(0)/l = 2V_1''(l)/l \\
&= -8\pi\rho\kappa\{V(l) + lV'(l)\}
\end{aligned} \tag{6.19}
$$

where the empirical form is given on the first line and three equivalent theoretical forms are given on the following lines. Thus, the experiments determine the curvature of the planar channel potential at the channel centre directly, independent of the potential form. Given data for a sufficient number of channels with distinct values of l, the potential itself can be evaluated by integration of the last of equations (6.19). For more limited data, the parameters of an assumed potential function can be evaluated. In either case, it is desirable to have a further test of the suitability of the resulting potential. This may be found by evaluating the random stopping power. If equation (6.14) is averaged over the channel cross section using equal weighting for all intervals, thus corresponding to a particle crossing the planar channel at a large angle, the random stopping power obtained is

$$(-dE/dz)_{\text{random}} \equiv \hat{S} = \alpha + (\beta/\pi l)[V_2(l) - V_2(0)]^{\frac{1}{2}} \tag{6.20}$$

where equations (6.16) and (6.18) have been used. The potential derived from the channeling data can be used to evaluate $\hat{S}$. The result can be compared with experimental observations, and be used to test the validity of the potential form assumed.[9]

Experimental curvature parameters are listed in Table 6.1 for each combination of ion, energy, and channel. The values are in no way dependent

---

* An equivalent theoretical treatment in which α and β are not separated and no assumptions concerning $S(x)$ as in equation (6.14) can be shown to give the same result.[11]

on assumptions about the potential function. The values of y are strongly dependent upon the ion and its energy, showing that the planar channel potential must also depend on these variables. However, the ratio of the values for the two channels appears to be essentially independent of the ion and its energy within experimental error $\pm 10\%$. If the interaction potential between the channeled ion and the individual lattice atoms is of the screened Coulomb type, the curvature parameter is

$$y = -8\pi\rho Z_1 Z_2 e^2 b\phi'(bl), \qquad (6.21)$$

where $Z_1 e$ and $Z_2 e$ are the (effective) nuclear charges of the two particles, $\phi(bl)$ is the screening function, and b is a constant. The ratio of the values for two channels determines the value of the screening constant, b. The constancy of the ratio y_{111}/y_{100} for Au in Table 6.1 implies the similar constancy of b, which is thus shown to be a property of the gold crystal. The values of b shown in Table 6.1 are therefore derived from the weighted mean values of the curvature parameter ratios of all probing ions. Physically, the implication is that the nuclear charge of the ion is always screened by the same number of electrons, whatever its position in the channel. As it approaches a target atom, it never gets so close that this screening is reduced significantly. Thus, it acts always as a simple charged particle which does not much perturb the medium, that is, all three ions may be considered as test charges. The strong variation of the experimental curvature parameters with the ion and its energy results from changes in the magnitude of the ionic charge with these variables.

Robinson considered several physically reasonable potentials which have screening functions which reduce to a single exponential in the range of distances near the centre of the two gold channels. For gold, the observed screening length 0·31 Å is nearly one-fifth smaller than the theoretically expected value 0·36 Å from the Molière potential. It is, however, in excellent agreement with the value expected from a machine calculation discussed by Tucker *et al.*[12] They describe a relativistic self-consistent-field program for the evaluation of atomic wave functions. The electrostatic potential of an isolated gold atom in its ground state is compared in Figure 6.12 with some related potentials calculated by the same program, as well as with the Thomas–Fermi and Molière functions, and with the empirically deduced potential. The last is simply an exponential screening function with the empirical value of b, passed through the Hartree potential midway between the centres of the two channels. It is clear from the figure that confining the Au atom to a spherical Wigner–Seitz cell to simulate solid state effects does not make an important alteration in the potential for the interatomic distances sensed by the planar channeling measurements. Like the Molière potential, the Hartree potential may be represented rather accurately by a sum of three exponentials.[12] Similar results are obtained with Ag and Si.

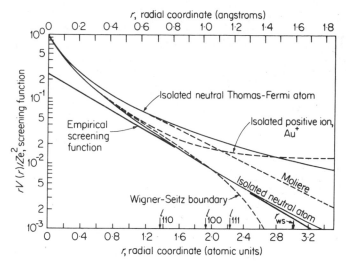

Figure 6.12. Comparison of a screening function deduced from planar channeling experiments in Au with Thomas Fermi and relativistic Hartree calculations

In addition to these two, other potential forms including the Bohr, the Thomas–Fermi, and the Born–Mayer were considered and evaluated in terms of the predicted random stopping power (equation (6.20)) and ionic charge (equation (6.21)). From these considerations it was concluded that the Hartree is clearly superior. The predicted ionic charges (last column, Table 6.1) agree quite well with the nuclear charges for He and H ions and with the equilibrium charge state for channeled 10 MeV O ions[13] but are 1·5 to 2 times larger than those observed for I ions in Au and Ag.[14] Hence, even though the I ions may act essentially as test charges when near the centre of a channel, as they approach closely to a lattice atom, screening of their nuclear charge by their own electrons must decrease and the interaction between the ions and the gold atoms must increase more rapidly than described by any of the potentials used.

From equation (6.18) it can be seen that once a form for $V_2(x)$ is assumed, the conjugate stopping power function $\sigma(x)$ is uniquely determined. This would be true no matter what the form of the $(-dE/dz)$ against ω. The ubiquitous linearity of this relation remains to be explained: the more so since the linearity applies in all regions of the stopping power (i.e. in the velocity proportional region (15–60 MeV I), the v independent region (10 MeV O), and the $\ln E/E$ region (3–6 MeV He)) and for both totally stripped ions (He and H) and ions which retain electrons which themselves contribute to inelastic processes.

For any electrostatic potential it is certainly true that the electron density function $\rho(x)$ is related to the potential $V(x)$ through the Poisson relation, i.e. $\rho(x) = \nabla^2 V(x)$. If all electrons, independent of their binding energies, were to interact with equal probability, then it would be expected that $s(x) \propto \rho(x)$. This would predict that using the $V(x) = V_0 \cosh(bx)$ potential, $s(x) \propto \cosh(bx)$, whereas the data all indicate that for this potential $s(x) = \cosh(\frac{1}{2}bx)$; a much more slowly rising function. This reflects the fact that with the ion velocities used in the experiment nonadiabatic interactions with more strongly bound electrons are improbable.

Ohtsuki and Kitagawa[15] have recently utilized Firsov's theory relating inelastic loss to impact parameters to calculate the planar stopping function $s(x)$ for low velocity ions (60 MeV I in Au). They conclude that, in the low velocity region where the stopping power is proportional to velocity, good agreement is obtained with the $\cosh(\frac{1}{2}bx)$ function and that small differences at larger x can be interpreted in terms of the reduction of screening. This result therefore gives a theoretical rationalization for the observed linearity of ω against $(-dE/dz)$ for the I–Au system.

Calculations by the same authors[16] for 3 MeV He ions in Au $\{100\}$ planar channels do not agree with the derived stopping power. In both the semi-classical and quantum mechanical calculations the stopping power function rises more rapidly than the local electron density. However, in a recent paper on 400 keV H^+ in Si, Ohtsuki *et al.*[17] take into account exchange terms for inner electrons and find that a weaker position dependence of the stopping power is obtained which resembles more closely the experimentally observed relation.[7]

6.2.4 Potentials from Emergent Angular Distributions

In the method just discussed, the planar potential was determined from measurements of the stopping powers of the various groups against wavelength, and the potential function was shown to be uniquely related to its conjugate stopping power function. A major advantage of this type of determination is that it is insensitive to the mosaic defect structure and thickness variations which are present to varying degrees in all the thin single crystals used in such studies. An alternative method of investigating the planar interaction potential is to make detailed measurements of the populations of the various energy loss groups as a function of emergence angle. Since it is possible to correlate the emergence of an ion to its coordinates on entering the planar channel through the interaction potential, it is conversely possible by doing detailed angular distribution measurements to obtain information on the planar potential parameters. The advantage of this method is that the potential determination does not intimately involve a stopping power function. However, the experimental measurements are now quite sensitive to mosaic spread and thickness variations in the crystals.[3,18]

This method was employed by Lutz *et al.*[18] to analyse emergent angular distributions of He ions transmitted along the {001} planar channels of Au. A well collimated beam (0·002°) of 2 MeV He$^+$ ions was incident on a 1300 Å ($\pm 10\%$) thick crystal which could be tilted to vary the pathlength from 1600 to 3000 Å without changing the angle of incidence (α_0). The emergent ions were detected in a position-sensitive detector with an angular resolution of 0·07°. The Au crystal used had a mosaic spread of 0·11° FWHM. Figure 6.13(a) shows some normalized emergent intensity distributions measured as a function of channel length. These spectra show a rapid variation of intensity and structure with crystal thickness, l. A set of theoretical angular distributions (Figure 6.13(b)) was calculated and compared to these measured results using

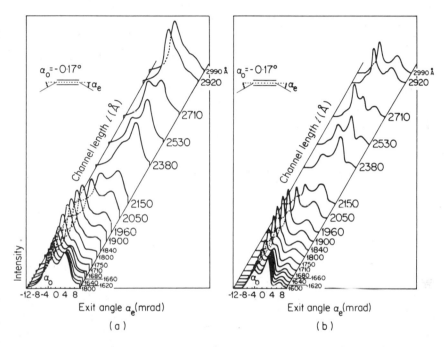

Figure 6.13. Angular distributions of He ions emerging from Au {001} planar channels as a function of channel length l. Incident energy 2 MeV, angle of incidence $\alpha_0 = -0·17°$. (a) Experimental; (b) calculated with $V_0 = 11$ eV, $b = 3·33$ Å^{-1} and $f = 0·11°$

the potential $V(x) = V_0 \cosh(bx)$ discussed previously. The influence of the inelastic energy loss on the trajectory was not taken into account since the path length was fairly short, less than or equal to 3λ. The angle of exit, α_e, was calculated numerically for each entrance coordinate, x_0. The entrance coordinate, x_0, was varied step by step to simulate a large number of particles penetrating the crystal (up to 4000). The mosaic spread of the crystal was introduced into the calculation by varying α_0 and α_e accordingly. The probability distribution of the mosaic spread was assumed to be of gaussian shape with f the FWHM. The finite experimental angular resolution was simulated by smoothing the calculated angular distributions three times over five channels (5 channels correspond to 1 mrad).

A good overall fit between the experimental and theoretical results of Figure 6.13 was obtained with the parameters $V_0 = 11$ eV, $b = 3.33$ Å^{-1}, and $f = 0.11°$. These parameters are the same as those derived by the stopping power method for 3 MeV He ions in Au $\{001\}$ planar channels.[2,3] However, the accuracy of these fitting parameters was apparently limited by the mosaic structure of the crystals.[18] The effect on the angular distributions when mosaic structure was included in the calculations was to both broaden and shift the peak positions. It was shown, for example, that a comparable fit to the data could be obtained using the parameters $V_0 = 15$ eV, $b = 2.94$ Å^{-1} and $f = 0.14°$.[18]

6.2.5 Potentials from Angular Dependence of Group Populations

A third method recently reported by Gibson and Golovchenko[19] combines measurements of the wavelengths (frequencies) and energies of transverse motion of planar oscillations to determine uniquely the planar continuum potential, (cf. equation (6.7)). The measurements were made for 1.8 MeV He ions channeled along the $\{111\}$ planar channels of a Au single crystal. The experimental arrangement is the same as that of Figure 6.1 where the transmitted He ions were measured in a solid state detector with an angular resolution of approximately 2.5×10^{-7} steradians, positioned in line with the incident beam.

The wavelength for a particular group was measured in the same manner as described in Section 6.2. The transverse energy of motion for a particular group was determined by measuring the energy-loss group population as a function of incidence angle. The results of such a measurement are given in Figure 6.14 which shows the population of the A_0 (best channeled) group (dashed wave), and a group with $\lambda \sim 920$ Å (solid curve) which makes $n = 3$ complete oscillations in transversing the (111) planar channel. The authors show[19] that the angle of incidence at which the $n = 3$ group is a maximum, $\psi_{\alpha3}$, has an energy of transverse motion $E_\perp$ given by

$$E_\perp = E\psi_{\alpha3} \tag{6.22}$$

where E is the energy of the incident ions. From a series of measurements such as these, the wavelength of oscillation was determined as a function of transverse energy.

The method outlined for extracting the potential function from these data proceeded as follows: (i) A wavelength function $\lambda(E_\perp, V)$ was chosen which

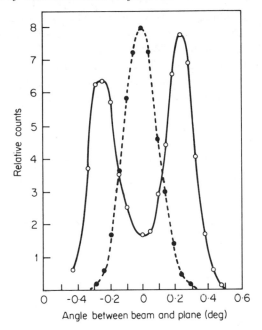

Figure 6.14. An angular scan of $n = 3$ group in the (111) plane of Au (pathlength $= 2750$ Å) for incident $1\cdot8$ MeV He ions. The dashed curve indicates the A_0 population distribution (intensity $\times$ 1/3)

best fitted the data. The functional dependence of λ on the interaction potential was represented by the parameters a_1, a_2—(i.e. $\lambda(E_\perp, a_1, a_2 \ldots)$). (ii) The inverse of the potential was then obtained via the transformation

$$x(V) = (8\pi^2 M v_f^2)^{-\frac{1}{2}} \int_0^v \frac{\lambda(E_\perp)\,dE_\perp}{(V - E_\perp)^{\frac{1}{2}}} \tag{6.23}$$

where v_f is the velocity of the ions along the channel direction.

The λ function chosen was

$$\lambda = \lambda_0(2/\pi)(1 + 4\Gamma B)^{-\frac{1}{4}}\kappa[\tfrac{1}{2} - \tfrac{1}{2}(1 + 4\Gamma B)]^{-\frac{1}{2}} \tag{6.24}$$

with $\lambda_0 = 2\pi v_f(M_1/a_1)^{\frac{1}{2}}$, $\Gamma = E_\perp/E_{\max}$, $B = E_{\max}a_2/a_1^2$, κ the complete elliptic integral of the first kind, and $E_{\max}$ a normalization parameter. Transformation of this function resulted in the potential function

$$V = \tfrac{1}{2}a_1 x^2 + \tfrac{1}{4}a_2 x^4 \tag{6.25}$$

A plot of λ/λ_0 (equation (6.24)) against $E_\perp/E_{\max}$ is shown in Figure 6.15 for various values of $B = E_{\max}a_2/a_1^2$. Also shown on the same graph are the experimental measurements of $E_\perp$ against λ. The fit to these data chosen by the authors is the solid curve in Figure 6.15 which corresponds to the parameters $\lambda_0 = 1408$ Å and $B = E_{max}a_2/a_1^2 = 3\cdot61$ with $E_{max} = 103\cdot5$ eV. The coefficients of the potential which result from this fit are $a_1 = 71\cdot6 \pm 4\cdot6$ and $a_2 = 179 \pm 26$. The uncertainties are attributed only to uncertainty in the crystal thickness and transverse energy determinations.

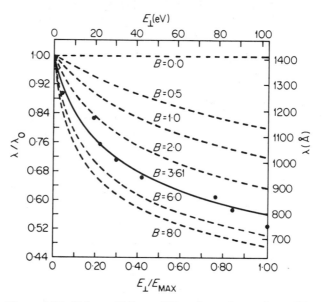

Figure 6.15. Values of λ/λ_0 vs. E/E_{max} for various values of B.
Data points shown are those of reference 19

The strength of this method for determining the planar interaction potential is that the experimental data of wavelength as a function of transverse energy provides enough information to determine uniquely the continuum potential. This is to be contrasted with the method described in Section 6.2.3 where the experimental data (curvature parameters) uniquely determine the connection between the potential and its stopping power function but not the potential function itself. The weakness of the group population method is that the accuracy of the experimental data is quite dependent on the same sample imperfections mentioned in connection with the previous method (i.e. thickness nonuniformity, mosaic spread, etc.). A comparison between the values of the potentials derived by these two methods for He ions in Au (111) shows satisfactory agreement. For example, the values of V (equation (6.25)) and $V_2(x) - V_0 = \{V_0 \cosh(bx - V_0)\}$, where $V_0 = 7$ eV and $b = 3\cdot22$ Å^{-1}, evaluated at $x = 0\cdot5$ and 1 Å are $V = 11\cdot8, 80\cdot6$ eV and $V_2(x) - V_0 = 11\cdot2, 79\cdot1$ eV, respectively. Also, comparing the curvature parameters derived by Robinson[9] $y_{111} = V''_{(0)}/l = a_1/l = 60.8 \, (\text{eV Å}^{-3})$ from the data of Gibson and Golovchenko[19] and $y_{111} = V_0 b^2/l = 64 \, (\text{eV Å}^{-3})$ from Appleton et al.[5]

6.3 AXIAL POTENTIALS FROM HYPERCHANNELING MEASUREMENTS

Since solid state effects on the ion–atom interaction potentials occur at large distances from the atoms, it is desirable to investigate the largest impact parameter events possible. In the planar channeling experiments discussed so far, the maximum impact parameters are determined by the planar spacings. It is possible by doing experiments in an axial channeling geometry to probe even larger impact parameter interactions provided one can study ions confined to only one particular axial channel. In planar channeling, a channeled ion travels in a single channel between adjacent atomic planes, whereas in ordinary axial channeling the majority of channeled ions wander from one axial channel to another. Axially channeled ions of a limited class, however, are expected to have transverse energies less than the potential barrier between adjacent rows and accordingly should be constrained to travel through the crystal in a single axial channel. This phenomenon, called hyperchanneling or proper axial channeling, was recognized in the early computer calculations of Robinson and Oen[20] and was observed by Eisen[21] as a low energy loss 'tail' in axial energy loss distributions. In recent experiments[22,23] hyperchanneling was observed as a distinct well-separated group and systematic investigations of the effect indicate that solid state effects may in fact be present.

Hyperchanneled ions are expected to have at least two distinctive characteristics, a smaller acceptance angle than ordinary axial channeling and lower energy losses because they sample larger average impact parameters than ions which wander from channel to channel. This effect was studied by Appleton et al[22,23] using a beam of 21.6 MeV I ions collimated to less than or equal to 0.01° full width onto a 0.60 μm thick [001] Ag crystal which had a mosaic spread of 0.05° FWHM. The energy spectra of those ions transmitted within $\pm 0.012°$ of the incident beam direction were recorded with an energy resolution of 185 keV FWHM and normalized by a beam monitor system. Energy spectra measured with the beam incident at various angles to [011] in ($1\bar{1}1$) are shown in Figure 6.16. The prominent group on the right, lying within a narrow angular range and centred on a dE/dx value of 3.15 MeV/μm, is due to hyperchanneling. It should be emphasized that the angular extent of this series of spectra ($\pm 0.45°$) was well within the calculated critical angles for ordinary axial channeling (i.e. $\psi_{\frac{1}{2}} \simeq 1.2°$ from Barrett[23] and $\Psi_1 \simeq 1.9°$ from Lindhard).[25] The major peaks in the energy loss distributions in Figure 6.16 taken at $\pm 0.45°$ relative to [011] are nearly identical to those obtained at

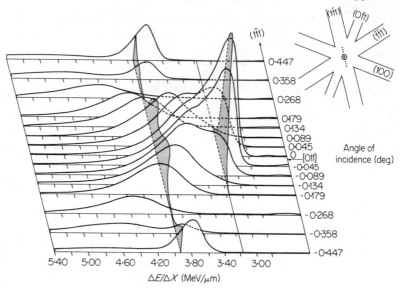

Figure 6.16. Normalized energy loss distributions at various incidence angles to [011] for 21·6 MeV I ions transmitted through 0·85 μm of Ag. The insert shows the angular extent of the measurements (dotted line) and the hyperchanneling region (shaded central spot). The widths shown for the planar channels are twice the half angles calculated from reference 7

angles 5 to 10° from [011] in (1$\bar{1}$1). This indicates that when the beam is incident parallel to a low index plane many of the transmitted ions have distributions characteristic of planar channeling even at angles of approximately 0·25 ψ_1. As the incidence angle with respect to [011] is decreased, the energy loss distributions in Figure 6.16 lose the characteristic {111} shape at about 0·30°, and a new group of ions with significantly lower energy losses is dominant within approximately 0·12° of [011]. This low loss group has the two characteristics expected of hyperchanneled ions. Related emergent patterns recorded photographically agreed well with these spectra. Where planar energy losses were evident, a characteristic elongated planar spot centred on the incident beam was observed superimposed on a ring of intensity centred on [011]. Within the range of hyperchanneling the ring had collapsed into an intense circular spot centred on [011].

These data were analysed to determine a characteristic angle and absolute fraction for hyperchanneling. It was estimated from the shapes of the population curves at various energy losses, such as those for 3·15 and 3·95 MeV/μm in Figure 6.16 that only ions with loss rates less than or equal to 3·7 MeV/μm were hyperchanneled. The fraction of all ions recorded with energy losses below a specified value was extracted as a function of incidence angle and the results are the data points in Figure 6.17.

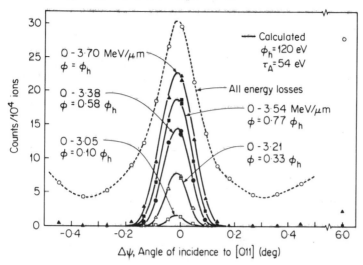

Figure 6.17. Comparison of measured (data points) and calculated (solid lines) fractions of the beam reaching the detector as a function of angle of incidence to [011]. The dashed curve is not calculated but can be understood qualitatively in terms of emergent intensity distribution and detector geometry

The solid curves in Figure 6.17 were calculated using the following model. As an ion of energy E entered the crystal at an angle ψ_i to the axial direction, it possessed a transverse kinetic energy $\tau_i = E\psi_i^2$. In addition, its entry point gives it a transverse potential energy ϕ_i chosen in the contours of Figure 6.18. These were calculated using a potential determined by Robinson[10] from planar channeling experiments for 21·6–60 MeV I in Ag.[6] As the ion penetrated the crystal, it tended to transverse the entire space within the contour corresponding to its total transverse energy. It was assumed that the crystal used was of sufficient thickness for the ion to achieve statistical equilibrium over this contour, which implied that the probability of finding it was constant inside the contour and zero outside.[25] In moving through the crystal, the ion acquired additional transverse energy ε_a from multiple scattering by electrons, lattice vibrations, and defects. The values of ε_a for the trajectories bounded by a particular contour were assumed to have an exponential distribution with mean value ε_A.

Those ions which had total transverse energies less than the critical contour, $\phi_h \simeq 60$ eV, in Figure 6.3 were treated as hyperchanneled. The fraction of the area in Figure 6.18 enclosed within ϕ_h is $f_h = 0.30$. Calculations using these values gave curves which agreed qualitatively with the experimental data in Figure 6.17 but had peaks about three times too high and were about 25 % too narrow at the base. Varying ϕ_h influenced only base

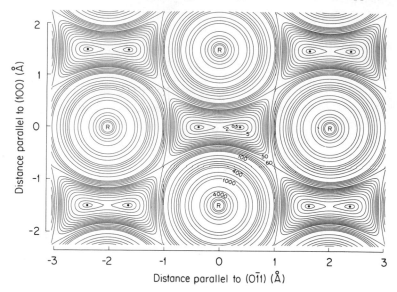

Figure 6.18. Potential energy contours for I ions in the continuum potentials of ⟨011⟩ rows (R) of Ag atoms. All values are in eV and are relative to the minimum (●), which has an absolute value of 72 eV

width and ε_A only the peak height, while ψ_M, the mosaic spread, influenced both. Values of 120 eV for ϕ_h and 54 eV for ε_A provided the best fit to the 0–3·70 MeV μm data and are shown as the solid lines in Figure 6.17. Since all combinations of parameters which gave a good fit of theory to experiment had values of ε_A several times the estimated value, it was concluded that there was considerably more multiple scattering than could be estimated for the mechanisms believed to be important. Such extra multiple scattering should also affect other experiments involving dechanneling rates. The results also indicated that hyperchanneling investigations provide information on potentials in solids at impact parameters not accessible to other channeling experiments. The value of ϕ_h used in calculating the solid curves in Figure 6.17 was twice as large as the value obtained from the contours of Figure 6.18. This may be an indication of the effects of lattice structure on the ion–atom potential. A simple way of taking this into account would be the application of Wigner–Seitz rather than free atom boundary conditions to the electron wave functions.[22,23] Robinson[9,10] has examined the resulting potentials and finds that they differ only near the cell boundary where the Wigner–Seitz potential drops more rapidly (cf. Figure 6.12). It appears that the Wigner–Seitz potential will give better agreement with the hyperchanneling experiments than the potential used in Figure 6.18 and at the same time preserve agreement with earlier planar channeling experiments.[2,5,7,9]

6.4 SUMMARY

The ultimate test of the channeling concept lies in the applicability of the continuum potential model. In the work presented in this chapter, it has been demonstrated that with experimental results obtained from channeling experiments on oscillation frequency against stopping power one is capable of deriving directly detailed information on the separate interatomic potentials which make up the continuum planar and row potentials, even to the extent of selecting between reasonable interatomic potential forms.

The most apt atomic potential appears to be the Hartree potential derivable from first principle[12] which can be expressed as the sum of three exponential terms. Given this atomic potential the planar potentials can be evaluated from equations (6.11) and (6.12) for any lattice plane of any material when fast light ions are used. The method of Gibson and Golovchenko[19] is apparently capable of giving empirical planar continuum potential parameters for a given lattice plane. The first results reported using this method are in accord with the predictions based on atomic potentials. For heavier ions, the effective charge on the ion is energy dependent and must be determined from the measurements to give the complete interaction potential. Moreover, for close distances of approach the screening owing to electrons remaining on the penetrating ion can not be neglected.

For the hyperchanneling experiments the same atomic potentials must still be applicable but correction must be made for the decline in the potential at the Wigner–Seitz cell boundary at the larger internuclear distances involved. Experiments demonstrating qualitative accord with this picture have been carried out but detailed testing remains to be done.

Finally, the conjugate planar stopping power functions obtained from the stopping power against oscillation frequency measurements (equation (6.18)) can be understood for slow ions in terms of Firsov's theory, but for higher velocity ions theoretical fits are still lacking.

References

1. Lutz, H. O., Datz, S., Moak, C. D. and Noggle, T. S., *Phys. Rev. Lett.*, **17**, 285 (1966).
2. Datz, S., Moak, C. D., Noggle, T. S., Appleton, B. R. and Lutz, H. O., *Phys. Rev.*, **179**, 315 (1969).
3. Robinson, M. T., *Phys. Rev.*, **179**, 327 (1969).
4. Datz, S., Moak, C. D., Appleton, B. R., Robinson, M. T. and Oen, O. S., *Atomic Collision Phenomena in Solids*, Ed. D. W. Palmer, M. W. Thompson and P. D. Townsend, Amsterdam, North Holland, p. 374 (1970).
5. Appleton, B. R., Datz, S., Moak, C. D. and Robinson, M. T., *Phys. Rev. B*, **4**, 1452 (1971).
6. Noggle, T. S., Moak, C. D., Datz, S. and Appleton, B. R., unpublished data.
7. Eisen, F. H. and Robinson, M. T., *Phys. Rev. B*, **4**, 1457 (1971).
8. Moak, C. D. and Brown, M. D., *Phys. Rev.*, **149**, 244 (1966).

10. Robinson, M. T., *Proc. Battelle Colloquium on Interatomic Potentials and Simulation of Lattice Defects, June 14–19, 1971* (to be published).
11. Robinson, M. T., private communication.
12. Tucker, T. C., Roberts, L. D., Nestor, C. W., Jr. and Carlson, T. A., *Phys. Rev.*, **178**, 998 (1969).
13. Datz, S., Martin, F. W., Moak, C. D., Appleton, B. R. and Bridwell, L. B., *Radiation Effects*, **12**, 163 (1972).
14. Datz, S., Moak, C. D., Lutz, H. O., Northcliffe, L. C. and Bridwell, L. B., *Atomic Data*, **2**, 273 (1971).
15. Ohtsuki, Y. H. and Kitagawa, M., *Phys. Lett.* **40A**, 313 (1972).
16. Kitagawa, M. and Ohtsuki, Y. H., *Phys. Rev.* B, **5**, 3418 (1972).
17. Ohtsuki, Y. H., Kitagawa, M. and Mizuno, M., *Phys. Stat. Solidi* (*b*), in press.
18. Lutz, H. O., Ambros, R., Mayer-Böricke, C., Reichelt, J. and Rogge, M., *Z. Naturforschung*, **26A**, 1105 (1971).
19. Gibson, W. M. and Golovchenko, J., *Phys. Rev. Lett.*, **28**, 1301 (1972).
20. Robinson, M. T. and Oen, O. S., *Phys. Rev.*, **132**, 2385 (1963).
21. Eisen, F. H., *Phys. Lett.*, **23**, 401 (1966); *Can. J. Phys.*, **46**, 561 (1968).
22. Appleton, B. R., Moak, C. D., Noggle, T. S. and Barrett, J. H., *Phys. Rev. Lett.*, **28**, 1307 (1972).
23. Appleton, B. R., Barrett, J. H., Noggle, T. S. and Moak, C. D., *Radiation Effects*, **13**, 171 (1972).
24. Barrett, J. H., *Phys. Rev.* B, **3**, 1527 (1971).
25. Lindhard, J., *Dan. Vid. Selsk., Mat. Fys. Medd.*, **34**, No. 14 (1965).

CHAPTER VII
Dechanneling

F. GRASSO

7.1 INTRODUCTION

A channeling experiment is performed by allowing a parallel and uniformly distributed ion beam to enter a single crystal in the direction of a low index axis or plane. Under this condition a large reduction occurs in the yield χ of the processes which require a close encounter with an atom of the crystal, for example nuclear reactions or large angle Rutherford scattering.

To describe such an experiment, the ion–crystal interaction may be treated as an elastic interaction with a lattice of rigid atomic potentials which has crystal symmetry. The details of the potential are unimportant, provided that it is screened.

For a fast moving ion in this scheme only two kinds of stationary motion are possible:[1] channeled and random motions. An ion which hits the first atom in the crystal row with an impact parameter b_1, which is larger than the potential screening radius, will suffer a small deflection and thereafter gentle correlated collision, without approaching closer than b_1 to any other atom; such ions are called channeled ions. However, ions which hit the first atom with a small impact parameter will be scattered through large angles and will then have a uniform probability of hitting any of the atoms in the crystal as occurs in an amorphous target. A uniform and parallel ion beam entering a lattice at an angle ψ_e with respect to a low index row or plane, less than a critical value $\psi_{\frac{1}{2}}$ is divided into two components, an aligned and a random beam. In the above scheme no particle transition can occur between the two beams.

If one does a backscattering experiment in a thick single crystal the yield, χ, measured when the beam direction is parallel to a low index axis or plane and normalized to the random incidence yield, increases as the ion beam penetrates deeper into the crystal, that is

$$\chi = \chi(z)$$

where z is the traversed thickness which is related to the measured energy of backscattered particles.

This means that the probability of a close encounter with a nucleus changes along the ion's path inside the crystal; i.e. the ion undergoes a transition from the channeled beam, where the probability of a close encounter is roughly zero, to the random beam, where this is roughly one. This transition, *dechanneling*, can be justified only by a more detailed description of the interactions between the moving ion and the crystal. In this respect even a perfect crystal differs from the rigid lattice description both because of thermal vibrations and the electron gas.

The thermal vibrations of the nuclei destroy the crystal symmetry and hence destroy the correlation between the successive collisions of a channeled ion. Its interaction with the electrons is not described completely by the screened atomic potential because of the inelastic processes involving excitations of the core or valence electrons. One should consider then the collisions between the moving ion and electrons and, in particular, the close ones[2,3,4] which produce appreciable changes in the ion momentum. These two contributions to the scattering of a channeled ion were considered theoretically by Lindhard[1,2] and termed nuclear and electronic reduced multiple scatterings.

Experiments on dechanneling were started by Davies *et al.*,[5] Ellegaard and Lassen,[6] Appleton *et al.*,[7] soon after the discovery of channeling in 1965. A systematic study of the comparison between dechanneling experiments and the Lindhard theory is, however, more recent.[8-14] This study is important as a quantitative interpretation of dechanneling experiments provides a significant test of the theory of elementary interactions between a channeled ion and a crystal. Conversely, if these are known, dechanneling may be used as a tool for studying the crystal,[12] i.e. for studying the potential at distances greater than the screening distance where it is difficult to investigate by other methods or the electronic densities and stopping powers in ordered structures. Moreover the results of this study are not only useful in that they add to our understanding of directional phenomena in thick crystals, for example channeling,[15] blocking,[16] flux peaking;[17] but also for their applications in for example lattice location of impurities, nuclear half-life measurements, and ion implantation.

In the following, we will refer to the more relevant experimental results in Section 7.3, comparing them with the quantitative estimates which can be obtained, in terms of elementary interactions, from Lindhard's theory in (Section 7.4). The aim is to investigate the role played by the different parameters used to describe the ion–crystal interaction. The case in which the incident beam is not parallel to the channel will be considered to justify the depth and temperature dependence of angular dips (Section 7.5). A few remarks on the experimental methods and techniques are included in Section 7.2.

The discussion has been limited to perfect crystals and light ions, since in this case, more experiments have been performed and a fuller understanding has been achieved. A treatment of the defect contribution to dechanneling is beyond the scope of the present work; however, a few remarks have been inserted on this topic.

The theoretical treatment is developed on the basis of Lindhard's theory[1] (i.e. in the continuum model approximation) the significance of which has been treated extensively in the previous chapters. This theory provides a complete description of the experiments. However, a few minor effects, such as feeding-in (the ion transition from a random trajectory to a channeled one) and the energy loss of the moving ion, have been neglected.

7.2 EXPERIMENTAL TECHNIQUES

To perform a dechanneling experiment one needs to measure the fraction χ of the incoming beam which has been dechanneled at a depth z inside the crystal. A schematic diagram of a typical set up is shown in Figure 7.1. An ion beam collimated within 0·01 to 0·1° is allowed to strike a single crystal which can be oriented with respect to the beam direction. The experimental techniques so far used for dechanneling studies are basically the same as those developed for channeling; they differ, however, because of the role played by

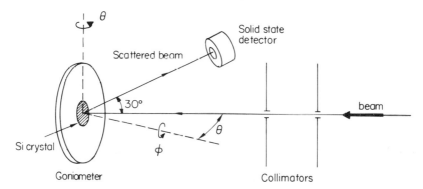

Figure 7.1. Schematic diagram of the apparatus used for dechanneling experiments

different factors. It should be mentioned that beam collimation, for example, is much more important in dechanneling than in channeling. A channeling dip, is, in fact, unaffected by the beam divergence provided that it is 5 to 10 times smaller than the critical angle $\psi_{\frac{1}{2}}$. Moreover, the finite angular resolution can be easily taken into account. In dechanneling however, one studies the change of the angular dip with the thickness traversed inside the crystal, and this change is related to the angular spreading of the beam with penetration. Because of the nonlinearity of the latter process the initial collimation plays an important role, and it is difficult to separate the contribution of the beam divergence from the results. The effect of the beam collimation is illustrated in Figure 7.2, where the angular dips measured at different depths are shown and the effect of a poor beam collimation on the dechanneled fraction is simulated by increasing the angle of incidence relative to the axis. The same argument holds for a single crystal target; bending and mosaic spread should be carefully avoided especially when thin crystals are used. Crystal surfaces should be carefully prepared and maintained during the experiment, to avoid amorphous layers or contamination affecting the beam collimation.

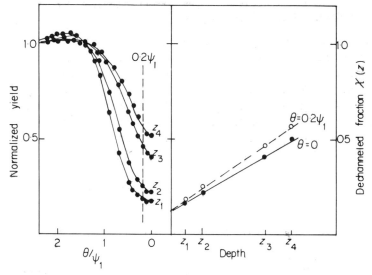

Figure 7.2. Normalized yields of backscattered particles (schematic) as a function of the incident angle measured at different depths (left hand side) and dechanneled fraction against depth (right hand side) for perfect alignment, and for a slight misalignment to simulate the effect of poor beam collimation

Even thicknesses of a few atomic layers are unacceptable. The goniometer used to orient the single crystal with respect to the beam, should have a sensitivity and reproducibility comparable with or better than the beam collimation. Solid state detectors are used in dechanneling experiments, the only requirement being an energy resolution of the order of a few per cent.

A classification of the experimental techniques can be made depending on whether the detector is lying inside or outside the critical angle ψ_1 with respect to the beam direction; the two extreme cases are referred to as transmission and backscattering techniques. Some attention should be paid to the analysis of the experimental data in order to obtain the dechanneled fraction $\chi(z)$ from the energy spectra determined by these two techniques.

7.2.1 Interpretation of Backscattering Experiments

To be scattered through a large angle and reach the detector, an ion from the incoming beam must experience a close collision with a target atom. The relative probability of such an interaction can be assumed to be either zero or one for ions moving in a channeled or random direction respectively; consequently, only nonchanneled or dechanneled ions may reach the detector. If the angular resolution of the detector is poor enough ($\Delta\theta \simeq 5°$), the directional effects along the outcoming path of the ion can be neglected. The backscattering technique appears to be ideal as it involves the definition of channeled and dechanneled particles and, therefore, the dechanneled fraction as well. In a more detailed description, some difficulties arise because the close encounter probability does not change as sharply as stated above, being finite even for a channeled ion, mainly because of the thermal motion of the target atoms. It is also different from one for ions moving with angles slightly larger than ψ_1, because of the compensation shoulders. However the contribution of these corrections are negligible because an ion spends only a small fraction of its trajectory in this transitional region. Two typical backscattering energy spectra corresponding to aligned and random incidence are shown in Figure 7.3.

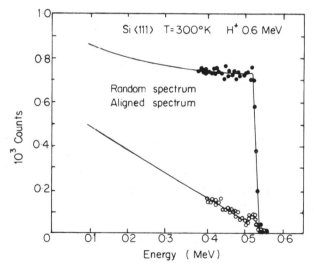

Figure 7.3. Energy spectra of 0·6 MeV protons backscattered from a Si target at room temperature: ● random incidence; ○ incidence aligned with the $\langle 111 \rangle$ axes

The backscattering technique involves a strong and intrinsic limitation as it concerns the determination of the depth z where the dechanneling process occurs. The penetration depth can be determined only through the energy loss by the moving ion. However, since the stopping power is different along the channeled and random parts of the incoming path the energy spectra of the backscattered ions should be interpreted both for an ion incidence in (a) a random direction, and (b) a channeling direction (see Figure 7.4).

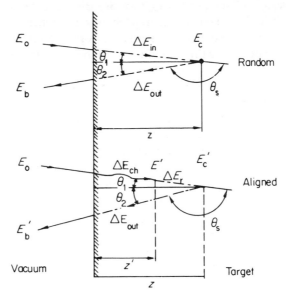

Figure 7.4. Backscattering of ions incident onto a crystal along a random (upper part) and an aligned direction (lower part schmatic)

(a) If a beam of ions of mass M_1 and energy E_0 strikes a crystal in a random direction, the number of backscattered particles detected in an interval ΔE_b at energy E_b is given by the incident* flux n times the differential scattering cross section $d\sigma/d\omega$ times the number $N.\,\Delta z$ of nuclei of mass M_2 available in a layer Δz at depth z; that is:

$$\frac{dN(E_b)}{dE_b}\Delta E_b = n\frac{d\sigma}{d\omega}\Delta\omega N\,\Delta z \qquad (7.1)$$

$\Delta\omega$ being the solid angle subtended by the detector and E_c the collision energy at the scattering point: $E_c = E_0 - \Delta E_{in} = E_0 - \int_0^{z/\cos\theta_1} S(E)\,dz$, that is the beam energy E_0 minus the loss ΔE_{in} along the incoming path because of the stopping power $S(E)$[18] (see Figure 7.4). The relation connecting the observed

---

* The flux of particles can be assumed to be constant inside the crystal because of the small scattering cross section.

energy E_b and scattering depth in equation (7.1) is:

$$E_b = k^2 E_c - \Delta E_{out} = k^2 \left(E_0 - \int_0^{z/\cos\theta_1} S(E)\,dz \right) - \int_{z/\cos\theta_2}^0 S(E)\,dz \qquad (7.2)$$

k being the kinematic factor: $k = (M_1 \cos\theta_S + M_2)/(M_1 + M_2)$, for $M_1 \ll M_2$ and ΔE_{out} the energy loss along the outcoming path. Differentiating equation (7.2) one obtains from equation (7.1) the random energy spectrum[18]

$$\frac{dN(E_b)}{dE_b}\,\Delta E_b = \frac{n\sigma(E_c)N}{k^2 S(E_c) + (\cos\theta_1/\cos\theta_2)S(k^2 E_c)}\,\Delta E_b. \qquad (7.3)$$

It should be mentioned that, by neglecting the straggling, the energy-to-penetration relation is unique for random incidence because of the single-valued stopping power function $S(E)$.[19]

(b) An ion which enters the crystal in a channeling direction does not contribute to the backscattering spectrum. If at a depth z' the ion is dechanneled, it behaves thereafter as a random particle which entered the crystal a depth z' with an energy $E' = E_0 - \Delta F_{ch}$, ΔF_{ch} being the energy loss along the channeled path. This ion can suffer a scattering at any depth $z > z'$, which means that the number of ions scattered from a layer Δz at z depends on the total flux $n'(z)$ of random particles or particles dechanneled before z.

The stopping power $S^*(E) = \beta S(E)$ along a channeling trajectory is reduced by a factor β with respect to that $S(E)$ of a random path (cf. Chapter 4); a particle scattered at depth z can be found then in a wide range of energies depending on the depth z' where it has been dechanneled.

The collision and backscattering energies are defined in this case by: (see Figure 7.4)

$$E_c' = E_0 - \Delta E_{ch} - \Delta E_r = E_0 - \int_0^{z'/\cos\theta_1} \beta S(E)\,dz - \int_{z'/\cos\theta_1}^{z/\cos\theta_1} S(E)\,dz$$

and

$$E_b' = k^2 E_c' - \Delta E_{out} = k^2 \left(E_0 - \int_0^{z'/\cos\theta_1} \beta S(E)\,dz - \int_{z'/\cos\theta_1}^{z/\cos\theta_1} S(E)\,dz \right)$$

$$- \int_{z/\cos\theta_2}^0 S(E)\,dz. \qquad (7.4)$$

These equations show that the relation between E_b', E_c', z and z' is not unique in this case, but can be fulfilled for a range of values ranging from $z' = 0$ to $z' = z$, i.e. for dechanneling in any z' point between the crystal surface and the backscattering point. The energy spectrum becomes in this case:

$$\frac{dN'(E_b')}{dE_b'}\,\Delta E_b' = \int_{z'=0}^{z'=z} \frac{N\,dn'/dz'\sigma(E_c')\,dz'}{k^2 S(E_c') + (\cos\theta_1/\cos\theta_2)S(kE_c')}\,\Delta E_b' \qquad (7.5)$$

The dechanneled fraction, i.e. the ratio between the flux of ions dechanneled within a depth z and the total flux of ions can therefore be obtained from the two energy spectra corresponding to aligned and random incidence as:

$$\chi(z) = \frac{\int_0^z (dn'/dz)\,dz'}{n} = \frac{dN'/dE'_b}{dN/dE_b}\Lambda(z) \qquad (7.6)$$

where $\Lambda(z)$ is a function which can be determined[9] by an iterative procedure solving equations (7.1), (7.5) and (7.6) with respect to the flux, $n'(z)$. Since the energy losses and the ion energy E_c and E'_c at a given depth are different for the two spectra, the function $\Lambda(z)$ also accounts for the energy dependence of the scattering cross section, and for the fact that the relation between E'_b and z', is not unique. The relevance of this correction was first discussed by Ellegaard and Lassen[6] with the following example: 'if a layer of the target were a perfect crystal, whereas for greater depths the atoms were randomly arranged, then the random and aligned spectra would never meet, but show a 'channeling effect' for any depth, simply because the particles having passed the crystalline layer in the channeling direction have the higher energy'.

The function $\Lambda(z)$ ranges[9] between one at small depths, where the difference in the energy loss and hence in the cross section can be neglected, and approximately 1·2 at the higher penetration of interest, i.e. when the ion loses approximately $\frac{1}{3}$ of its initial energy E_0 along the incoming path (see Figure 7.5). Average values $S^*(E) \simeq \frac{1}{2}S(E)$ have been recently used for protons in evaluating the absolute dechanneled fraction to be compared with theoreti-

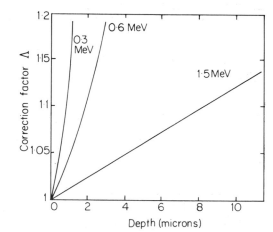

Figure 7.5. Correction factor $\Lambda(z)$ required to compute the dechanneled fraction $\chi(z)$ from the ratio of the aligned to random spectra; the figure refers to protons of different energies in Si $\langle 111 \rangle$

cal predictions. It should be noted that the assumption of a reduced stopping power along the channeled trajectory reduces the measured channeled fraction, thus compensating in part for the effect of the $\sigma(E_c)$ correction.

It should be mentioned that the correction factor is very close to one when the stopping power decreases steeply with an increase in beam energy, while it can introduce a factor of two when the stopping power is roughly energy independent. This is because in the first case the main part of the energy loss occurs along the outgoing path, which is always a random one.

When the absolute dechanneled fraction is unimportant, as it is when comparing different experimental results, the problems above can be overcome by simply assuming that the stopping power is independent of the ion trajectory[20] i.e. $S^*(E) = S(E)$. In this case the factors $\Lambda(z)$ reduces to one and the approximate dechanneled fraction is given simply by the ratio of the aligned to random spectra.

7.2.2 Interpretation of Transmission Experiments

A particle moving in a single crystal has a lower stopping power along a channeled trajectory than along a random one. By means of this difference the fraction χ of particles dechanneled at depth z can be determined from the energy spectrum of the transmitted ion beam.

The energy of a transmitted particle, dechanneled at depth z in a crystal of thickness t is

$$E_t = E_0 - \int_0^z \beta S(E)\, dz' - \int_z^t S(E)\, dz'$$

Here, in contrast with the backscattering case the relation between E_t and z is unique if the stopping powers $S(E)$ and $S^*(E) = \beta S(E)$ are known and single-valued functions. However β depends on the trajectory followed by the channeled particle. Moreover, the dechanneling depth is determined in this case by the difference between the stopping powers, so that a small uncertainty in $S^*(E)$ causes a much larger error in z than in the backscattering case. Even if the dechanneling depth could be determined correctly the dechanneled fraction could only be determined by subtracting from the transmitted-particles energy spectrum the contributions given by the particles which remain either nonchanneled or channeled along the whole crystal. The latter cannot be determined independently and, in the case of a crystal thick enough to exhibit an appreciable dechanneling, the beam spreading is comparable with the transmission spectrum itself (see Figure 7.6). The determination of the dechanneled fraction implies the analysis[21] of the transmission spectrum under specific assumptions. It is therefore difficult and limited by the experimental energy resolution. Finally thin target preparation is much more difficult, and strong masking effects can be easily introduced into the

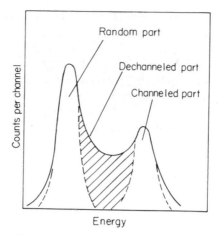

Figure 7.6. Energy spectrum (schematic) of an ion beam transmitted through a single crystal, showing how the dechanneled fraction should be determined by subtracting both the channeled and random spectra

experiment because of lattice imperfections. The above difficulties have limited the use of the transmission technique to a few experiments.

Larger uncertainties in the conversion of energy to penetration scale arise, in both the experimental methods described above, because of the uncertainties involved in the evaluation of $S^*(E)$. This depends in fact on the electron density sampled by the ion along its path.

The stopping power $S^*(E)$ has been measured by transmission experiments in thin single crystal for different incident angles (cf. Chapter 4). In a thick crystal however the beam is spread along its path because of the interaction with electrons and nuclei; and in this case the reduction factor β in the channeled stopping power changes along the ion path and the actual energy loss can be evaluated only through an average stopping power, which cannot be simply related to that measured by transmission.

7.3 EXPERIMENTAL RESULTS

In the following two Sections, 7.3.1 and 7.3.2, we will discuss a few of the more relevant experiments on planar and axial dechanneling. The most striking results will be presented and interpreted qualitatively with the aid of simple physical concepts. The significance of the results will be discussed in some cases in connection with the preceding section on the treatment of experimental data, and taking into account the possible masking effects.

7.3.1 Planar Dechanneling

Planar dechanneling, (i.e. the change of the minimum yield, $\chi(z)$, with penetration depth, z), for an ion beam incident parallel to a close packed plane has been studied for different ions and crystal targets. Experimental results have been reported by Appleton *et al.*[7] Davies *et al.*[5] and Campisano *et al.*[22]

The interpretation of planar dechanneling experiments from the backscattered energy spectra is relatively simple because the aligned spectrum rises steeply in the planar case, approaching the random spectrum when the energy loss along the ingoing path is only $0 \cdot 1 \, E_0$ (see Figure 7.7) In computing

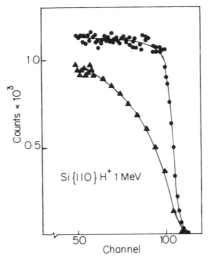

Figure 7.7. Energy spectrum of 1 MeV protons backscattered from a Si target at room temperature: ● random incidence; ▲ incidence aligned with the {110} plane

$\chi(z)$ from the energy spectra the correction factor can then be neglected by assuming a stopping power which is independent of the trajectory. The error introduced by this approximation remains in the limit of a few per cent, which is of the order of the experimental accuracy. Only for crystals like W where an appreciable fraction of the beam remains channeled up to higher penetration depths does the difference in the stopping powers need to be considered.

The minimum yield at the crystal surface, $\chi(0)$ is of the order of $0 \cdot 2$ for the lower order planes and it is of the order of $2a/d_p$, a being the potential screening radius and d_p the distance between the atomic planes. $\chi(0)$ exhibits a very small dependence on the beam energy and on the crystal temperature as shown[5] in Figure 7.8. With increasing depth z the channeled fraction $1 - \chi(z)$ decreases because of transitions of particles from the aligned to the random beam. Apart for a small initial region where the depth dependence is smoother, the decrease is exponential with respect to z in all the cases investigated,

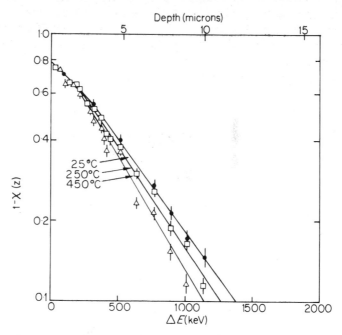

Figure 7.8. Depth dependence of the channeled fraction $(1 - \chi(z))$ in the $\{110\}$ plane of W for different target temperatures[5]

which suggests the introduction of a half depth, $z_{\frac{1}{2}}$ for channeling.[7] The energy dependence of $z_{\frac{1}{2}}$ has been investigated in detail for protons in $Si\{110\}$ in the range 0·3 to 9 MeV[7,22] and for protons in $W\{110\}$ in the range 0·5 to 12 MeV[23]; in both cases $z_{\frac{1}{2}}$ increases linearly with the beam energy as shown in Figure 7.9, and suggest the parametrization of the dechanneled fraction by a single parameter, $z_{\frac{1}{2}}/E_0$.

The effect of the lattice vibration on $\chi(z)$ has been investigated for the two cases above and for $Ge\{110\}$.[22] The change in $\chi(z)$ was found to be very small and just outside the experimental errors; the increase of the dechanneled fraction with crystal temperature is only appreciable for low beam energies and high penetration depths. The corresponding changes in $z_{\frac{1}{2}}$ are of the order of 5 to 10% when the mean square vibrational amplitude changes by a factor of two (see, for example, Figure 7.8).

A qualitative interpretation of the planar dechanneling experiments in terms of the ion beam spreading has been proposed by several authors.[7]

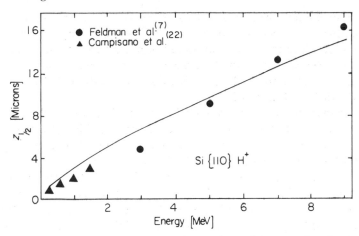

Figure 7.9. Half thickness of planar channeling $z_{\frac{1}{2}}$ against beam energy for protons in Si $\{110\}$. Comparison between experimental (▲ reference 22 and ● reference 7) and calculated values (cf. Section 7.4)

The small temperature dependence suggested that the interaction occurs mainly with the temperature-independent electron gas and this was consistent with the strong reduction in the rate of interaction with nuclei as revealed by channeling dips. Moreover, the exponential dependence of $\chi(z)$ on the penetration z suggested a diffusion mechanism, which could be attributed to a statistical process such as multiple scattering by the electron gas.[7,10] This interpretation was supported by the energy dependence which is in agreement with that expected from the multiple scattering theory.

The quantitative interpretation of planar dechanneling,[22,24,25] which will be treated in detail in Section 7.4.3 in terms of the elementary interaction processes[1] between the moving ion and the crystal, confirms the above description.

7.3.2 Axial Dechanneling

When an ion beam enters a single crystal parallel to a low order axis the energy spectrum of the backscattered particles starts from a very small value on the high energy side, and increases smoothly with decreasing energy, still remaining significantly different from the random spectrum, as shown in Figure 7.3. In this case a significant part of the beam remains channeled up to high penetrations. The difference in the stopping powers, cannot be neglected therefore, and the correction factor $\Lambda(z)$ should be calculated from the stopping powers $S(E)$ and $S^*(E)$ and from the Rutherford cross section $\sigma(E)$. The interpretation of the axial dechanneling results is more difficult than in the planar case, because of the average estimates of $S^*(E)$ involved.

Moreover, the minimum yield at the crystal surface, $\chi(0)$, can be estimated from the forbidden cross sectional area[1] $(a/\rho_{ch})^2 = 0.02$, ρ_{ch} being the channel radius. The result in this case is an order of magnitude smaller than in the planar case. Hence even a small surface contamination or crystal imperfection can change appreciably the $\chi(0)$ value[25] and especially the $\chi(z)$ curve at large depth[26] (cf. Figure 7.2).

The agreement between the axial dechanneling data obtained by different authors is therefore not as good as in the planar case. This discrepancy can be attributed partly to the experimental conditions, but it arises mainly from the different ways of deriving the results from the experimental spectra.

The axial case was investigated for a series of ion–crystal combinations by Davies *et al.*[5] and Ellegaard *et al.*[6] It was found that in this case the channeled fraction $1 - \chi(z)$ decreases with energy E_0 and with increasing depth z. By analogy with the planar case it was suggested that the depth dependence was roughly exponential. However the range of variation of $\chi(z)$ was too small to determine the complete dependence and significant deviations from the simple exponential law occurred. The most striking result of these experiments was, in contrast with the planar case, a strong dependence of the dechanneled fraction on the crystal temperature. This result was surprising because the nuclear interaction was expected to have only small influence on a channeled beam which is prevented from close encounters.

A systematic investigation of the temperature and energy dependence of the dechanneling of protons and deuterons in Si and Ge was carried out by the Catania group[8-12] which showed that at large penetration $\chi(z)$ was closer to a linear function of the penetration depth than to a simple exponential. This is shown in Figure 7.10. They found, moreover, that the dechanneled

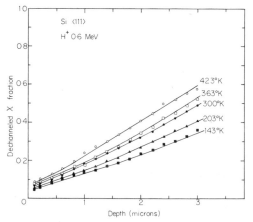

Figure 7.10. Depth dependence of the dechanneled fraction $\chi(z)$ for 0·6 MeV protons along the $\langle 111 \rangle$ axes of Si at different temperatures

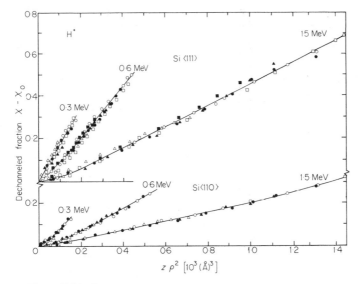

Figure 7.11. Dependence of the proton dechanneled fraction $\chi(z) - \chi(0)$ on $z\rho^2$, along the $\langle 111 \rangle$ and $\langle 110 \rangle$ axes of Si at different temperatures:[9] △ 80 K, ■ 143 K, ▲ 203 K, ● 300 K, □ 363 K, ○ 423 K

fraction $\chi(z) - \chi(0)$ along the main axis of Si and Ge can be expressed as a function of the depth z, of the beam energy E_0 and of the mean square vibrational amplitude ρ^2 through a single scaling parameter $z\rho^2/E_0$. This result can be seen by comparing Figures 7.10, 7.11 and 7.12 which show how the

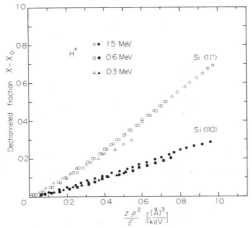

Figure 7.12. Dependence of the proton dechanneled fraction $\chi(z) - \chi(0)$ on $z p^2/E_0$ along the $\langle 111 \rangle$ and $\langle 110 \rangle$ axes of Si[9]

results taken at different temperature and energies collapse into a single curve if plotted against $z\rho^2/E_0$. The same functional dependence on the target temperature was found to be valid for the results previously published by Davies *et al.*[5] for 3 MeV protons in W$\langle 100 \rangle$.

The simple proportionality of $\chi(z)$ to the mean square vibrational amplitude suggested that the main contribution to the beam scattering in the axial case came from the vibrating nuclei and that the contribution from the temperature independent electron gas was negligible. The strong temperature dependence of axial dechanneling in comparison with the planar case has, however, a simple physical interpretation. The planar channeling can be described by a simple model; namely, as the motion of a charged particle in the electric field produced by a continuous planar distribution of charges*. The field produced by such a distribution is independent of the distance, so that any fluctuation of the plane does not change the force acting on the charged particle. The axial channeling however is described as the motion of a charged particle in the field produced by a continuously charged string; the field in this case is a function of the distance and it fluctuates at the particle's position if the string fluctuates because of thermal vibrations. Ions moving along an axial channel then experience a time dependent field so that they suffer random momentum changes according to the force fluctuations and, therefore, undergo angular dispersion. Moreover, in the axial case the critical angle, i.e. the maximum angle allowed for a channeled ion, decreases strongly with increasing temperature so that an ion beam which has been spread by a given angle is dechanneled more easily at higher temperatures. The combination of these two effects gives, in the case of Si and Ge, the simple ρ^2 scaling found in experiments. Furthermore, in this case the qualitative description outlined above was confirmed, quantitatively, by detailed calculation based on Lindhard's theory.

The simple $z\rho^2/E_0$ parametrization found for the axial dechanneling in Si and Ge stimulated several other experiments both to confirm this result and to extend them to other ion–target combinations.[14,23,27–30] In particular Fujimoto *et al.*[14] repeated the experiments on Si and Ge and found that the ratio of the channeled to the random spectrum, i.e. the noncorrected dechanneled fraction, can be fitted by an exponential function depending on three parameters χ_0, L and Δz. The plots of the experimental data they reported agree roughly with the $z\rho^2$ parametrization except for the curve obtained at liquid nitrogen temperature where a strong surface contamination occurred.[14] The authors tried to attribute a simple physical significance to the L parameter as a diffusion pathlength on the basis of a diffusion model proposed by Ellegaard[6] and found that L could be expressed as:

$$L \propto (A\rho^2 + B)^{-1}.$$

* We neglect the finite thickness ($\simeq a$) of the charge distribution.

They concluded that A and B represent the nuclear and electronic contribution to dechanneling in the case investigated. The physical meaning of this interpretation is not clear however, since the other two parameter χ and Δz are also temperature dependent, and because of the many assumptions involved in the application of the diffusion model.

A more substantial disagreement has been reported recently by Davies *et al.*[23] Their experimental results show that the proton dechanneled fraction $\chi(z)$ in W$\langle 111 \rangle$ is roughly independent of the beam energy in the range from 0·5 to 12 MeV. Furthermore $\chi(z)$ in W$\langle 111 \rangle$ is less dependent on the vibrational amplitude than in Si, Ge or W$\langle 100 \rangle$. The ρ^2 dependence found in the previous cases is, as stated above, the net result of two different contributions which change from case to case and is not taken as a general law. However, the energy dependence of the dechanneled fraction results from the energy dependence of the multiple scattering, which depends inversely on the beam energy.[1] A tentative interpretation[31] of the W$\langle 111 \rangle$ results is based on the effect of a misalignment between the ion beam and the crystal axis; this could arise either because of poor collimation of the beam; or because of the mosaic spread in the crystal. The effect of a misalignment on the dechanneled fraction increases with increasing beam energy because of the critical angle reduction. It has been estimated[31] that a misalignment of the order of 0·1° could fit the experimental results; though this value is reasonable, no data are available at present to support this interpretation.

The relevance of crystal imperfections or contamination in dechanneling requires detailed treatment which is beyond the scope of the present review. It should be mentioned however that several investigations on this point have been reported. The experiments were performed either by covering the crystal surface with an amorphous layer[26,32] or by introducing disorder into the crystal itself.[33] In both cases, the dechanneled fraction increases strongly with an increase in the number of atoms displaced from lattice sites. The effect is not limited to the damage zone but extends and increases with penetration depth. The defect contribution to dechanneling is similar to that sketched in Figure 7.2 and can be related to the beam scattering from defects.

7.4 DECHANNELING THEORY

Several discussions have been reported on the validity and on the physical significance of the classical description of channeling phenomena.[1,34,35] Although each of the many processes involved need to be treated quantum mechanically, the whole channeling process may be described by the simple and elegant theory developed by Lindhard.[1] This classical theory has been shown to be valid when the ion mass and energy are high, while for low masses, quantum effects should be expected.[36,37] This theory has been discussed in

Chapter 2, where the mathematical framework has been developed in some detail.

The following theory of dechanneling is developed, starting from Lindhard's point of view and using the relationships he derived to describe the interaction of a moving ion with a single crystal target. Lindhard showed[1] that a fast ion moving with a small angle ψ from a close-packed atomic row (or plane) experiences a repulsive force which, as a first approximation, can be replaced by a continuum potential which depends only on the distance **r** from the row (or plane), and not on the position z along it, i.e.*

$$U(\mathbf{r}, z) \simeq U(\mathbf{r}) \simeq U_1(r). \tag{7.7}$$

This implies the conservation of a parallel component, $p_z \simeq p$, of the particle momentum and this in turn implies the possibility of describing completely the motion of an ion in the plane transverse to the string (or in the line transverse to the plane). By definition the total transverse energy is given by

$$E_\perp = \frac{\{p\psi(r)\}^2}{2M_1} + U(r) = E_0\psi^2(r_0) \tag{7.8}$$

and the transverse momentum

$$\mathbf{p}_\perp = \mathbf{p} \times \boldsymbol{\psi}(r) \tag{7.9}$$

where M_1, p, and E_0 are the ion mass, momentum and energy respectively, and r_0 is the point of minimum potential where $U(r_0) = 0$. Since the potential is independent of z, then $U(r)$ is independent of time ($t = zM_1/p_z$) and the ion transverse energy is conserved. However when the thermal vibrations of string atoms are considered $U(r)$ becomes time dependent, and the transverse energy changes with t, i.e. with depth z along the path. The same non-conservation of $E_\perp$ also results when close collisions with the crystal electrons spread in the channel are considered. In this scheme the dechanneling process is described by the angular dispersion of a collimated beam because of interactions both with the electrons and with the vibrating nuclei of the crystal. The momentum changes experienced by a channeled ion in a single elementary interaction is small and hence the overall dispersion can be described by a statistical treatment (i.e. as a diffusion in transverse momentum, under the proper boundary conditions and symmetries[1]). The boundary conditions are specified by the initial distribution in transverse momentum of the ions entering the channel, and by the probability that an ion of given transverse momentum suffers a close collision with a nucleus, and is dechanneled. The dechanneling problem then becomes that of studying (in transverse

---

* The last approximation means that, because of the screening, the potentials of two adjacent rows do not overlap appreciably and the potential coincides with that of a single row, which has a cyclindrical symmetry.

momentum space $\mathbf{p}_\perp$) the diffusion of a beam as a function of the time t or, because of the relation $t = zM_1/p$, as a function of the penetration depth z.

7.4.1 The Diffusion Equation

If the scattering of the moving ion is isotropic and involves only small changes $\Delta\mathbf{p}_\perp = \mathbf{p}'_\perp - \mathbf{p}_\perp$ in the transverse momentum, the scattering probability will be a function of the average transverse momentum $\{\mathbf{p}_\perp + (\Delta\mathbf{p}_\perp/2)\}$ of the modulus $\Delta\mathbf{p}_\perp$, and of the position $\mathbf{r}$ occupied by the ion in the transverse space, i.e.

$$P(\mathbf{p}_\perp, \mathbf{p}'_\perp, \mathbf{r}) = P\{\mathbf{p}_\perp + (\Delta\mathbf{p}_\perp/2), \Delta\mathbf{p}_\perp, \mathbf{r}\}.$$

However, if statistical equilibrium in the transverse phase space (cf. Chapter 2) has been reached, a particle which has a transverse momentum $\mathbf{p}_\perp(r_0)$ at r_0, will be distributed in the allowed $\mathbf{r}$ region according to a known function $P_0(E_\perp, \mathbf{r})$ (cf. Chapter 2). It is then possible to average the scattering probability over the allowed $\mathbf{r}$ space and obtain an average scattering probability $P\{\mathbf{p}_\perp(r_0) + (\Delta\mathbf{p}_\perp/2), \Delta\mathbf{p}_\perp\}$ which depends only on the transverse momentum measured at the point r_0 and on the $\Delta\mathbf{p}_\perp$. If the scattering processes do not modify the statistical equilibrium appreciably, we can work out the distribution in transverse momentum of the particles from the above scheme, bearing in mind that it should be measured at the channel axis, r_0.

The diffusion equation is:

$$\frac{\partial}{\partial z} n(\mathbf{p}_\perp(r_0), z) = \mathrm{div}_{\mathbf{p}_\perp(r_0)} D(\mathbf{p}_\perp(r_0))\, \mathrm{grad}_{\mathbf{p}_\perp(r_0)} n(\mathbf{p}_\perp(r_0), z) \qquad (7.10)$$

and in the general case it can only be solved by numerical methods.[38] For axial and planar symmetry in the $\mathbf{p}_\perp$ space the diffusion equations become respectively:

$$\frac{\partial}{\partial z} n(p_\perp, z) = \frac{1}{p_\perp} \frac{\partial}{\partial p_\perp} \left\{ D(p_\perp) \frac{\partial}{\partial p_\perp} n(p_\perp, z) \right\} \qquad (7.11)$$

and

$$\frac{\partial}{\partial z} n(p_\perp, z) = \frac{\partial}{\partial p_\perp} \left\{ D(p_\perp) \frac{\partial}{\partial p_\perp} n(p_\perp, z) \right\} \qquad (7.12)$$

where $p_\perp \geqslant 0$ is the modulus of the transverse momentum measured at r_0.

A better insight into the physical significance of the diffusion equation may be gained by studying along a path length δz (see Figure 7.13), the behaviour of a beam of particles having at depth z a well defined $p_\perp(r_0)$ or $E_\perp$:

$$n(\mathbf{p}_\perp(r_0), z) = \delta\{p'_\perp(r_0) - p_\perp(r_0)\}. \qquad (7.13)$$

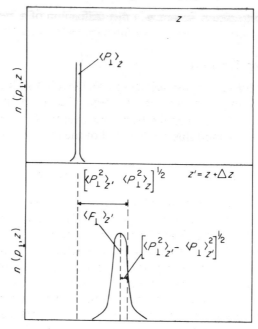

Figure 7.13. Schematic diagram of the distribution in transverse momentum at depth $z + \Delta z$ (lower part) of a beam collimated like a δ function at depth z (upper part). The shift of the rms $p_\perp$ values between z and $z + \Delta z$ is compared with the rms spreading

The changes in the mean square momentum after δz, at $z' = z + \delta z$ is:[1]

$$\frac{\delta \langle p_\perp^2 \rangle}{\delta z} = \frac{\langle p_\perp^2 \rangle_{z'} - \langle p_\perp^2 \rangle_z}{\delta z} = 2jD(p_\perp) + j\frac{\mathrm{d}D(p_\perp)}{\mathrm{d}p_\perp} \qquad (7.14)$$

and in terms of transverse energy $E_\perp$:

$$\frac{\delta \langle E_\perp \rangle_{E_\perp}}{\delta z} = \frac{j}{M_1}D(p_\perp) + \frac{j}{2M_1}p_\perp\frac{\mathrm{d}D(p_\perp)}{\mathrm{d}p_\perp} \qquad (7.15)$$

where j is the number of dimensions in the transverse momentum space ($j = 1, 2$ in the planar and axial case respectively). At depth $z' = z + \delta z$ the mean square spreading in transverse momentum of a beam, which was collimated at z becomes:

$$\frac{\langle (p_\perp - \langle p_\perp \rangle)^2 \rangle_{z+\delta z} - \langle (p_\perp - \langle p_\perp \rangle)^2 \rangle_z}{\delta z} = \frac{\langle p_\perp^2 \rangle_{z+\delta z} - \langle p_\perp \rangle_{z+\delta z}^2}{\delta z}$$

$$= jD(p_\perp)$$

which for a straightforward comparison with equation (7.15) can be written as

$$\frac{1}{2M_1}\frac{\langle p_\perp^2\rangle_{z+\delta z} - \langle p_\perp\rangle_{z+\delta z}^2}{\delta z} = \frac{j}{2M_1}D(p_\perp) \qquad (7.16)$$

By comparing equations (7.15) and (7.16) one finds that the spreading of a collimated beam see equation (7.13) after it has traversed a path length δz is smaller than the average energy change, provided that $D(p_\perp)$ is a nondecreasing function of $p_\perp$ (see Figure 7.13). This condition always holds of course because with increasing $p_\perp$ the ion approaches closer to the atomic string (or plane) and hence the scattering probability increases. The same equation suggests two extreme and opposite approximations suitable for solving the diffusion equation.

(a) When the diffusion coefficient is constant, $D(p_\perp) = D_0$, the mean square spreading of the beam in transverse momentum at z' is $\frac{1}{2}$ of its mean square change along δz, so that the diffusion is relevant. In this case however the diffusion equation can be solved analytically[39] with the proper boundary conditions.

(b) When the diffusion function, $D(p_\perp)$, changes rapidly with $p_\perp$, the change in transverse energy is determined by the term $(p_\perp/M_1)(\mathrm{d}D/\mathrm{d}p_\perp)$ of equation (7.15). The other term D/M_1, which is small, determines the beam spreading and is negligible with respect to the change in transverse energy $\delta\langle E_\perp\rangle$. In this case a beam of particles which has a well defined transverse energy or momentum at depth z will have also a well defined energy at any other depth. We can therefore take $\delta\langle E_\perp\rangle = \mathrm{d}E_\perp$ and avoid solving the diffusion equation by following the energy changes of each particle of the beam, (steady increase approximation).[1,9]

To solve the diffusion equation, either in its complete form or in the approximations described above, we shall determine, in terms of the elementary scattering processes, either the diffusion function $D(p_\perp)$ or the change in the average transverse energy $\delta\langle E_\perp\rangle$. These are related by equation (7.15) i.e. by

$$\frac{\delta\langle E_\perp\rangle}{\delta z} = \frac{1}{M_1}\operatorname{div}_{\mathbf{p}_\perp}\{\mathbf{p}_\perp D(\mathbf{p}_\perp)\}. \qquad (7.17)$$

According to Lindhard[1] (cf. Chapter 2) we can separate the contribution of the vibrating nuclei, from that of the close collisions with electrons. Assuming that statistical equilibrium has been reached, the average change in the transverse energy per unit path length, owing to the lattice vibrations is:

$$\frac{\delta\langle E_\perp\rangle_n}{\delta z} = \frac{d}{4E}\langle(\delta\mathbf{k}(\mathbf{r}))^2\rangle \qquad (7.18)$$

where $\mathbf{k} = -\nabla_r U(r)$ is the force acting, $\delta\mathbf{k}(\mathbf{r}) = \boldsymbol{\rho}_r \nabla_r^2 U(\mathbf{r})$ is its fluctuation, $\boldsymbol{\rho}_r$ the projection in the $\mathbf{r}$ space of the atomic displacement caused by thermal vibrations, and the symbol $\langle \ \rangle$ now means the average of the allowed positions of the moving ion in the $\mathbf{r}$ space. The electronic scattering produces an average change in transverse energy given by[1]

$$\frac{\delta\langle E_\perp \rangle_e}{\delta z} = \frac{m}{2M_1} < \rho_e(\mathbf{r}) > S_e \qquad (7.19)$$

where m and $\rho_e(\mathbf{r})$ are the electronic mass and density respectively, the latter being related to the potential by the Poisson equation. S_e is the stopping power per electron, which for a fast ion of velocity v can be written as [1,4]

$$S_e = \frac{4\pi Z_1^2 e^4}{mv^2} L_e = \frac{4Z_1^2 e^4}{mv^2} \ln\frac{2mv^2}{I} \qquad (7.20)$$

$I \simeq Z_2 \times 10 \text{ eV}$ which is the average excitation energy per electron.

7.4.2 The Axial Case

In the axial case the spatial distribution $P_0(E_\perp, \mathbf{r})$ of a channeled ion beam is uniform[1] (cf. Chapter 2) over the allowed region of the transverse plane, i.e. in the region between r_0 and $r_{\min}$ where the ion's total transverse energy is larger than the potential, $E_\perp = U(r_{\min})$, i.e.

$$P_0(E_\perp, \mathbf{r}) = \frac{1}{\pi(r_0^2 - r_{\min}^2)} \simeq Nd \quad \text{for } r_{\min} \ll r_0 \qquad (7.21)$$

where N is the atomic density of the crystal and d the mean atomic spacing along the string. Using the string potential $U_1(r)$ one has:

$$U(r) = U_1(r) - U_1(r_0) = \frac{Z_1 Z_2 C^2}{d} \frac{1}{A}\left(\frac{C^2 a^2}{r^2} + 1\right) \qquad (7.22)$$

where $A = \{(C^2 a^2/r_0^2) + 1\}$, Z_1 and Z_2 are the atomic numbers of the ion and of the crystal, $C^2 = 3$, and $a = 0.46\ (Z_1^{\frac{2}{3}} + Z_2^{\frac{2}{3}})^{-\frac{1}{2}}$ Å, is the Thomas–Fermi screening distance.

The scattering of a channeled ion interacting with vibrating nuclei can now be evaluated[9] by performing, in axial symmetry, the averages in equation (7.18):

$$\delta\langle E_\perp \rangle_n = \delta z \frac{d}{4E_0}\langle\{\delta\mathbf{k}(\mathbf{r})\}^2\rangle = \delta z \frac{d}{4E_0}\langle\{\boldsymbol{\rho}\nabla_r^2 U(r)\}^2\rangle$$

$$= \delta z \frac{d}{4E_0}\frac{\rho_{ax}^2}{2}\langle k^2 r^{-2} - k'^2\rangle$$

$$= \delta z \frac{d\rho_{ax}^2}{8E_0}\int_{r_{\min}}^{r_0} P_0(E_\perp, \mathbf{r})\{(k^2 r^{-2} - k'^2)\}2\pi r\, dr.$$

Therefore

$$\delta\langle E_\perp \rangle_n = \delta z \frac{\pi N d^2 \rho_{ax}^2}{E_0 C^2 a^2} \left(\frac{Z_1 Z_2 e^2}{d} \right)^2 \frac{(Ae^{\varepsilon_\perp} - 1)}{A(e^{\varepsilon_\perp} - 1)} \left\{ (Ae^{\varepsilon_\perp} + \tfrac{2}{3}) \left(1 + \frac{e^{-\varepsilon_\perp}}{A} \right)^3 \right\} \quad (7.23)$$

where

$$k' = d^2 U/dr^2 \quad \text{and} \quad \rho_{ax}^2 = \tfrac{2}{3}\langle \rho^2 \rangle.$$

The dimensionless quantity $\varepsilon_\perp$ (reduced transverse energy) is given by

$$\varepsilon_\perp = 2E_\perp/E_0 \psi_1^2 = E_\perp/E_0 Z_1 Z_2 c^2/d \quad (7.24)$$

and has been introduced[11] as a scaling factor for the ion transverse energy using the Lindhard angle $\psi_1 = (2Z_1 Z_2 e^2/E_0 d)^{\frac{1}{2}}$. For a channeled ion $\psi(r_0) < \psi_1$ and then $0 \leqslant \varepsilon_\perp \lesssim 2$. Equation (7.23) can be written in terms of $\varepsilon_\perp$ by separating the factors A_n, which depend on the ion mass, atomic number and energy, and on the crystal atomic number, density, axis and temperature, from those factors, f_n, which depend only on the reduced transverse energy $\varepsilon_\perp$.

Rewriting equation (7.23) in these terms gives

$$\delta\langle \varepsilon_\perp \rangle_n = \delta z A_n(E_0, T) f_n(\varepsilon_\perp). \quad (7.25)$$

The electronic contribution to the beam scattering can be evaluated from the sampled electronic density:

$$\langle \rho_e(r) \rangle = \frac{Nd}{4\pi Z_1 e^2} \int_{r_{min}}^{r_0} P_0(E_\perp, \mathbf{r}) \left(\frac{1}{r} \frac{d}{dr} \left(r \frac{dU_1}{dr} \right) \right) 2Mr \, dr$$

$$= \frac{Nd}{2Z_1 e^2} \left(r \frac{dU}{dr} \right)_{r_{min}}.$$

Therefore

$$\delta\langle E_\perp \rangle_e = \delta z \frac{\pi Z_1 e^2 Nd}{E_0} \left(\frac{Z_1 Z_2 e^2}{d} \right)^2 L_e \frac{Ae^{\varepsilon_\perp} - 1}{A(e^{\varepsilon_\perp} - 1)} \left\{ \frac{e^{-\varepsilon_\perp}}{A} \left(1 - \frac{e^{-\varepsilon_\perp}}{A} \right) \right\}$$

$$\delta\langle \varepsilon_\perp \rangle_e = \delta z B_e(E_0, v) f_e(\varepsilon_\perp) \quad (7.26)$$

in which, once again, the part depending on the transverse energy has been separated from the other. It is relevant that the B_e factor depends on the ion energy and, through L_e it also depends on the ion velocity, v (cf. equation (7.20)). The function $f_n(\varepsilon_\perp)$ and $f_e(\varepsilon_\perp)$ are plotted in Figure 7.14 while the factors A_n and B_e are given in Table 1 for some typical cases.

From the above figure and table it is evident that the main contribution to the beam scattering in axes comes from the nuclear term which is larger for transverse energies, $\varepsilon_\perp \gtrsim 0.7$. The ions which enter the channel with high

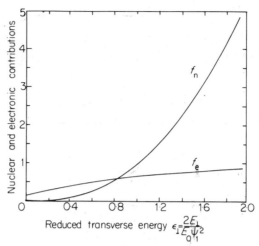

Figure 7.14. Changes of the nuclear f_n and
electronic f_e scattering as a function of reduced
transverse energy for protons in Si $\langle 111 \rangle$ (see
equations (7.23) and (7.25) for f_n and equation
(7.26) for f_e respectively)

transverse energy and are then dechanneled at small depths (i.e. after a few scatterings) interact mainly with the vibrating nuclei. On the other hand, ions which enter with low transverse energy are diffused by the electrons, which lift them to a higher transverse energy region, where they interact with the vibrating string. This is consistent with the strong temperature dependence found in axial dechanneling experiments. Moreover, the rapid change of f_n with respect to $\varepsilon_\perp$ causes a rapid change in $\delta\langle\varepsilon_\perp\rangle/\delta z$ with respect to $\varepsilon_\perp$, as shown in Figure 7.15. We therefore expect a negligible contribution from the beam diffusion with respect to its shift in transverse energy. The two terms are compared in the same figure where D/M_1 has been obtained by integration of equation (7.17)[14] which for axial symmetry gives:

$$\frac{\delta\langle\varepsilon_\perp\rangle}{\delta z} = \frac{\delta\langle\varepsilon\rangle_n + \delta\langle\varepsilon_\perp\rangle_e}{\delta z} = \{A_n f_n + B_e f_e\} \tag{7.27}$$

and

$$\frac{D}{M_1} = \frac{1}{p_\perp^2}\int_0^{p_\perp} (A_n f_n + B_e f_e)p_\perp' \, dp_\perp. \tag{7.28}$$

Since the beam diffusion is small in the axial case it can be neglected, and hence under these conditions the steady increase approximation described in Section 7.4.1(a) applies. This allows us to determine the depth $z(E_{\perp 0})$ where a particle which entered with transverse energy $E_{\perp 0}$ reaches the

Table 7.1 Values of the coefficients $A_n(E_0, T)$ and $B_e(E_0, v)$ in μm^{-1}

Beam and target	A_n at 1 MeV against crystal temperature (K)						B_e against beam energy (MeV)					
	80	140	200	300	420	720	0.3	0.6	1.0	1.5	2.0	3.0
H$^+$ in Si ⟨111⟩	0.11	0.15	0.18	0.21	0.24	—	0.55	0.40	—	0.22	—	—
D$^+$ in Si ⟨111⟩	0.11	0.15	0.18	0.21	0.24	—	0.30	0.28	—	—	—	—
He$^+$ in Si ⟨111⟩	—	—	—	0.44	—	—	—	—	—	0.24	—	—
H$^+$ in Ge ⟨111⟩	0.34	0.44	—	0.58	0.61	—	0.24	0.23	0.19	0.15	—	—
H$^+$ in W ⟨100⟩	—	—	—	1.35	—	2.93	—	—	—	—	—	0.07
H$^+$ in W ⟨111⟩	0.31	—	—	0.89	—	2.02	—	—	0.09	—	0.07	—

The mean square vibrational amplitude has been calculated from the Debye model.

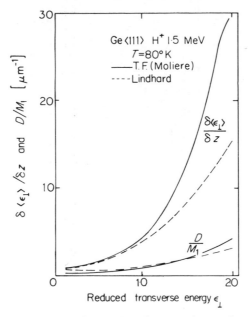

Figure 7.15. Comparison between the steady increase, $\delta\langle\varepsilon_\perp\rangle/\delta z$, and diffusion, D/M_1, terms for axial dechanneling

maximum value $E_\perp^*$ allowed for a channeled ion. By integration of equation (7.27) one obtains in terms of $\varepsilon_\perp$:

$$z(\varepsilon_{\perp 0}) = \zeta(\varepsilon_{\perp 0}) - \zeta(\varepsilon_\perp^*) \qquad (7.29)$$

where

$$\zeta(\varepsilon_\perp) = \int (A_n f_n + B_e f_e)^{-1} \, \delta\varepsilon_\perp.$$

The fraction, χ, of particles dechanneled between zero depth and the depth $z(\varepsilon_{\perp 0})$ is given, on the other hand, by the fraction $P(\varepsilon_{\perp 0})$ of the incoming beam which entered with transverse energy larger than $\varepsilon_{\perp 0}$. The dechanneled fraction $\chi(z)$ can then be calculated by eliminating $\varepsilon_{\perp 0}$ from equations (7.21) and (7.29) as shown in Figure 7.16.

The function $P(\varepsilon_{\perp 0})$ can be computed from the surface transmission.[10,11] An ion which enters a single crystal parallel to a row at a distance r from it, passes from a region where the potential is zero to a region where it is $U(r)$; the ion is then deflected and it gains a transverse energy $E_{\perp 0} = U(r)$. The fraction dN of the incident beam having a transverse energy within $dE_{\perp 0}$ at $E_{\perp 0}$ is equal to the fraction which enters the channel within dr at a distance *r* from the string, i.e.

$$dN = \frac{dN}{dE_{\perp 0}} dE_{\perp 0} = n(E_{\perp 0})\, dE_{\perp 0} = \frac{2\pi r\, dr}{\pi r_0^2}$$

and then:

$$n(\varepsilon_{\perp 0}) = \frac{C^2 a^2}{r_0^2} \frac{A e^{\varepsilon_{\perp 0}}}{(A e^{\varepsilon_{\perp 0}} - 1)^2} \tag{7.30}$$

$$P(\varepsilon_{\perp 0}) = \int_0^{\varepsilon_{\perp 0}} n(\varepsilon'_{\perp 0})\, d\varepsilon'_{\perp 0} = \frac{A}{A e^{\varepsilon_{\perp 0}} - 1} \tag{7.31}$$

The functions $P(\varepsilon_{\perp 0})$ and $n(\varepsilon_{\perp 0})$ for the Si$\langle 111 \rangle$ channel are plotted in Figures (7.16) and (7.17) in terms of $\varepsilon_{\perp 0} = 2E_{\perp 0}/E\psi_1^2$.

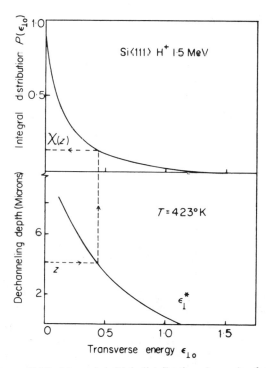

Figure 7.16. Integral initial distribution (upper) of 1·5 MeV protons entering along the $\langle 111 \rangle$ axis of Si. The depth at which (lower) a particle having a reduced initial transverse energy reaches the critical value to be dechanneled. The dechanneled fraction is obtained by eliminating $\varepsilon_{\perp 0}$ between the lower and the upper parts. For the case shown in the figure $E_{\perp}^* \simeq 48$ eV

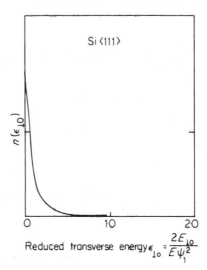

Figure 7.17. Differential transverse energy distribution, $n(\varepsilon_{\perp 0})$, for ions which entered the Si $\langle 111 \rangle$ axial channel

When the crystal surface is covered by an amorphous layer, and this is always the case experimentally, the scattering caused by the amorphous layer should be considered.[11] A different treatment will be adopted for this depending on whether the average number of close collisions that the ion suffers in the amorphous region is small, comparable with or much larger than one. A detailed treatment of this effect is beyond the scope of the present work; the reader is referred however to the literature on single, plural[40,41] and multiple[42] scattering in amorphous materials. As discussed in Section 7.2 and shown in Figure 7.2, the spreading in the amorphous layer contributes appreciably to axial dechanneling, so that it cannot be neglected in a quantitative treatment.

A final point to be dealt with concerns the significance of a maximum allowed transverse energy, $E_\perp^*$, for channeling. In the limit of the above description of axial dechanneling in terms of the steady increase approximation one can assume that the probability of a close encounter between an ion and the nuclei is either zero or one depending on whether the ion's transverse energy is lower or larger than $E_\perp^* = E_0 \psi_1^2$. We know, however, that this step function is an approximation to a continuous[1] scattering probability $\pi(E_\perp)$ which, because of the thermal vibration of the string atoms, becomes smoother with increasing temperature. The step function approximation introduces an error which increases with the temperature. This can be reduced by considering a temperature dependent critical angle $\psi_{\frac{1}{2}}(T)$ and taking

$$E_\perp^* = E_0 \psi_{\frac{1}{2}}(T)^2 \tag{7.32}$$

this temperature dependent critical angle can be calculated in a consistent way, as will be shown in Section 7.5; and this calculated result is in agreement with the experimental value.

7.4.3 The Planar Case

In contrast with the axial case the spatial distribution in statistical equilibrium[1] of an ion moving in a planar direction is not uniform as its motion in the transverse plane is similar to a classical one-dimensional oscillator. In the continuum approximation it depends on the distance y from the plane as $1/p_\perp(y)$. If the potential relative to its minimum if $Y(y)$, the equilibrium probability distribution is then

$$P_0(E_\perp, y) = \frac{g(E_\perp)}{d_p} \left(\frac{E_\perp}{E_\perp - Y(y)} \right)^{\frac{1}{2}} \tag{7.33}$$

d_p being the distance between the planes surrounding the channel and $g(E_\perp)$ a normalization factor.

The planar potential is a smooth function of distance, hence the contributions of both planes must be considered. If one considers only the contribution of the two atomic planes surrounding the channel the potential acting on the ion is:

$$Y(y) = Y_1(y) + Y_1(d_p - y) - 2Y_1(d_p/2) \tag{7.34}$$

using the Lindhard continuum potential

$$Y_1(y) = 2\pi Z_1 Z_2 e^2 N d_p Ca \left[\left\{ \left(\frac{y}{Ca} \right)^2 + 1 \right\}^{\frac{1}{2}} - \frac{y}{Ca} \right]$$

$$\equiv F \left[\left\{ \left(\frac{y}{Ca} \right)^2 + 1 \right\}^{\frac{1}{2}} - \frac{y}{Ca} \right] \tag{7.35}$$

in which N is the atomic density of the crystal and $C = \sqrt{3}$. As shown in Figure 7.18, $Y(y)$ can be approximated near the centre of the channel by an harmonic potential[14]

$$Y_2(y) = F(2a/d_p)^3 C(y/a - d_p/2a)^2 \tag{7.36}$$

while higher order terms should be considered near the planes (see Y_4 in Figure 7.18).

Knowing both the potential, $Y(y)$ and the probability distribution $P_0(E_\perp, y)$, the change of transverse energy per unit length averaged over the sampled

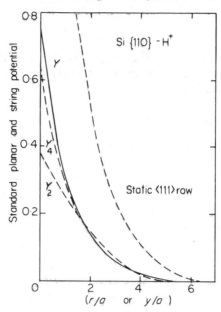

Figure 7.18. Comparison between Lindhard axial (Si $\langle 111 \rangle$ dashed line) and planar (Si $\{110\}$ full line) potentials measured in $E_0 \psi_1 / 2$ and F units respectively. The harmonic (Y_2) and quartic (Y_4) fits of the planar potential are shown

region can be evaluated. The contribution from the vibrating nuclei yields:

$$\delta \langle E_\perp \rangle = \delta z \frac{d}{4E_0} \langle \{\delta \mathbf{k}(\mathbf{r})\}^2 \rangle = \delta z \frac{(Nd_p)^{\frac{1}{2}}}{4E_0} \frac{\rho_p^2}{2} \langle \{k'(y)\}^2 \rangle \qquad (7.37)$$

where $k = -dY/dy$ and $k' = -d^2Y/dy^2$, and $\rho_p^2 = \frac{1}{2}\rho_{ax}^2 = \frac{1}{3}\langle \rho^2 \rangle$ is the component of the lattice mean square vibrational amplitude normal to the plane. In the harmonic approximation $k' = $ constant, and equation (7.37) becomes

$$\delta \langle E_\perp \rangle_n = \delta z \frac{(Nd_p)^{\frac{1}{2}}}{8E_0} \rho_p^2 \{F(2/d_p)^3 Ca\}^2. \qquad (7.38)$$

The contribution from electron scattering is

$$\delta \langle E_\perp \rangle_e = \delta z \frac{1}{2} \frac{m}{2M_1} S_e \langle \rho_e(\mathbf{r}) \rangle = \delta z \frac{\pi Z_1^2 e^4}{2E_0} L_e \langle \rho_e(y) \rangle \qquad (7.39)$$

where the factor $\frac{1}{2}$ has been introduced because the spreading parallel to the plane has no effect on planar dechanneling. In the harmonic approximation we get:

$$\delta\langle E_\perp\rangle_e = \delta z \frac{\pi Z_1^2 e^4}{2E_0} L_e \{F(2/dp)^3 Ca\}^2. \tag{7.40}$$

The average change per unit path length caused by both processes can be written in terms of a dimensionless reduced transverse energy $\varepsilon_\perp^p = E_\perp/F$, and separating the parts dependent on the crystal and beam parameters:

$$\delta\langle\varepsilon_\perp^p\rangle = \delta\langle\varepsilon_\perp^p\rangle_n + \delta\langle\varepsilon_\perp^p\rangle_e = A_n^p(E_0, T)f_n^p(\varepsilon_\perp^p) + B_e^p(E_0, v)f_e^p(\varepsilon_\perp^p).$$

The functions $f_n(\varepsilon_\perp^p)$ and $f_e(\varepsilon_\perp^p)$ are plotted in Figure 7.19 against $\varepsilon_\perp^p$, while the coefficients A_n^p and B_e^p are reported in Table 2 for a typical case.

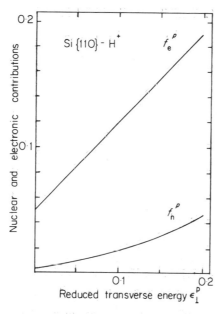

Figure 7.19. Changes of the nuclear f_n^p and electronic f_e^p scattering as a function of reduced transverse energy $\varepsilon_\perp^p$ for protons in Si {110}. (See equations (7.38), (7.39) and (7.41) respectively)

As expected from the small temperature dependence found experimentally, and from the qualitative description given in terms of the electric field (cf Section 7.3.2), the comparison between Figure 7.14 and 7.19 and Tables 1 and 2 show that the relative importance of nuclear to electronic contribution is much smaller for planes than for axes. Moreover, $\delta\langle\varepsilon_\perp^p\rangle/\delta z$ is a smooth

Table 7.2 Values of the coefficients $A_n^p(E_0, T)$ and $B_e^p(E_0, v)$ in μm^{-1}

Beam and target	A_n^p at 1 MeV against crystal temperature (K)				B_e^p against beam energy (MeV)			
	80	140	300	420	0·3	0·6	1·0	1·5
H$^+$ in Si {110}	0·12	0·15	0·21	0·24	1·04	0·73	0·53	0·41
D$^+$ in Si {110}	0·12	0·15	0·21	0·24	0·60	0·52	0·40	0·32
H$^+$ in Ge {110}	0·48	0·62	0·80	0·84	0·72	0·70	0·58	0·46

The mean square vibrational amplitude has been calculated from the Debye model.

function of $\varepsilon_\perp^p$; the diffusion of the beam then becomes relevant, as can be verified by integration of equation (7.17) for planes:

$$\frac{\delta\langle E_\perp \rangle}{\delta z} = \frac{\delta\langle E_\perp \rangle_n + \delta\langle E_\perp \rangle_e}{\delta z} = \frac{1}{M_1}\frac{d}{dp_1}\{p_\perp D(p_\perp)\}$$

$$\frac{D(p_\perp)}{M_1} = \frac{F}{p_\perp}\int_0^{p_\perp} (A_n^p f_n^p + B_e^p f_e^p)\, dp'. \tag{7.41}$$

For the low index planes of Si and Ge the electronic density is of the order of three electrons per atom. The scattering is therefore caused by valence electrons, which are described much better as nearly free than by the Thomas–Fermi model. In this assumption the electronic density should be nearly constant, and approximations (7.38) and (7.40) hold. The diffusion coefficient (7.41) therefore becomes independent of transverse momentum, $D(p_\perp) = D_0$. This allows the analytical solution of the diffusion equation (7.12), under the proper boundary conditions:

$$n(p_\perp, 0) = n(p_{\perp 0}); n(p_\perp^*, z) = 0 \tag{7.42}$$

where $p_\perp^*$ is the maximum transverse momentum beyond which a channeled ion experiences close nuclear interactions and is dechanneled. As in the axial case this can be evaluated from the critical angle $\phi_{\frac{1}{2}}$ for planar channeling, i.e. $p_\perp^* = \sqrt{2M_1}\phi_{\frac{1}{2}}\sqrt{E_0}$. The sink at $p_\perp^*$ implies a neglect of the second order process, i.e. the possibility that a dechanneled ion feeds into the channel later on.[24,43] $n(p_{\perp 0})$ is the distribution in transverse momentum of the ions which are deflected by the planar potential while crossing the crystal surface. To calculate $n(p_{\perp 0})$ we again make use of the conservation of energy. The fraction dN of the incident beam having a transverse energy within $dE_{\perp 0}$ of $E_{\perp 0}$ is equal to the fraction which enters the channel within dy of the distance y from the plane, where $E_{\perp 0} = Y(y)$, i.e.

$$dN = \frac{dN}{dE_{\perp 0}}dE_{\perp 0} = n(E_{\perp 0}) = \frac{dy}{d_p/2}$$

and in terms of transverse momentum

$$dN = n(p_{\perp 0})\, dp_{\perp 0}; \quad n(p_{\perp 0}) = \frac{2}{d_p}\left(\frac{dp_\perp}{dy}\right)^{-1} \tag{7.43}$$

which in the harmonic potential approximation results in a constant n_0 normalized to the incident flux:

$$n_0 = \frac{2a}{d_p}\left(2M_1 CF\left(\frac{2a}{d_p}\right)^3\right)^{-\frac{1}{2}}; \quad \int_0^{p_{\max}} n(p_{\perp 0})\, dp_{\perp 0} = 1 \tag{7.44}$$

where $p_{\perp\max}$ is the transverse momentum gained by a particle which enters at the point of maximum potential, $p^2_{\perp max}/2M_1 = Y(0)$. If higher order terms are considered in the potential, $n(p_{\perp 0})$ becomes a slowly decreasing function of $p_{\perp 0}$ as shown in Figure 7.20.

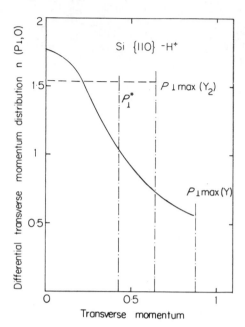

Figure 7.20. Differential transverse momentum distribution for protons entering the Si {110} planar channel calculated from the Lindhard potential Y (full time) and from the harmonic approximation Y_2 (dashed line). Refer to Figure 7.18 for the potentials

The diffusion equation for planes with the condition $D(p_{\perp}) = D_0$ is

$$\frac{\partial}{\partial z} n(p_{\perp}, z) = D_0 \frac{\partial^2}{\partial p_{\perp}^2} n(p_{\perp}, z)$$

and its solution with the above boundary conditions is

$$n(p_{\perp}, z) = \sum_{n=1}^{\infty} a_n \sin n\pi \left(\frac{p_{\perp} + p_{\perp}^*}{2p_{\perp}^*} \right) \exp\left(-\frac{\pi^2 n^2}{8M_1 F} \frac{D_0 z}{p_{\perp}^{*2}} \right) \qquad (7.45)$$

where the coefficients:

$$a_n = \frac{1}{p_{\perp}^*} \int_0^{p_{\perp}^*} n(p_{\perp}, 0) \sin n\pi \left(\frac{p_{\perp} + p_{\perp}^*}{2p_{\perp}^*} \right) dp_{\perp} \qquad (7.46)$$

are fixed by the initial distribution $n(p_{\perp}, 0)$.

The channeled fraction at depth z becomes

$$1 - \chi(z) = N(z) = \int_0^{p^*_\perp} n(p_\perp, z)\, dp_\perp \qquad (7.47)$$

which in the harmonic approximation with $z = 0$ gives:

$$1 - \chi(0) = N(0) = \int_0^{p^*_\perp} n_0\, dp_\perp = \frac{p^*_\perp}{p_{\perp max}} \simeq \frac{2}{3} \qquad (7.48)$$

and is shown in Figure 7.20. For large z where only the first term of the series is important, it gives:

$$1 - \chi(z) = N(z) \propto \exp(-D_0 z / p^{*2}_\perp) \qquad (7.49)$$

where D_0 decreases with increasing beam energy E_0, since the multiple scattering decreases (i.e. neglecting the small energy dependence of L_e cf. equation (7.37) and (7.39)) as

$$D_0 \propto 1/E_0 \qquad (7.50)$$

7.4.4 Comparison with Experiment

The dechanneled fraction $\chi(z)$ has been calculated from the above theory for a few ions and targets and along different axes or planes.[8–12,17,22,38,44] For the axial case the calculated results agree to within 30% of the experimental data and reproduce both the linear variation of the $\chi(z)$ against z curve and its parametrization in terms of zp^2/E_0. A better agreement has been found by a fuller treatment which included the scattering in the amorphous layer covering the crystal surface, the finite resolution of the incident beam and the small overlap between the string potentials at the channel axis.[8–12,17] It is evident that these three factors affect the initial distribution in transverse energy, suggesting that a more direct investigation of this would be relevant today. Some typical results including the above refinements on the initial distribution are shown in Figures 7.21 and 7.22. The latter is of particular interest as the different roles of nuclear and electronic scattering in different crystals (Si and W) is clearly shown.[10] In the case considered, even if the nuclear contributions are the same at the specified beam energies, the larger electronic density in the Si channel causes a higher dechanneling fraction in agreement with experiments. The same figure shows that the dechanneling at small depths is determined only by the nuclear term (cf Section 7.4.2). Figure 7.23 shows how it is possible to measure the electronic term by taking advantage of the fact that the electronic cross section is dependent on both the ion energy and velocity and independent of the ion mass. The difference in the dechanneled fraction between protons and deuterons of the same energy can then be ascribed to the differences in the electronic scattering.[11]

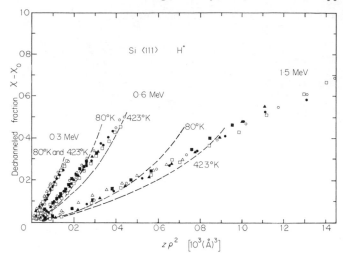

Figure 7.21. Experimental and calculated dechanneled fraction against $z\rho^2$ for protons along the $\langle 111 \rangle$ axes of Si

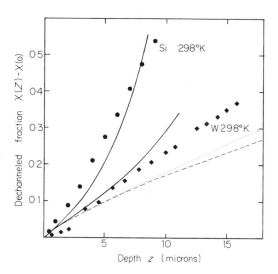

Figure 7.22. Dechanneled fractions of ($\bullet$) 1·5 MeV protons in Si $\langle 111 \rangle$ at 298 K (reference 10) and of ($\blacklozenge$) 3·0 MeV protons in W $\langle 100 \rangle$ (reference 5) compared with theory: full lines calculated by including both nuclear and electronic scatterings; dotted (Si) and dashed (W) lines calculated by considering the nuclear scattering only

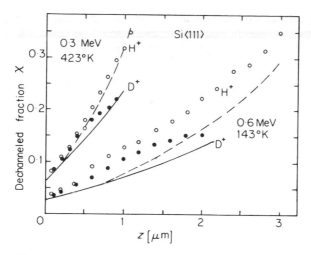

Figure 7.23. Measured dechanneled fractions for protons (○), and deuterons (●), in Si ⟨111⟩, compared with theoretical results

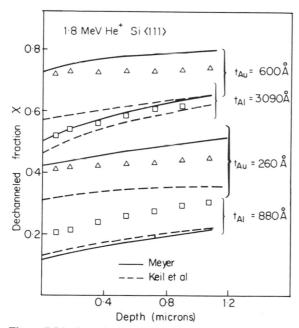

Figure 7.24. Experimental and calculated dechanneled fraction for 1·8 MeV He[+] in Si ⟨111⟩ covered with amorphous layers of Al and Au.[32] In the calculation the spreading introduced by the amorphous layer is described according to Keil *et al.*[40], and to Meyer[41], treatment respectively

Finally, Figure 7.24 shows,[32] within the limits of the scattering theory,[37,38] the results of dechanneling experiment in crystals covered by thick amorphous layers. The theory of axial dechanneling gives a good description of the processes involved and justifies the available experimental data. The main discrepancy, between theory and experiment is illustrated by the upward curvature of the calculated $\chi(z)$ curve against z at large z values. This can be attributed to neglect of the diffusion contribution in the steady increase approximation. Here the ion's transverse energy can only increase, while if diffusion is considered there is a finite probability of a reduction in the ion's transverse energy and then in dechanneling. Moreover, it is expected that this effect increases with time t, spent by the moving ion in the channel, and hence with z.

The effect of a small diffusion term is that of increasing slightly $\chi(z)$ at small z, and of reducing it at large z values.[12] This qualitative interpretation

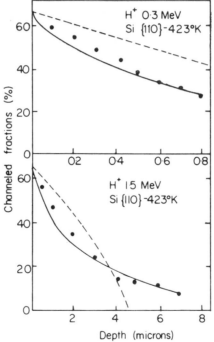

Figure 7.25. Experimental and calculated channeled fraction for protons in the {110} plane of Si. The full line refers to the diffusion approximation, while the dashed line to the steady increase approximation; as expected the last fails in the planar case because of the large diffusion term[22]

has been confirmed recently both by numerical solutions of the complete diffusion equation[38] and by considering the diffusion as a small perturbation in the analytical treatment of axial dechanneling.[45] For planar dechanneling the approximations involved in the present theory are not so crude as in the axial case; moreover, the initial distribution is very smooth in this case, and it cannot be affected appreciably by slight contamination on the crystal surface. The agreement between the value of $(1 - \chi(z))$ calculated from the above theory and the experimental results is good, as it confirms both the exponential trend of $1 - \chi(z)$ against z and the $1/E_0$ scaling. Figures 7.25 and 7.9 show the results obtained for Si $\{110\}$.[22]

The present comparison confirms that the two extreme approximations to the diffusion equation discussed in Section 7.4.1 give a good description of axial and planar dechanneling and suggest the possibility of better approximations.[45]

7.5 ANGULAR DIPS

In the preceding section the problem of dechanneling has been restricted to the case of an incident beam parallel to axes or planes. The extension to a nonzero incident angle ψ_e is however straightforward.

The yield of a close encounter process is related to the probability that a particle in the beam hits a vibrating lattice nucleus. It is then given by the rate at which a particle reaches a distance **r** from the atomic row multiplied by the probability that a nucleus is displaced to **r** by thermal vibrations.[1] Because of the relationship between the transverse energy $E_\perp$ of the particle and the potential $U(r_{min})$, one can integrate over the probability distributions in term of $E_\perp$.

The angular yield at a depth z and for an entrance angle ψ_e is then given by[17]

$$Y(\psi_e, z) = \int n'(E_{\perp z})\Pi(E_{\perp z}) \, dE_{\perp z} \tag{7.51}$$

where $E_{\perp z}$ is the transverse energy measured at a depth z; $n'(E_{\perp z})$ is the differential distribution of the channeled particles at the same depth and $\Pi(E_{\perp z})$ is the probability that a particle of reduced transverse energy $E_{\perp z}$ hits a nucleus. We know the initial distribution $n(E_{\perp 0})$ which has been evaluated in Section 7.4.2 (cf equation (7.30)). If the entrance angle ψ_e is different from zero, the transverse energy of the particles just beneath the crystal surface will be given by the potential at the entrance point plus the transverse kinetic energy which they have by virtue of the nonzero incident angle.

$$E'_{\perp 0} = U(r) + E_0\psi_e^2. \tag{7.52}$$

Hence the differential transverse energy will be shifted to a higher transverse energy by an amount $E_0\psi_e^2$.

We shall now determine the changes in transverse energy distribution along the particle's path. We already know, from the steady increase approximation, the relationship between the transverse energy $E_{\perp z}$ at a depth z and the corresponding $E'_{\perp 0}$. This relationship is obtained by solving equation (7.29) with respect to $E_\perp$.

The differential distribution $n'(E_{\perp z})$ in transverse energy at a depth z is then related to the above $n(E_{\perp 0})$ at zero depth through the conservation of the particle flux, and through the relationship between $E_{\perp z}$ and $E_{\perp 0}$:

$$n'(E_{\perp z})\, \mathrm{d}E_{\perp z} = n(E_{\perp 0})\, \mathrm{d}E_{\perp 0} \qquad (7.53)$$

which holds if the beam enters parallel to the atomic row. This relationship means that the particles which have an energy between $E_{\perp 0}$ and $(E_{\perp 0} + \mathrm{d}E_{\perp 0})$ at zero depth, will have energies between $E_{\perp z}$ and $(E_{\perp z} + \mathrm{d}E_{\perp z})$ at depth z.

The last point to be dealt with, refers to the probability that an ion of a given transverse energy $E_\perp$ hits a vibrating lattice nucleus. This probability in the continuum approximation is given by[1]

$$\Pi(E_\perp) = \int \mathrm{d}P(r)\bigg|_{U(r) < E_\perp} \qquad (7.54)$$

where $\mathrm{d}P(r) = \{\exp(-r^2/\rho^2)\, \mathrm{d}r^2/\rho^2\}$ is the spatial probability distribution of vibrating nuclei in the plane normal to the row, and corresponds to the assumption that atoms vibrate with an isotropic gaussian distribution and that their vibrations are uncorrelated.

The integration of equation (7.54) must be performed over the permitted region of the transverse plane (i.e. in the region where the transverse energy $E_\perp$ is higher than the continuum potential for a vibrating string). This potential is related to the standard string potential by

$$U(r) = \int_0^\infty \int_0^{2\pi} U\{(r'^2 + r^2 - 2rr' \cos\theta)^{\frac{1}{2}}\}\, \mathrm{d}P(r')\, \mathrm{d}\theta. \qquad (7.55)$$

The probability $\Pi(E_\perp)$ as given by equation (7.54) is plotted in Figure 7.26 for different crystal temperatures.[44]

Knowing the initial transverse energy distribution, its change along the path inside the crystal, and the probability of hitting nuclei, one can compute the yield profile by means of equation (7.51).

The Figures 7.27, 7.28 and 7.29 show the results of these calculations[44] for the angular yields $Y(\psi_e, z)$ and for the half width at half minimum $\psi_{\frac{1}{2}}(z, T)$ against z, and $\psi_{\frac{1}{2}}(z, T)$ against ρ^2 at $z = 0$.

It must be remarked that because of the continuum approximation one cannot obtain shoulders in the calculated yield profiles. This approximation however, should not affect appreciably the half width at half minimum $\psi_{\frac{1}{2}}$ as is confirmed by the agreement with the experimental values resulting from Figures 7.28 and 7.29.

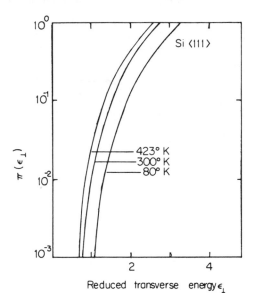

Figure 7.26. Collision probability $\Pi(\varepsilon_\perp)$ against $\varepsilon_\perp$, of a channeled proton with a Si crystal atom at different temperatures

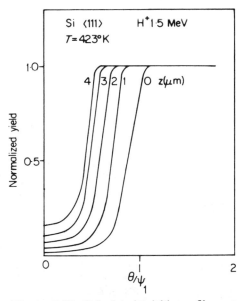

Figure 7.27. Calculated yield profiles, at different depths, in the continuum model approximation including the nonconservation of transverse energy

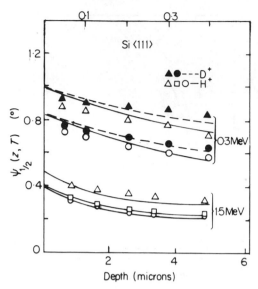

Figure 7.28. Comparison between experimental and calculated[46] against penetration depth for protons and deuterons in Si ⟨111⟩

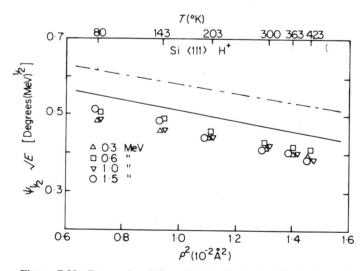

Figure 7.29. Extrapolated $\psi_{\frac{1}{2}}$ at zero depth from experimental values compared with theory (reference 46, full line). The dashed line represents the values calculated by J. U. Andersen[47]

References

1. Lindhard, J., *Dan. Vid. Selsk., Mat. Fys. Medd.*, **34**, 14 (1965).
2. Lindhard, J., *Dan. Vid. Selsk., Mat. Fys. Medd.*, **28**, 8 (1954); 'Studies in Penetration of Charged Particles in Matter' *Nuclear Science Series Report Number 39*. National Academy of Sciences—National Research Council, Washington, D.C. p. 4 (1964).
3. Erginsoy, C., Conf. on Solid State Physics Research with Accelerators, Brookhaven National Laboratory Upton, New York, 1967 BNL 50083, 30 (1967).
4. Bethe, H., *Ann. Physik*, **5**, 325 (1930); *Handbuch d. Physik*, **24/1**, 491, Berlin, Springer (1933).
5. Davies, J. A., Denhartog, J. and Whitton, J. L., *Phys. Rev.*, **165**, 345 (1968).
6. Ellegaard, C. L. and Lassen, N. O., *Dan. Vid. Selsk., Mat. Fys. Medd.*, **35**, 16 (1967).
7. Appleton, B. R., Feldman, L. C., and Brown, W. L., Conference on Solid State Physic with Accelerators, Brookhaven, National Laboratory, Upton, New York, September 1967, BNL 50083, 45 (1967).
8. Foti, G., Grasso, F. and Rimini, E., *Lett. Nuovo Cimento*, **1**, 941 (1969).
9. Foti, G., Grasso, F., Quattrocchi, R. and Rimini, E., *Phys. Rev. B*, **3**, 2169 (1971).
10. Foti, G., Grasso, F., Quattrocchi, R., Quercia, I. F. and Rimini, E., *Lett., Nuovo Cimento*, **4**, 707 (1970).
11. Campisano, S. U., Foti, G., Grasso, F., Quattrocchi, R. and Rimini, E. *Phys. Lett.*, A, **33**, 433 (1970).
12. Campisano, S. U., Grasso, F. and Rimini, E., *Rad. Effects*, **9**, 153 (1971).
13. Morita, K. and Itoh, N., *Phys. Lett.*, A, **34**, 359 (1971); *J. Phys. Soc. Japan*, **31**, 313 (1971).
14. Fujimoto, F., Komaki, K., Nakayama, H. and Ishii, M., *Phys. Lett.*, A, **33**, 432 (1970); *Phys. Stat. Sol.*, **6**, 623 (1971); *Rad. Effects*, **13**, 43 (1972).
15. Campisano, S. U., Foti, G., Grasso, F. and Rimini, E., *Phys. Lett.*, A, **35**, 119 (1971).
16. Appleton, B. R. and Feldman, L. C., *Atomic Collisions in Solids*, Ed. D. W. Palmer, M. W. Thompson and P. D. Townsend, Amsterdam, North-Holland, p. 417 (1970).
17. Andersen, J. U., Andreasen, O., Davies, J. A. and Uggerhøj, E., *Rad. Effects*, **7**, 25 (1971); Eisen, F. H. and Uggerhøj, E., *Rad. Effects*, **12**, 233 (1971).
18. Gorodetzky, S., Pape, A., Cooperman, E. L., Chevallier, A., Sens, J. C. and Armbruster, R., *Nucl. Instrum. Methods*, **70**, 11 (1969); experimental data and Williamson, C., Boujot, J. P. et Picard, J., Rapport CEA-R 3042 Saclay (1966), calculated data.
19. For a more complete treatment see: K. Bardin, Thesis, Cal. Inst. Techn. Pasadena, California 1961.
20. This convention was agreed on during an informal Symposium on dechanneling, held at Aarhus University in Sept. 1970.
21. Altman, M. R., Feldman, L. C. and Gibson, W. M., *Phys. Rev. Lett.*, **24**, 464 (1970).
22. Campisano, S. U., Foti, G., Grasso, F., Lo Savio, M. and Rimini, E., *Rad. Effects*, **13**, 157 (1972).
23. Davies, J. A., Howe, L. M., Marsden, D. A. and Whitton, J. L., *Rad. Effects*, **12**, 247 (1972).
24. Feldman, L. C., Appleton, B. R. and Brown, W. L., Conf. on Solid State Physics with Accelerators, Brookhaven National Laboratory, September 1967, BNL 50083 58 (1967).
25. Van Vliet, D., *AERE Report* 6395 (1970).

26. Bøgh, E., *Can. J. Phys.*, **46**, 653 (1968); Fujimoto, F., Komaki, K. and Nakayama, H., *Phys. Stat. Sol.*, **5**, 725 (1971).
27. Fontell, A., Arminen, E. and Leminen, E., *Rad. Effects*, **12**, 255 (1972).
28. Jack, H. E., *Rad. Effects*, **3**, 101 (1972).
29. Morita, K., Itoh, N., Tachibana, K., Miyagawa, S. and Natsunami, N., Int. Conf. Atomic Collisions in Solids, Gausdal, Norway, 1971.
30. Pedersen, M. J., Andersen, J. U., Elliot, D. J. and Laegsgaard, E., Int. Conf. Atomic Collisions in Solids, Gausdal, Norway, 1971.
31. Foti, G. and Grasso, F., unpublished results, Catania 1971.
32. Rimini, E., Lugujjo, E. and Mayer, J. W., *Phys. Rev.*, B, **6**, 618 (1972).
33. Mory, J., Quéré, Y., *Rad. Effects*, **13**, 57 (1972).
34. Lerving, P., Lindhard, J. and Nielsen, V., *Nucl. Phys.*, A, **95**, 481 (1967).
35. DeWames, R. E., Hall, W. F. and Lehman, G. W., *Phys. Rev.*, **174**, 392 (1968).
36. Andersen, J. W., Augustiniak, W. M. and Uggerhøj, E., *Phys. Rev.*, B, **3**, 705 (1971).
37. Grasso, F., Lo Savio, M. and Rimini, E., *Rad. Effects*, **12**, 149 (1972).
38. Bonderup, E., Esbensen, H., Anderson, J. U. and Schiøtt, H. E., *Rad. Effects*, **12**, 261 (1972).
39. Crank, J., *The Mathematics of Diffusion*, Oxford 1956.
40. Keil, E., Zeitler, E. and Zinn, W., *Z. Für Naturforshung*, **15**, 1031 (1960).
41. Meyer, L., *Phys. Stat. Sol.*, B, **44**, 253 (1971).
42. Scott, W. T., *Rev. Mod. Phys.* **35**, 231 (1963).
43. Appleton, B. R. and Feldman, L. C., *Rad. Effects*, **2**, 65 (1969).
44. Björkquist, K., Cartling, B. and Domeij, B., *Rad. Effects*, **12**, 267 (1972).
45. Campisano, S. U., Foti, G., Grasso, F., unpublished results, Catania, 1971.
46. Campisano, S. U., Foti, G., Grasso, F., Quercia, I. F. and Rimini, E., *Rad. Effects*, **13**, 23 (1972).
47. Andersen, J. U., *Dan. Vid. Selsk.*, *Mat. Fys. Medd.*, **36**, 7 (1967).
48. Andersen, J. U. and Uggerhøj, E., *Can. J. Phys.*, **46**, 517 (1968).
49. Anderson, J. U., and Feldman, L. C., *Phys. Rev.*, B, **1**, 2063 (1970).

CHAPTER VIII

Range Profiles

J. L. WHITTON

8.1 INTRODUCTION

When energetic ions are injected into amorphous material, the resulting distribution or range profile is approximately gaussian and may be derived from the theory of Lindhard, Scharff and Schiøtt.[1] This theory, however, does not take into account any departure from a gaussian, owing either to some order in the target material, as in polycrystalline or monocrystalline targets, or to movement of the implanted atoms after injection.

The only way, therefore, to predict accurately the range profiles of injected ions into ordered targets, is to make the appropriate physical measurements and to make them in a wide variety of materials and experimental conditions until sufficient data are collected to allow safe extrapolation to other systems. There is now a collection of useful data which will allow us to take into account such diverse effects on the range profile as incident ion type and energy, target type, orientation, temperature and so on.

This chapter will be devoted to a detailed description and appraisal of the sensitive measuring techniques developed specifically for this type of experiment. Then, an assessment will be made of the effect of various parameters on the final range profile of implanted ions (in the energy range 5–500 keV) in polcrystalline and monocrystalline targets of metals, semiconductors and other materials. Finally, the possibility of predicting profiles will be considered and examples given, where appropriate, throughout the text.

8.2 MEASUREMENT OF RANGE PROFILES

8.2.1 Sectioning Techniques

It is not an easy matter to measure range profiles which extend only a few hundred ångström units into a target. Obviously, to obtain sufficient data points over the range profile, measurements must be made in depth increments of only tens of ångström units, i.e. a few atom layers.

Of the many techniques developed to make such measurements, those involving the removal of thin sections from the surface of a target, coupled with a measurement of the implanted atoms removed, or remaining, have been most successful. Figure 8.1 shows, schematically, the sectioning and measuring techniques using, in this case, radioactive isotopes. First, the target is injected with energetic radioactive ions; then the depth distribution, or range profile, is measured by removing thin uniform layers from the surface and measuring the residual activity after each layer is removed.

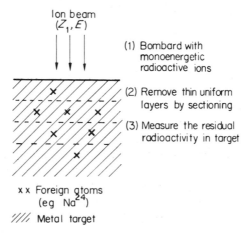

Figure 8.1. Schematic diagram of sectioning
and measuring technique

The sectioning and the measuring may be done in a variety of ways; various possibilities are shown in Table 8.1.

Since none of the techniques shown in Table 8.1 is universally applicable or acceptable, an appreciation of their limitations is necessary for proper application to various systems. These limitations are discussed below.

Table 8.1. The various techniques for removing thin sections from materials and for measuring the range profile of implanted atoms

Removal Technique	Detection Technique
Chemical dissolution	Radiotracers
Electrolytic dissolution	Photoluminescence
Anodic oxidation/dissolution	Auger spectroscopy
Anodic oxidation/mechanical stripping	Ion–ion
Corrosive film/dissolution	Ion–X-ray
Low energy sputtering	Nuclear reactions
Vibratory polishing	
Ultramicrotomy	α-spectroscopy
	β-spectroscopy
	Momentum analysis
	(Rutherford backscattering)

(1) *Chemical Dissolution*

This has been used with a fair degree of success in the study of range profiles in Ge,[2] in NaCl and KBr[3] and in GaAs.[4] The major disadvantage lies in the preferential etching rates on different crystallographic faces of single crystals, at grain boundaries, dislocations, and in fact at any point where the local structure differs from that of the matrix as, for example, in ion-implanted zones. This disadvantage may disappear when the specimen material is amorphous, as shown by Pringle[5] in the chemical dissolution of Ta_2O_5, although even in this case the rate of chemical dissolution differs between the implanted and nonimplanted areas. Jech[6] has exploited the fact that amorphous materials dissolve faster than crystalline, by implanting single crystal insulators and semiconductors, thus deliberately creating an amorphous layer by low energy bombardment, then dissolving the damaged layer. The difference in rate of dissolution between the disordered layer and the undamaged substrate is such that the substrate is hardly affected by the solvent. However, although Jech has shown a novel method of exploiting what is normally a disadvantage of chemical polishing, the disadvantage remains in the general application of chemical dissolution.

(2) *Electrolytic Dissolution*

The disadvantages of chemical dissolution apply equally to this method, with the additional effect of nonuniformity of the electric field between cathode and anode resulting in a faster rate of attack at the edge than at the centre of the specimen.

(3) *Anodic Oxidation/Chemical Dissolution*

In this technique the specimen is made the anode of an electrolytic cell and anodized to some thickness of oxide which is controlled by the applied voltage. The technique depends for its operation on the uniform anodic oxidation of the target and subsequent selective dissolution of the oxide. This technique is undoubtedly the most sensitive and reproducible. Unfortunately, it is presently limited to silicon,[7,8,9] aluminium,[10] tungsten,[11] gold,[12] molybdenum[13] and indium antimonide.[14] The limitation is caused not so much by the difficulty of anodizing the targets but by that of finding solvents which will dissolve the oxide without attacking the substrate.

(4) *Anodic Oxidation/Mechanical Stripping*

This technique circumvents the difficulty of finding preferential solvents for anodic oxides by utilizing the fact that anodic oxides formed in solutions containing HF may be very weakly bonded at the oxide–metal interface. Pawel and Lundy[15] have applied this to tantalum and niobium. After oxidation a piece of adhesive tape is affixed to the oxide which is then forcibly detached from the metal. This technique is of limited application with the additional disadvantage that small sections of oxide could be left adhering to the metal.

(5) *Corrosive Film/Chemical Dissolution*

This is based on the formation of a corrosive film on 'soft' metals when immersed in an organic solvent containing a halogen; the film can then be selectively dissolved by chemicals which form complexes with the metal halide. Andersen and Sørensen[16] have successfully applied this technique to range studies in copper, silver and gold. It is suggested by these authors that the metals tin and lead may also be treated in this way.

(6) *Low Energy Sputtering*

In theory this method is applicable to all materials but in practice, as with the chemical dissolution method, it is very sensitive to changes in crystal structure. Sputtering rates are different for different metals, so there is a possibility of implanted atoms being sputtered at a faster or slower rate than the matrix. Sputtering rates measured using a commercial ion gun vary from 300 Å/min for niobium to 3000 Å/min for silver. Sputtering rates also changes very markedly between crystalline and amorphous targets, by as much as a factor of two as shown by Farren and Scaife[17] in a study of sputtering yields from gallium arsenide. Of more concern in range profile studies is that in materials damaged by ion implantation, where the damage profile changes with depth in the crystal, a continuous but nonuniform change in sputtering rate through the range profile is likely. It is also essential to ensure that any surface structure resulting from the ion sputtering is small compared to the range profile. Finally, the possibility of a change in range profile owing to collision events cannot be disregarded. However, notwithstanding these objections, the technique has been used by Lutz and Sizmann[18] and by Pöhlau et al.[19] for the measurement of range profiles in copper and gallium arsenide respectively. In both cases very high implantation doses were used. The technique is particularly useful when it is desirable to measure a range profile without removing the target from the implantation environment, as shown by Hermann et al.[20] in the measurement of krypton in tungsten at liquid nitrogen temperature.

(7) *Vibratory polishing*

In this technique the specimen is held in a stainless steel holder and laid face down in an aqueous slurry of $0.05 \mu m$ Al_2O_3 on a soft polishing cloth in a stainless steel bowl. The bowl vibrates by means of an electromagnet and leaf springs, causing the specimens to move around the periphery of the bowl in such small steps that they appear to have a smooth flowing motion. Directional effects are eliminated since the specimen rotates around its own vertical axis while moving round the bowl. Thin sections are thus removed by an abrasion process; the thickness of the sections is controlled by time. The technique has been used[21] to measure range profiles in a wide variety of materials. These are listed in Table 8.2 along with approximate removal

Table 8.2 The removal rate of surface layers from various crystals by vibratory polishing

Material	Type	Removal rate* Å/min of vibratory polishing	Calibration method
Cu	Single crystal	150	Indent measurement
Ag	Single crystal	35	Indent measurement
Mo	Single crystal	30	Indent measurement
Ta	Single crystal	30	Indent measurement
W	Single crystal	20	Radiotracer
V	Single crystal	25	Indent measurement
Nb	Single crystal	30	Indent measurement
UO_2	Single crystal	20	Weight loss
Si	Single crystal	50	Indent measurement
Ge	Single crystal	100	Weight loss
GaP	Single crystal	150	Weight loss
GaAs	Single crystal	100	Weight loss
MgO	Single crystal	90	Weight loss
SiO_2	Single crystal	90	Weight loss
Steel	Polycrystal	95	Indent measurement
ZrO_2	Amorphous	10	Spectrophotometer
Ta_2O_5	Amorphous	100	Spectrophotometer

* Since these rates vary with, for example, specimen size, the approximate figures are given mainly as a guide to the removal rates which can be achieved. Most of these rates were obtained from disc specimens of approximately 1 cm diameter and 1 mm thick.

rates determined by weight loss. Disadvantages of the technique are (i) some deformation is introduced in specimens which could result in anomalously shortened range profiles (ii) the polished surface is not completely flat and (iii) following from the nonflatness of the surface, weight loss measurements are not as accurate as one could expect from a flat surface. Nonetheless, the almost universal application of the technique often compensates for these objections, which can be circumvented (i) by treating the vibratory polished surface to a very short chemical polish before implantation; enough to dissolve the damaged region but not to alter the surface contour, (ii) by ensuring that the implant zone is always centrally located and (iii) by making use of a recently tested microhardness indenter depth gauge[22] instead of making weight loss measurements.

Although, to the author's knowledge, no one has applied the technique of ultramicrotomy to the sectioning of metals for range profile measurements it appears, at least, to be feasible. Reimer[23] has shown how sections of approximately 200 Å thick may be removed from metal of approximately 1 mm diameter, with little distortion taking place. One possible disadvantage is that smearing may occur as the section is being removed, resulting in a lateral movement of the matrix and implanted atoms. The specimen diameter is limited to approximately 1 mm, otherwise the microtome blade will not make a clean cut.

8.2.2 Techniques for the Detection of Implanted Atoms—Destructive

Some of the detection techniques listed in Table 8.1 and used in conjunction with the removal of layers are very specific, as, for example, the photo-luminescence study by Arnold[24] to measure the range profile of 200 keV protons in silver phosphor/glass. On the other hand, Auger spectroscopy,[25] ion–ion interactions[26] and the ion X-ray studies of Cairns *et al.*[27] should be capable of wider application, although results to date suggest that a very high concentration of implanted ions is necessary to provide an adequate reaction yield.

The radiotracer method is undoubtedly the simplest and most sensitive technique. Radioactive atoms may be injected directly from an isotope separator[28] or, in some instances,[29] it is possible to implant stable ions and later neutron-activate the matrix and implanted atoms. The direct injection of radioactive atoms is preferable, in that a much lower implant dose is required and there are no competing radiotracers from the matrix atoms.

Where there are no suitable radioactive isotopes of an element, as for example, with aluminium and oxygen, nuclear reactions can be used in conjunction with the sectioning technique. These reactions are specific to any one nuclide and there is a wide enough choice to allow detection of most of the light elements. A comprehensive survey of such reactions has been compiled by Amsel *et al.*[30]

8.2.3 Techniques for the Detection of Implanted Atoms—Nondestructive

Several nondestructive techniques have also been developed for the measure-ment of range profiles. Some of these are, however, very specific, as in the case of energy loss measured from implanted α-emitters,[31] and of β-ray spectroscopy,[32] while the more general technique of momentum analysis by Rutherford backscattering is, in the main, limited to the measurement of implanted atoms which are considerably heavier than the matrix and of high ($> 10^{14}$–10^{16} atoms/cm^2) concentration.[33] The latter condition is particularly limiting in the easily damaged semiconductors.

Nuclear reactions may also be used in a nondestructive way, as has recently been shown[34] in the measurement of range profiles of 18-oxygen in gallium phosphide. As with the momentum analysis, however, rather high concentrations of implanted atoms are often required to give adequate reaction yields and the mathematical treatment of data is complex.

While all of the techniques mentioned may be applied to metals and non-metals, a few techniques have been developed which apply only to semiconductors. These depend in one way or another on the measurement of electrically active atoms in the matrix and utilize the phenomena of Hall effect and conductivity, capacitance-voltage and the preferential stain-etching of junctions. (See, for example, Mayer et al.[35]) The main criticism of these methods is that the measured profile is caused by the electrically active atoms, which may make up only a small fraction of the number implanted and also to electrically active damage centres. However, in the study of ion-implanted semiconductors, where the electrical activity profile is of as great interest as the ion profile, these are undoubtedly valuable techniques.

So, from a consideration of all the aforementioned techniques, we conclude that some form of sectioning combined with the implantation of radiotracers is the most versatile and sensitive. Certainly most of the available range profile data have been obtained by these techniques and except for the specific cases mentioned it will undoubtedly continue to be the preferred method.

Most of the described techniques provide us with integral depth distributions from which profiles may be extracted by differentiation. The data to be presented here are mostly in the form of integral depth distributions; only in cases in which the removed radioactivity, rather than the remaining radioactivity, is measured[8] do we get differential distributions directly.

8.3 RANGE PROFILES IN AMORPHOUS AND POLYCRYSTALLINE TARGETS

In any study of range profiles, a clear distinction may be made between those obtained from injection into amorphous, (random or completely disordered) polycrystalline (having some degree of order), and monocrystalline (well ordered) targets. An example of these different shapes of range profile is given in Figure 8.2 which shows the profiles obtained, by a sequence of layer removal and residual activity measurements, when 5 keV ^{85}Kr is injected into these three types of tungsten targets (Kornelsen et al.[36]).

The curve labelled amorphous (tungsten) was derived from range profile measurements in anodically formed amorphous tungsten oxide (WO_3) with the abscissa scale adjusted for the different stopping power of the oxygen. This corrected curve and others of Na, Ar and Xe of incident energies up to 160 keV fit well (within 10%) to the computer program of Schiøtt[37] based on the theory of Lindhard, Scharff and Schiøtt.[1] A great deal of theoretical work has been done for the 'amorphous' case, since the range and stopping powers are statistical concepts and are therefore tractable. Sanders[38] and Sigmund and Sanders[39] have, along with Schiøtt,[37] shown that their theories fit well with range profiles in Al_2O_3, Ta_2O_5 and WO_3. The most recent work by Winterbon *et al.*,[40] Sigmund *et al.*[41] and Winterbon[42] allows ready calculation of range profiles in all amorphous media.

It is unfortunate that, whereas we have a good theoretical treatment for range profiles in amorphous elemental materials, we have very little experimental data. The only range profile experiments in amorphous systems are in the three oxides mentioned. This is not a completely satisfactory situation since the use of oxides, because of the different stopping powers of metal and oxygen, makes any range or stopping power calculation more difficult.

The only experimental data available for ranges in amorphous targets, other than the oxides mentioned, are for Si[43] and GaAs.[44] In these cases good agreement is found with theory. Other attempts to compare experimental data with theory by using polcrystalline targets, instead of truly amorphous ones, have been unsatisfactory owing to a much deeper penetration than was expected from theory.

Reference to Figure 8.2 shows how markedly the range profile of 5 keV ^{85}Kr in polycrystalline tungsten diverges from that in the amorphous tungsten oxide. This phenomenon was noticed in some very early measurements of range profiles in polycrystalline germanium, aluminium and tungsten[8,45,46] and, in fact, provided one of the first indications of channeling. The most surprising feature, however, lies in the much deeper *overall* penetration

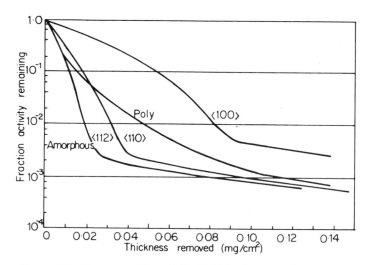

Figure 8.2. Range profiles of 5 keV ^{85}Kr in amorphous, poly-crystalline and single crystal tungesten (from Kornelsen *et al.*[36] 1964)

when compared with the theoretical gaussian profile. One might expect some small fraction of the incident beam to become channeled, but (Figure 8.2), the penetration depth of 20% of the injected atoms lies at about a factor of two deeper than the 20% level in the amorphous case. (Kornelsen[36] reported that the tungsten specimen showed no sign of preferred orientation). This behaviour has been seen in both aluminum and tungsten and recent work by Santry and Sitter[47] indicates that the median range of 40 keV ^{85}Kr ions in polycrystalline tungsten and gold is deeper by a factor of 1·7 than was calculated from the theory of Lindhard, Scharff and Schiøtt.[1] These data serve to emphasize the necessity of using amorphous targets when it is desired to make careful comparison of experimental data with theory.

In the present circumstances, the nearest that one may hope to come to an amorphous metal, is a polycrystalline foil of small grain size having no preferred orientation, although recent reports[43] of the formation of amorphous gold are encouraging. Other possible means of forming amorphous metals are by splat cooling,[44] by high dose self-ion implantation, e.g. W into W and by evaporation of Ta on to Ta or W substrates. These latter possibilities are being explored currently at the author's laboratory.

For a detailed discussion of range profiles and range straggling in amorphous media, the reader is referred to the book by Mayer *et al.*[35] and to the more recent articles cited here.

8.4 RANGE PROFILES IN SINGLE CRYSTALS

8.4.1 Introduction

Unlike the polycrystalline materials which are extremely difficult, if not impossible, to characterize, the single crystal, with its well ordered lattice and easily characterized condition, provides an excellent target material for the investigation of range profiles and of the many parameters which can affect the final distribution of implanted ions.

All implantations at energies of 5–500 keV considered here involve two basic types of energy loss: nuclear and electronic. The two processes are separated into (i) the ions which are channeled (electronic stopping) and (ii) those which are prevented from doing so by large angle collisions (nuclear stopping). However, an overlap of the two always occurs at energies of 5–500 keV.

Although, for channeled ions, electronic stopping is usually the dominant slowing down process, for nonchanneled ions it only dominates at high energies. The channeling phenomenon discriminates against nuclear collisions and allows us to study electronic stopping at energies very much lower than is possible in amorphous media, i.e. the crystal lattice serves as a very convenient and efficient filter for eliminating violent (nuclear) collisions.

In the following sections, after a brief discussion of the principles of, and requirements for, channeling the effect of various parameters on the channeling phenomenon and on the range profiles will be considered in detail.

8.4.2 The Principles of Channeling

When an energetic ion moves through a crystal lattice within a certain critical angle of a low index direction as shown schematically in Figure 8.3, each time it approaches one of these low index rows of atoms, the gradually increasing repulsive interaction between the screened Coulomb fields of incident ion and lattice row steers the ion away from the row, thus preventing the occurrence of large angle nuclear collisions.

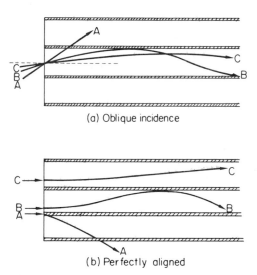

(a) Oblique incidence

(b) Perfectly aligned

Figure 8.3. Schematic of channeled and non-channeled ions (from Mayer *et al.*[35] 1970)

For channeling to occur, the critical angle ψ_{crit} within which the incident ion must approach the crystal lattice is, for particles of energy E and angle of incidence ψ relative to a low index row of atoms, given by[50] equations (2.34) and (2.35).

We can see from these two equations that ψ_{crit} depends only weakly on energy and atomic number, so for medium to heavy atoms of up to a few hundred keV, incident along low index rows, the critical angle is between 3 and 5°.

From this we can see that it is not too difficult to channel particles of these energies; conversely, we can see why it is difficult to suppress channeling completely, as shown earlier in the case of polycrystalline targets, cf Figure 8.2.

The energy lost by the incident ion is obviously much less for the channeled trajectory than for the nonchanneled, so it follows that the channeled ions will penetrate much deeper into the crystal lattice. We have seen this already in Figure 8.2 in the comparison of range profiles in amorphous, polycrystalline and single crystal tungsten. An even more marked illustration of the channeling effect is seen in Figure 8.4 which shows the differential distribution of

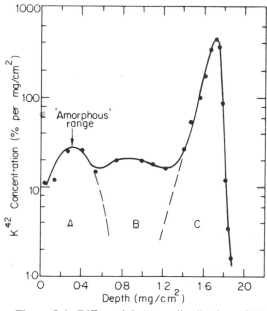

Figure 8.4. Differential range distribution of 500 keV ^{42}K ions along the $\langle 111 \rangle$ direction of W (from Eriksson[51] 1967)

500 keV ^{42}K ions incident along the $\langle 111 \rangle$ directions of tungsten.[51] Here we see the division of the incident beam into three distinct groups (A, B and C) as shown in Figures 8.3 and 8.4.

In group A the particles are those which enter the lattice at angles greater than the acceptance angle and are therefore unaffected by the lattice. These have a range distribution similar to particles in an amorphous material. In group B, the particles are those which are initially channeled, although poorly. These have large oscillations and are scattered out of the aligned direction, i.e. become dechanneled, long before what would be their normal channeled range and so do not penetrate as deeply as those in Group C. In group C, the particles are properly channeled and are, therefore, constrained to stay in the axial channels throughout their entire trajectory.

In a simple consideration, we can relate group A to nuclear stopping and

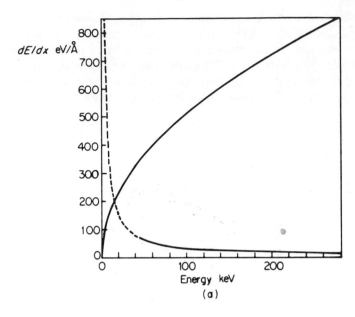

(a)

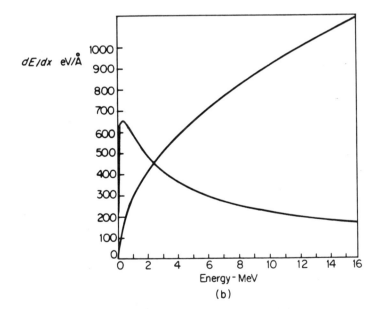

(b)

Figure 8.5. (a) The rate of energy loss for a well channeled Xe atom travelling down the $\langle 100 \rangle$ channel in a W lattice (b) The rate of energy loss for a random trajectory in W (from Winterbon[52] 1970). The monotonically increasing curves are for electronic stopping, the others for nuclear stopping

group C to electronic stopping with group B being a mixture of both. Figure 8.5(a) and (b) from Winterbon[52] show the relative importance of nuclear and electronic stopping in channeled and nonchanneled trajectories of Xe ions in a W lattice. Winterbon shows (Figure 8.5(a)) that in the $\langle 100 \rangle$ channeled case, electronic stopping starts dominating at an energy of approximately 20 keV. For the nonchanneled particle, however, (Figure 8.5(b)) electronic stopping is not the dominant process until the incident energy of the particle is 2·5 MeV. The experiments on channeled ranges at energies of only a few keV therefore, allow us to study inelastic stopping in a region which is normally completely masked by nuclear collisions. Much of the data to be presented here on electronic stopping processes have been measured with incident energies of only 40 keV.

The range–energy relationships for the two components are now well established. For channeled particles $R \propto E^{0.5}$ while for nonchanneled particles $R \propto E$ (in the energy range of 5–500 keV considered here). This will be discussed later in connection with the prediction of range profiles.

8.4.3 The Effect of Various Parameters on Range Profiles

Now to an examination of the effect of various parameters on the final range profiles of implanted ions. Since the influence of an ordered lattice on the depth of penetration is felt in any material, whether metal, semi-conductor or insulator, and since some effects are more clearly shown in one than the other, examples will be given from various systems. In the final section, particular attention will be paid to semiconductors and to the implantation phenomena peculiar to them, since these are currently attracting a great deal of interest.

The parameters to be considered are: (a) crystal orientation, (b) incident ion energy, (c) lattice type (fcc and bcc), (d) target temperature, (e) change in Z_1 (ion) and Z_2 (target), (f) crystal disorder, (g) bombardment dose.

(a) *Effect of Crystal Orientation*

The ease of penetration of ions into a crystal lattice is directly related to the openness of the lattice. Figure 8.6 shows that the relative ease of penetration of 40 keV Xe in gold varies with the size of these channels, i.e. in the face-centered cubic lattice $\langle 110 \rangle > \langle 112 \rangle > \langle 100 \rangle > \langle 111 \rangle$, while in the body centered cubic lattice the order is $\langle 111 \rangle \simeq \langle 100 \rangle > \langle 110 \rangle > \langle 112 \rangle$.

These relationships are true only when the crystal lattice is correctly aligned; any misalignment will effectively reduce the acceptance angle for channeling and will result in a fore-shortened profile. Davies *et al.*[53] have shown such an effect by tilting only 0·88° away from the $\langle 100 \rangle$ direction in tungsten (see Figure 8.7).

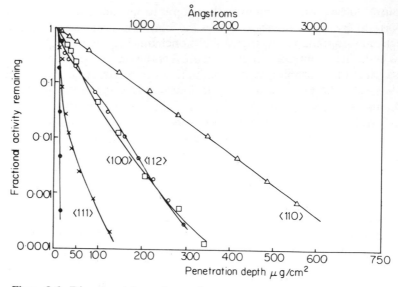

Figure 8.6. Directional dependence of penetration for 40 keV ^{133}Xe, in Au
(from Whitton[54] 1967)

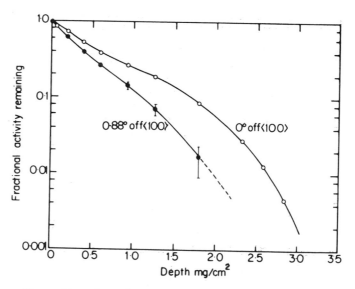

Figure 8.7. Effect of slight misorientation on the range profile
of 400 keV ^{133}Xe in ⟨100⟩ W (from Davies *et al.*[53] 1965)

(b) *Effect of Incident Energy*

The most obvious effect of change in energy is the corresponding change in depth of penetration of ions into the target. Figure 8.8 shows what happens when the incident energy of ^{133}Xe and ^{198}Au ions injected into gold is increased from 20 to 80 keV. The slope of the range–energy curve at the 0·001 fraction level corresponds to an energy exponent of 0·8.[54] This same energy exponent of 0·8 had earlier been reported by Piercy *et al.*[55] for various ions injected into aluminium.

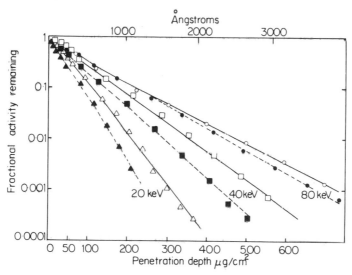

Figure 8.8. Energy dependence of ^{133}Xe (solid lines) and ^{198}Au (dashed lines) in $\langle 110 \rangle$ Au (from Whitton[54] (1967)

However, it turns out that the $R \propto E^{0·8}$ relationship arises from a combination of nuclear (E) and electronic ($E^{0·5}$) stopping processes.[55] Similar range–energy curves from ions in tungsten[51] (where the fraction of channeled ions is high) shows an $E^{0·5}$ exponent for channeled ions and an E for nonchanneled.

(c) *Effect of Lattice Type* (fcc and bcc)

The range profile of 40 keV Xe in the most open direction of gold, $\langle 110 \rangle$, is quite different from that of the same ions in the most open direction, $\langle 100 \rangle$, of tungsten. In gold, Figure 8.9, the profile is seen to decrease exponentially, while in tungsten, the number of channeled ions is greater and the profile is characterized by a well defined maximum range.

The maximum range, $R_{\max}$, is characteristic of best channeled particles and is the most reproducible parameter in range studies although in most cases it is not easily measured. Experimental data from a wide variety of

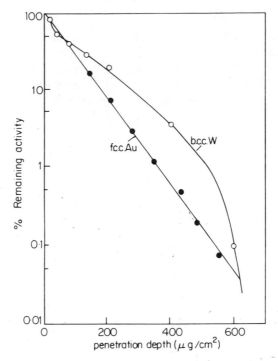

Figure 8.9. Range distributions of 40 keV ^{133}Xe along the most open directions of gold $\langle 110 \rangle$ and tungsten $\langle 100 \rangle$ (from Whitton[57] 1968)

studies in tungsten by Eriksson[51] and Davies *et al.*[56] have established that R_{max} depends only on crystal direction of the target and type and energy of incident ion. It does not change significantly with bombardment dose, crystal disorder, lattice temperature, surface disorder or small misalignment. However, the number of particles reaching R_{max} is sensitive to all these factors. Eriksson's studies of R_{max} in W have shown that, except for energies below a few keV, R_{max} shows an $E^{0.5}$ dependence characteristic of electronic stopping, indicating that, for a well channeled ion over long distances, electronic stopping is the only significant mechanism of energy loss.

Why then do we have a readily observable R_{max} in tungsten but not in gold, targets of almost equal density and atomic number? To answer this question, a series of range profiles was measured[57] in a number of targets, both fcc as for gold and bcc as for tungsten.

The results are summarized in Figure 8.10 which shows that the fcc targets Al, Cu and Au do not show maximum ranges but the bcc Ta and W do. This difference was attributed to the difference in atomic thermal

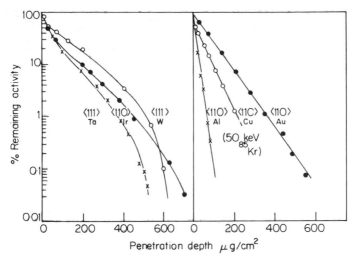

Figure 8.10. Range distributions of 40 keV ^{133}Xe along the most open directions of fcc and bcc crystals (from Whitton[57] 1968)

vibrational amplitude which is small for bcc metals and large for fcc. Confirmatory evidence of this is shown by the observable R_{max} in iridium which although a fcc metal, has a thermal vibrational amplitude close to that of the bcc metals. Some physical properties of fcc and bcc metals are given in Table 8.3.

Table 8.3 Physical properties of some fcc and bcc metals

Element	Lattice type	Unit cell length a in ångströms	Root mean square vibrational amplitude at 25°C in ångströms
W	bcc	3·158	0·085
Ta	bcc	3·296	0·11
Ir	fcc	3·831	0·09
Al	fcc	4·041	0·18
Cu	fcc	3·607	0·15
Au	fcc	4·070	0·15

It was concluded that, although all ions in any material should, by definition, have a maximum range, the rate of dechanneling in most fcc metals, owing to the high thermal vibrational amplitude, was sufficiently large to prevent a measurable number of particles reaching the maximum range. This is, therefore, primarily a temperature, rather than a lattice effect and the difference in range profiles shown in Figure 8.10 is true only for room temperature measurements. The depth distribution of ions implanted at low temperature are quite different to those implanted at room temperature. These effects are considered below.

(d) *Effect of Target Temperature*

Measurements of the depth distribution of 40 keV ^{133}Xe in $\langle 110 \rangle$ gold[58]—held at 20 K during the implantation—showed quite clearly an R_{max} to which about a 10^{-3} fraction of the incident ions penetrated. This profile is shown in Figure 8.11 compared with room temperature profiles of the same

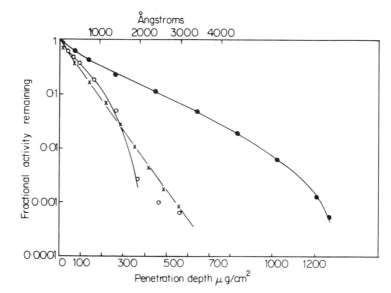

Figure 8.11. Effect of temperature on the depth distribution of 40 keV ^{133}Xe in gold. $\times$ $\langle 110 \rangle$ Au at 300 K; $\bigcirc$ $\langle 111 \rangle$ W at 300 K; $\bullet$ $\langle 110 \rangle$ Au at 20 K (from Whitton[57] 1968)

ions in gold and tungsten. An extrapolation of the room temperature ^{133}Xe in Au curve indicates that only about 1 in 10^7 of the incident ions would reach R_{max}, while 10^{-2} of the ^{133}Xe ions reach the R_{max} in tungsten—further evidence of the influence of the vibrating lattice atoms on the dechanneling rate.

More detailed measurements of range profiles have been made by Channing[59] and by Howe and Channing[60] in low temperature Al, W and Au. In all cases the dechanneling rate decreased with lowering of target temperature. The reverse is, of course, also true; when the lattice vibrations of tungsten are increased from 0·085 to 0·115 Å by heating the target to

$250\,°C,$[51] the number of initially channeled particles reaching the channeled peak is lower by a factor of 2 to 3 indicating that the dechanneling rate is correspondingly higher.

The existence of a two-fold difference between R_{max} of 40 keV ^{133}Xe in Au and in W is somewhat surprising since the two targets are of almost equal mass. We cannot attribute this to the difference in lattice temperatures because Davies *et al.*[56] have shown that the R_{max} in tungsten is completely independent of temperature. It was suggested[57] that the reason may be found in the electronic configurations of the lattice atoms. It follows that when the lattice has low stopping power, thus allowing deeper penetration of the incident ion, the possibility of dechanneling increases since dechanneling is a function of both pathlength and lattice vibrations.

This aspect of depth distribution and stopping processes will be considered in greater detail in the next section.

(e) *Effect of Changing Z_1 (ion) and Z_2 (target)*

Although the theoretical work of Firsov[61] and of Lindhard *et al.*[1] predicts the general trend of increasing electronic stopping cross section S_e with increase in atomic number Z_1 of incident ions at constant velocity, it fails to take into account the periodic oscillations in S_e observed in both amorphous and crystalline targets.

While a detailed account of the theoretical treatment of the oscillations of S_e with Z_1 is given in Chapter 4, for the sake of continuity, mention is made here of the experimental results leading to our present state of knowledge.

The oscillations in gaseous and amorphous media observed by Ormrod *et al.*[62] and by Fastrup *et al.*[63] showed variations of approximately 30% about the mean value derived from theory. In a measurement of R_{max} of various ions at a constant velocity of 1.5×10^{-8} cm s^{-1} along the $\langle 100 \rangle$ and $\langle 110 \rangle$ directions of W single crystals these oscillations were again observed[64] but were of much greater magnitude. The data of Eriksson *et al.* are shown in Figure 8.12 compared with the Firsov curve. Similar maxima and minima have been observed by Eisen[65] in his measurement of electronic stopping cross section of light ions $5 < Z_1 < 19$ channeled through silicon, and again by Bøttiger and Bason[66] in a similar study of various ions transmitted through Au single crystals (Figure 4.6).

Further theoretical work (described in detail in Chapter 4 by Poate) has established that the dependence of electronic stopping powers of channeled ions can be correlated with variations in the electronic density in the outer regions of the ion, i.e. in a simple sense on the size of the ion.

The Z_1 oscillations are particularly large only for channeled particles; compare the minima to maxima difference of approximately 30% in amorphous media, Fastrup *et al.*[63] with the approximately 15-fold variation

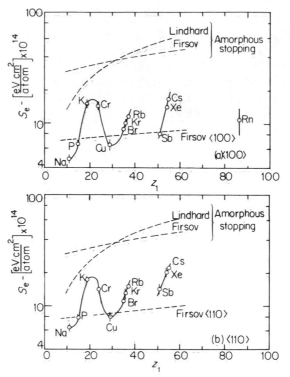

Figure 8.12. Electronic stopping cross section plotted as a function of atomic number of ions incident along the $\langle 100 \rangle$ and $\langle 110 \rangle$ directions of W. These measurements are taken from R_{max} (from Eriksson *et al.*[64] 1967)

obtained by Bøttiger and Bason.[66] These oscillations affect very strongly the range profile of channeled ions, since the greater S_e, the shorter the range and hence the closer R_{max} is to the surface of a target. We can see this in the range profile of 40 keV ^{35}S, ^{85}Kr and ^{24}Na implanted along the $\langle 110 \rangle$ direction of GaAs, Figure 8.13.[67] Reference to Figure 4.6 shows that of the three, S has the highest S_e, Kr the medium and Na the lowest. We would expect, therefore, penetrations to vary in reverse order, i.e. the highest S_e would have the shortest R_{max} and this is so, ^{35}S has an easily recognized R_{max} at 1·5 μm, ^{85}Kr at approximately 3 μm and the R_{max} for ^{24}Na is greater than 4 μm. Dearnaley *et al.*[43] have shown similar behaviour in silicon for ions having minima and maxima S_e.

The question now arises: does the electronic stopping cross section vary with Z_2, the target atom, as it does with Z_1? This is of considerable importance in the prediction of range profiles, since the effect on range profiles of

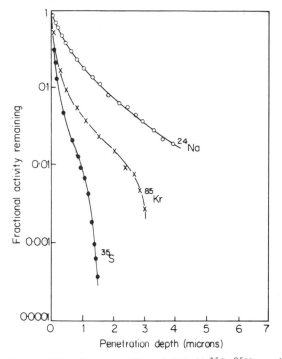

Figure 8.13. Range profiles of 40 keV ^{35}S, ^{85}Kr and ^{24}Na in $\langle 110 \rangle$ GaAs (from Whitton and Carter[67] 1970)

osillations even in Z_1 are not predictable; hence R_{max}, although the most reproducible of all parameters in range studies, can only be determined by experiment. Should the same oscillatory behaviour of S_e be equally strong in Z_2, there could easily be a compounding effect with both Z_1 and Z_2 at minima or maxima or a cancelling effect with either having minimum S_e and the other maximum S_e.

Measurement of R_{max} in different target materials suggests that the S_e oscillations apply also to Z_2. As mentioned earlier, we see a greater than two-fold difference in R_{max} of Xe in Au and W; the R_{max} of 40 keV ^{35}S is greater in GaAs than in GaP targets of similar lattice spacing by a factor of two (Ga and As have minimum S_e and P has $\frac{1}{2}$ maximum S_e (cf. Figure 4.6)); the maximum range of 40 keV ^{35}S is greater in Ge than in Si by a factor of two (again Si has a S_e greater than twice that of Ge).

Additional evidence of oscillatory S_e dependence on Z_2 comes from the work of Hvelplund[68] with the measurement of ions of $2 < Z_1 < 12$ in helium, air and neon and for $Z_1 = 2$ and $Z_1 = 8$ in hydrogen and oxygen targets. These measurements show oscillations similar to the Z_1 oscillations as do the results of like experiments by White and Mueller.[69] Similar studies of the S_e of Li (30 to 100 keV) in amorphous and polycrystalline solids by Bernhard *et al.*,[70] continued by Abel *et al.*[71], also support the earlier indications of oscillatory dependence of S_e on Z_2. However, as in the earlier Z_1 work in amorphous media, the oscillations are smaller than in single crystals, *viz.* about $\pm 50\%$ about the theoretical curves of Firsov[61] and of Lindhard *et al.*[1]

In an effort to amplify these oscillations, as was done in the Z_1 measurements by using single crystal targets, work is going on in the author's laboratory using single crystal targets of minimum and maximum S_e (assuming the minima and maxima S_e for Z_2 is the same as for Z_1 as suggested above). A study of Figure 4.6 shows that most targets of minimum S_e are fcc while those of maximum S_e are bcc. Targets of these crystal types which will cover a Z_2 range from Fe to W are being used in the present study. Preliminary results indicate enhanced oscillations in approximately the same positions as in the Z_1 experiments in W and Au.

(f) *Effect of Crystal Disorder*

The channeling effects described in the previous sections are particularly sensitive to various lattice and surface imperfections as, for example, oxide layers and other surface contaminants, disorder arising from poor surface preparation, dislocations, impurities, ion-bombardment-induced disorder etc.

It follows from this, that careful surface preparation of crystal targets is an essential in channeling studies and that equally careful characterization is mandatory. Some of the hazards and pitfalls likely to be encountered in routine preparation of single crystal surfaces have been described by Whitton[72] while for general surface preparation techniques and recipes for chemical and electropolishing, the reader is referred to the books of Samuels[73] and Tegart.[74] The channeling/Rutherford backscattering technique (cf Chapter 13) is probably the most useful analytical tool for the assessment of crystal perfection prior to range profile studies and should be used whenever possible.

The most direct effect of oxide, which generally forms an amorphous layer on the ordered surface of a crystal, is that of spreading the incident beam, thus causing many of the initially well aligned ions to enter the lattice at angles greater than ψ and so preventing channeling. (A similar effect is caused by ion-bombardment-induced disorder and by mechanically introduced surface disorder). An indirect effect of surface oxide is that of energy loss of the incident particle before it reaches the crystal lattice. Figure 8.14 from Eriksson[51] shows the decrease in penetration and reduction in channeling of K ions in $\langle 100 \rangle$ W owing to different thicknesses of anodically formed WO_3. In this case 1 V equals 10 Å of WO_3.

Although the presence of individual dislocations or dislocation networks

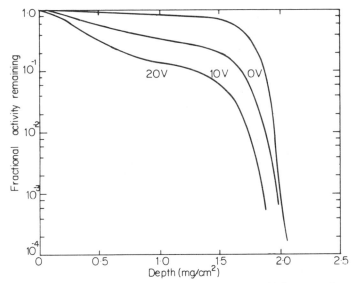

Figure 8.14. Effect on range profiles of different thicknesses of surface oxide. 500 keV ^{42}K in $\langle 100 \rangle$ W (from Eriksson[51] 1967) $1 \text{ V} \equiv 10 \text{ Å}, \text{WO}_3$

in metals has not been shown to influence low energy heavy ion range profiles, Quéré and Couve[75] have demonstrated the dechanneling of α-particles by grain boundaries in platinum. A similiar though lesser effect would be expected for low angle grain boundaries, of the type present in crystals having observable mosaic spread.

Dearnaley et al.[43] have shown how intrinsic defects in silicon can affect the range profile. However, these affect only the most penetrating few per cent. of incident particles, unlike the surface disorder which affects the entire range profile.

(g) *Effect of Bombardment Dose*

While all the listed parameters have an influence on the channeling behaviour of incident ions in metals and semiconductors, the effect of bombardment dose is particularly noticeable in the semiconductors. So far, it has not been possible to transform metal single crystals into the amorphous state by ion bombardment, but all semiconductors examined to date are completely disordered in the implanted region at relatively low incident ion doses. Typical crystalline/amorphous transition doses are listed in Table 8.4.

The influence of bombardment-induced disorder on range profiles in semiconductors is very marked. Figure 8.15 shows the change in range profile of 40 keV ^{85}Kr along the $\langle 111 \rangle$ direction of GaAs. A dose of 3×10^{11} Kr ion/cm^2 gives the profile expected in an undamaged crystal (lower doses

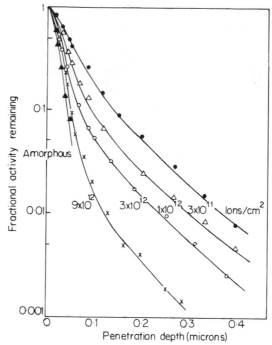

Figure 8.15. Effect of bombardment dose on the range profiles of 40 keV ^{85}Kr incident on $\langle 111 \rangle$ GaAs (from Whitton and Carter[67] 1970)

Table 8.4 Ion bombardment doses required, at room temperature, for the crystalline to amorphous transition in various materials.

Material	Incident ion	Incident ion energy (keV)	Crystalline to amorphous dose (ions/cm^2)
Si	P	40	7×10^{14}
Ge	S	40	5×10^{14}
GaAs	S	40	5×10^{14}
GaP	S	40	5×10^{14}
InSb	P	40	$10^{14}\text{–}10^{15}$
Au	Xe	40	$> 10^{16}$
W	Kr	40	$> 10^{16}$
NaCl	Xe	40	$> 2 \times 10^{16}$
KBr	Xe	40	$> 2 \times 10^{16}$
MgO	Xe	40	$> 2 \times 10^{16}$
SiO$_2$	Xe	40	$> 2 \times 10^{16}$

gave identical profiles) but further increase in incident ion dose results in a decrease in both the depth of penetration and of the channeled fraction.[44] Similar effects have been seen in Ge^{67} and in $Si.^{7,8}$

This effect is due to the crystal gradually becoming disordered, by large angle collisions of the incoming ions with the lattice atoms, resulting in a reduced possibility of later incident ions being channeled. The final result is that there is a limit to the number of ions which can become channeled; any further increase in dose affects only the distribution near the crystal surface. This is better illustrated when the data in Figure 8.15 are replotted to show number, rather than fraction, of incident ions channel (Figure 8.16). Now we see that an increase in dose from 3×10^{11} to 1×10^{12} ion/cm^2 increases the number channeled by a factor of two; further increase to 3×10^{12} gives only a 10% increase in the number channeled as does a further increase in dose to 10^{13} ion/cm^2. The damage effect, therefore, imposes an upper limit on the number of ions channeled in all semiconductor targets studied and this limit is fixed unless some form of disorder annealing is used.

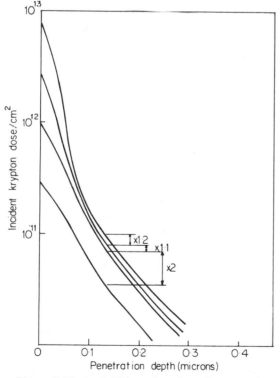

Figure 8.16. As 8.15 but plotted to show number rather than fraction of ions channeled (from Whitton and Carter[67] 1970)

(h) *Effect of Target Temperature in Semiconductors*

We have seen in Figure 8.11 that a lowering of metal target temperatures allows a greater number of incident ions to become channeled. This is also the case in semiconductors; both this and the reverse effect of reduction in the number of ions channeled owing to the crystal target being heated have been shown by Dearnaley *et al.*[8] for 40 keV ^{32}P ions implanted along the $\langle 110 \rangle$ direction of Si.

However, a more significant result of heating semiconductor targets arises from the annealing of ion-bombardment-induced disorder. Any re-ordering of the crystal lattice will enhance the chances of subsequent ions becoming channeled and this has been shown in the case of 40 keV ^{35}S ions along the $\langle 110 \rangle$ direction of GaAs and GaP.[76] The particular advantage of annealing disorder during implantation rather than after, lies in the much lower temperature required. Mayer *et al.*[77] and Nelson *et al.*[78] have shown that ion-bombarded silicon may be re-ordered at 300°C during implantation yet a temperature of 600°C is required for post bombardment annealing.

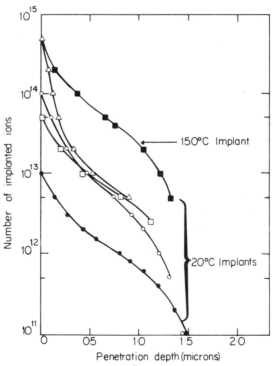

Figure 8.17. Effect of target temperature on the total number of ^{35}S ions channeled along the $\langle 110 \rangle$ direction of GaAs (from Whitton and Bellavance[76] 1971)

The result of target annealing, by heating during bombardment, is to increase the number of ions which can become channeled. Figure 8.17, a similar plot to that for Kr in GaAs (Figure 8.16), shows that the number of S ions channeled to a depth of approximately 1·5 μm in GaAs reaches a maximum at a dose of 5×10^{13} ion/cm^2; an additional order of magnitude increase in dose results in virtually no more channeling. However, when the target is heated to only 150°C during implantation, the number of ions reaching that depth is increased by a factor of six.

Finally, and briefly, although this description of channeling effects has been confined to metals and semiconductors, similar effects are seen in single crystals of the alkali halides NaCl and KBr and also in MgO and SiO$_2$.[3] The effect of bombardment dose on the range profiles in these materials is closer to that seen in semiconductors than in metals, although the dose required to cause a transformation of crystalline to amorphous in NaCl and KBr is at least an order of magnitude higher than in the Group IV and Group III–V semiconductors.

8.5 CHANGES IN RANGE PROFILE AFTER IMPLANTATION

Early investigations of range profiles in tungsten[36] showed a consistent, very deep penetration of the ions beyond their maximum range, involving typically 10^{-3} of the incident beam. Careful experiment by Davies and Jespergård[79] proved beyond reasonable doubt that this penetration was due to interstitial diffusion of perfectly channeled atoms, although previously the effect had been variously ascribed to rapid diffusion along defects (such as dislocations), to experimental artefacts such as incomplete anodic oxidation or dissolution and to macroscopic defects on the crystal surface. Typical profiles showing the deep penetration, or supertails, as they are also described are shown in Figure 8.18.

To date, conclusive evidence of this phenomenon has been observed only in tungsten, although the possibility of the effect has been advanced to explain apparently anomalously deep diffusion in silicon.[80]

A theoretical consideration of this enhanced penetration in tungsten has led McCaldin[81] to suggest that the following criteria be fulfilled before interstitial diffusion processes can take place:

(1) The damage ratio (i.e. the number of vacancies produced per incident ion) must be less than unity, otherwise each injected atom has a high probability of becoming trapped in one of the vacancies it has itself created.

(2) The interstitial atoms must have a reasonable lifetime in the crystal, and must not interchange readily with the lattice atoms.

(3) The activation energy for interstitial motion must be small enough (or the lattice temperature high enough) to permit measurable diffusion of the injected interstitials to occur.

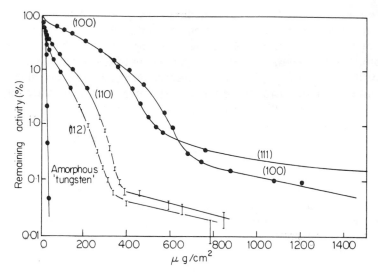

Figure 8.18. Range profiles of 40 keV ^{125}Xe
ions in various tungsten targets showing the
deep penetration caused by interstitial diffusion
(from Kornelsen *et al.*[36] 1964)

Additionally, Anderson and Sigmund[82] have pointed out that when
$Z_1 > Z_2$, even perfectly channeled ions can initiate collision cascades
and thus become trapped.

Experimental evidence for much of this has been provided in the many
cases of incident ions injected into tungsten: those with $Z_1 < Z_2$ (Na, K,
Ar, Cr, Kr, Rb, Xe and Cs) showed supertails, while Rn with $Z_1 > Z_2$ did
not.[64] However, the use of W as target material presents rather a biased
proof of the requirement that Z_1 must be less Z_2 in that there are few
available incident ion types heavier than W. It is hoped that the recently
developed anodic oxidation/dissolution sectioning technique for the bcc
Mo[13] will allow a more stringent test of this criterion.

Hermann *et al.*[20] have shown in another way that the enhanced penetration
results from a diffusion process. They observed that implantation at 80 K
followed by sectioning and measuring at the same temperature resulted in
no channeled ions diffusing but that subsequent warming to room temperature before sectioning did. The fraction of channeled ions taking part in the
diffusion process is typically 10^{-2}–10^{-4}; however, this fraction is strongly
dependent on lattice defect concentration, since a significant decrease in
this fraction is seen when the implantation is preceded by another of lighter
ions which create point defects at and beyond R_{max}.[56]

While this exceptionally deep penetration of ions in tungsten, and possibly
other metals, is undoubtedly of interest in the general study of range profiles,

the situation in semiconductors is quite different. In the latter case, because of the possible large effect on the electrical properties of solid state devices owing to deep penetration, or in fact any departure from a gaussian profile, of even a small fraction of the incident ions, the problem becomes of practical importance.

Departures from the gaussian profile in the form of a general broadening or of deeper penetration of a small fraction of the incident ions has been observed, not surprisingly, in semiconductors implanted at elevated temperatures. The more interesting observations, however, come from deep penetrations seen in semiconductors implanted at room temperature. Much attention has been given to this problem by Dearnaley et al.[80] but it is still far from satisfactorily explained whether the effect is caused by channeling or diffusion or a combination of both.

The main difficulty in interpreting the deep penetrations observed in, for example, silicon is that of completely suppressing the deep penetrations due only to channeling. Although careful tilting of a single crystal can present an apparently random array of atoms to the incoming beam, no one has yet succeeded in obtaining a gaussian profile in a single crystal, i.e. some fraction of the incident ions can always find its way into planes or axes and thus be channeled. Only when the single crystal nature of the target is destroyed by, for example, high implantation dose and rendered amorphous can the gaussian profile be achieved.

Dearnaley et al.[80] have shown that enhanced penetration takes place in silicon both in channeled and nonchanneled implants. It is not particularly surprising that the deep penetration takes place beyond the R_{max} since presumably the McCaldin criteria are satisfied (although the $Z_1 < Z_2$ requirement is not). However, it is more difficult to assign the enhanced penetration beyond the nonchanneled distribution (Figure 8.19) to the same mechanism, since the possibility of nonchanneled particles coming to rest in an undamaged lattice is very remote. Rutherford backscattering measurements[83] of silicon implanted with 400 keV phosphorus along the $\langle 110 \rangle$ direction show no detectable damage at R_{max} but a high level of damage near the surface. Dearnaley et al.[80] show evidence which suggests that both of these deep penetrations are caused by diffusion processes. However, it may be that the processes are different for channeled and nonchanneled particles. For the channeled particles the process may be an interstitial type diffusion as proposed for the damage free tungsten lattice and for the nonchanneled particles a radiation-enhanced diffusion in the heavily damaged silicon lattice.

Further evidence on this subject has recently been reported by Reddi and Sansbury who have measured, by a differential capacitance technique, the range profiles of 30 up to 600 keV phosphorus along the low index axes of silicon. They observe only slight evidence of deep penetration beyond the R_{max}.

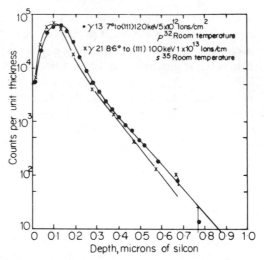

Figure 8.19. Differential range distributions of 120 keV ^{32}P and 100 keV ^{35}S implanted into misoriented Si crystals (from Dearnaley *et al.*[80] 1971)

Whatever the mechanism, or combination of mechanisms, there is no doubt that this is one of the least understood phenomena of ion implantation and channeling and one which deserves much more attention. For a full description of the very detailed research into this subject the reader is referred to the reports by Dearnaley *et al.*[80,85]

8.6 SUMMARY

From all of the foregoing evidence and cited references, it is now possible to make certain predictions about the form of range profiles as a function of some of the parameters listed. However, we can only indicate the trend as we cannot, as yet, give the final profile.

Of the seven parameters listed, only those of crystal orientation, incident ion energy, incident ion type and target type tend to affect the depth to which incoming particles penetrate, while the remaining three—target temperature, crystal disorder and bombardment dose—mainly affect the number of incoming particles reaching that penetration depth.

We can only predict these trends as almost all profiles are influenced throughout the entire range by the two processes of nuclear and electronic stopping. In cases in which nuclear stopping is mainly involved, as in low energy implants into amorphous materials, the profile may be readily forecast by application of current theory. In a similar way, for properly channeled particles slowed by electronic stopping, the measurement of R_{max} of one incident ion, at one energy, into one target material allows extrapolation of R_{max} to higher and lower energies by use of the $R_{max} \propto E^{0.5}$ relationship.

Our knowledge of the oscillating electronic stopping cross section with Z_1 allows an estimate to be made of change in R_{max} of any incident ion injected at one energy into one material but at this stage not with any accuracy. The trend in change of R_{max} when only Z_2 is changed can be shown with even less accuracy.

The trend is also clear when the target temperature is changed but, in metals, this affects only the number of incident ions penetrating deeply into the crystal while in semiconductors the additional effect of annealing has to be considered.

In the same way, an increase in crystal disorder and in bombardment dose leads to a decrease in the number of incident particles penetrating deeply, in short, the effect is shown in the dechanneling rate. This rate is the least predictable of any part of range profiles and because it can vary so much with even slight changes in, for example, defect density, surface oxide, atomic vibrational amplitude, beam misalignment, target misalignment *etc.*, it is likely to remain the least predictable part, although computer calculations of dechanneling rates (cf Chapter 7) are gradually showing the influence of some of these parameters.

The same arguments hold for the prediction of range profiles which alter after implanation, since, again, the number of ions taking part in diffusion processes is very heavily dependent on the defect density of the crystal.

8.7 FUTURE EXPERIMENTS

From the foregoing, it is fairly obvious that a careful assessment of the effects of crystal disorder (of any and all types) on dechanneling rate is necessary if some control over profiles is to be exercised. This is a formidable task, entailing very careful preparation and characterization of target material and equally careful control over implant conditions.

An extension of the study of electronic stopping cross section variation with Z_1 could well be made to include materials other than the tungsten, silicon and gold used to date. Such an extension has been curtailed in the past by the nonavailability of suitably thin single crystal targets. However, recent success in the growth of thin single crystals of W by reduction of WF_6[86] and the epitaxial growth of thin single crystals of Cu[87] and Ag[88,89] could allow an extended study. The provision of higher energy heavy ion isotope separators, such as the one being installed at the Chalk River Nuclear Laboratories, will allow the use of thicker and presumably more easily obtained targets.

The present investigation of change in electronic stopping cross section with $Z_2{}^{90}$ by measuring the R_{max} of one incident ion in many targets, may in the same way as above be extended to transmission studies through thin single crystal films.

Such experiments should certainly increase our knowledge of the effect of these parameters on range profiles and thus allow us greater facility in making predictions.

For further study of the phenomenon of profile change after implantation, the availability of other bcc metals such as Ta, Mo, V, Nb, presently being used in the S_e against Z_2 experiments could well confirm or otherwise the presently accepted criteria for interstitial diffusion in bcc targets.

Such experiments could go a long way to filling some of the gaps in our knowledge of channeling and dechanneling. The past decade has provided us with a wealth of information which, if properly applied, could well stimulate an equally fruitful next decade.

References

1. Lindhard, J., Scharff, M. and Schiøtt, H. E., *Dan. Vid. Selsk. Mat. Fys. Medd.*, **33**, No. 14 (1963).
2. Bredov, M. M. and Okuneva, N. M., *Doklady Akad. Nauk SSSR*, **113**, 795 (1957).
3. Whitton, J. L. and Matzke, Hj., *Can. J. Phys.*, **44**, 2905 (1966).
4. Sansbury, J. D. and Gibbons, J. F., *Rad. Effects*, **6**, 269 (1970).
5. Pringle, J. P. S., *J. Electrochem. Soc.*, **119**, 482 (1972).
6. Jech, C., *Phys. stat. sol.*, **27**, 573 (1968).
7. Davies, J. A., Ball, G. C., Brown, F. and Domeij, B., *Can. J. Phys.*, **42**, 1070 (1964).
8. Dearnaley, G., Freeman, J. H., Gard, G. A. and Wilkins, M. A., *Can. J. Phys.*, **46**, 587 (1968).
9. Pryborski, W., Roed, J., Lippart, J. and Sarholt-Kristensen, L. *Rad. Effects*, **1**, 33 (1969).
10. Davies, J. A., Friesen, J. and McIntyre, J. D. *Can. J. Chem.*, **38**, 1526 (1960).
11. McCargo, M., Davies, J. A. and Brown, F. *Can. J. Phys.*, **41**, 1231 (1963).
12. Whitton, J. L. and Davies, J. A., *J. Electrochem. Soc.*, **111**, 1347 (1964).
13. Arora, M. R. and Kelly, R., *J. Electrochem. Soc.*, **119**, 2 (1972).
14. Wilkins, M. A. and Dearnaley, G., *Proc. Conf. on Ion Implantation, Reading, Stevenage, England*, Peter Peregrinus, p. 193 (1970).
15. Pawel, R. E. and Lundy, T. S., *J. appl. Phys.*, **35**, No. 2, 435 (1964).
16. Andersen, T. and Sørensen, G., *Rad. Effects*, **2**, 111 (1969).
17. Farren, J. and Scaife, W. J., *Talanta*, **15**, 217 (1968).
18. Lutz, H. and Sizmann, R. *Phys. Lett.*, **5**, 113 (1963).
19. Pöhlau, C., Lutz, H. and Sizman, R., *Z. Angew. Physik*, **17**, 404 (1964).
20. Hermann, H., Lutz, H. and Sizmann, R., *Z. Naturforschung*, **21a**, 365 (1966).
21. Whitton, J. L., *J. appl. Phys.*, **36**, 3917 (1965).
22. Whitton, J. L., Plattner, H. H. and Quenneville, A. F., 1972, to be published.
23. Reimer, L., *Z. für Metallkunde* **50**, No. 1, 37 (1959).
24. Arnold, G. Sandia Labs. (1972), private communication.
25. Chang, C. C., *Surf. Sci.*, **25**, 53 (1971).
26. Socha, A. J., *Surf. Sci.*, **25**, 147 (1971).

27. Cairns, J. A., Holloway, D. F. and Nelson, R. S. *Proc. Conf. on Ion Implantation, Reading, England*, Stevenage, England (Peter Peregrinus (1970)).
28. Davies, J. A., Brown, F. and McCargo, M., *Can. J. Phys.*, **41**, 829 (1963).
29. Fairfield, F. and Crowder, B. J., *Trans. Met. Soc. of AIME*, **245**, 439 (1969).
30. Amsel, G., Nadal, J. P., D'Artemare, E., David, D., Girard, E. and Moulin, J., *Nucl. Instrum. and Meth.*, **92**, No. 4, 481 (1971).
31. Davies, J. A. and Domeij, B., *J. Electrochem. Soc.*, **110**, 849 (1963).
32. Graham, R. L., Brown, F., Davies, J. A. and Pringle, J. P. S., *Can. J. Phys.*, **41**, 1686 (1963).
33. Powers, D., Chu, W. K. and Bourland, P. D., *Phys. Rev.*, **165**, 376 (1968).
34. Whitton, J. L., Mitchell, I. V. and Winterbon, K. B., *Can. J. Phys.*, **49**, 10 (1971).
35. Mayer, J. W., Eriksson, L. and Davies, J. A., *Ion Implantation in Semiconductors*, New York, Academic Press (1970).
36. Kornelsen, E. V., Brown, F., Davies, J. A., Domeij, B. and Piercy, G. R., *Phys. Rev. A*, **136**, (1964).
37. Schiøtt, H. E., *Can. J. Phys.*, **46**, 449 (1968).
38. Sanders, J. B., *Can. J. Phys.*, **46**, 455 (1968).
39. Sigmund, P. and Sanders, J. B., *Proc. Int. Conf. on Application of Ion Beams to Semiconductor Technology, Grenoble*. Ed. P. Glotin. Editions Ophrys p. 215 (1967).
40. Winterbon, K. B., Sanders, J. B. and Sigmund, P., *Dansk. Vid. Selssk. Mat. Fys. Medd.*, **37**, 14 (1970).
41. Sigmund, P., Mathie, M. T. and Phillips, D. L., *Rad. Effects*, **11**, 39 (1971).
42. Winterbon, K. B., *Rad. Effects* (1972) in press.
43. Dearnaley, G., Wilkins, M. A., Goode, P. D., Freeman, J. H. and Gard, G. A. (1969) *Proc. Conf. on Atomic Collision Phenomena in Solids, Sussex*, Amsterdam, North-Holland, (1970).
44. Whitton, J. L. and Carter, G., *J. Mater. Sci.*, **4**, 208 (1969).
45. Davies, J. A. and Sims, G. A., *Can. J. Chem.*, **39**, 601 (1961).
46. McCargo, M., Davies, J. A. and Brown, F., *Can. J. Phys.*, **41**, 1231 (1963).
47. Santry, D. C. and Sitter, C. (1972), to be published.
48. Quéré, Y. (1971), Centre d'Etudes Nucleaires de Fontenay aux Roses, private communication.
49. Ramachandrarao, P. and Anatharaman, T. R. (1970), Proc. Symp. on Materials Science Research, India. (Dept. of Atomic Energy Government of India) and refs. within.
50. Lindhard, J., *Dan. Vid. Selsk. Mat. Fys. Medd.*, **34**, 14 (1965).
51. Eriksson, L. *Phys. Rev.*, **161**, 235 (1967).
52. Winterbon, K. B. (1972), Chalk River Nuclear Labs., private communication.
53. Davies, J. A. Eriksson, L. and Jespersgård, P., *Nucl. Instrum. and Meth.*, **38**, 245 (1965).
54. Whitton, J. L., *Can. J. Phys.*, **45**, 1947 (1967).
55. Piercy, G. R., Brown, F., McCargo, M. and Davies, J. A., *Can. J. Phys.*, **42**, 1070 (1964).
56. Davies, J. A., Eriksson, L. and Whitton, J. L., *Can. J. Phys.*, **46**, 573 (1968).
57. Whitton, J. L., *Can. J. Phys.*, **46**, 581 (1968).
58. Channing, D. A. and Whitton, J. L., *Phys. Lett.*, **13**, 27 (1964).
59. Channing, D. A., *Can. J. Phys.*, **45**, 2455 (1967).
60. Howe, L. M. and Channing, D. A., *Can. J. Phys.*, **45**, 2467 (1967).
61. Firsov, O. B., *Zh. Eksp. i. Teor. Fiz.*, **36**, 1517 (1959); *Sov. Phys. J.E.T.P.*, **9**, 1076 (1959).

62. Ormrod, J. H., MacDonald, J. R. and Duckworth, H. E., *Can. J. Phys.*, **43**, 175 (1966).
63. Fastrup, B., Borup, A. and Hvelplund, P., *Can. J. Phys.*, **46**, 489 (1968).
64. Eriksson, L., Davies, J. A. and Jespersgård, P., *Phys. Rev.*, **161**, 219 (1967).
65. Eisen, F. H., *Can. J. Phys.*, **46**, 561 (1968).
66. Bøttiger, J. and Bason, F., *Rad. Effects*, **2**, 105 (1969).
67. Whitton, J. L. and Carter, G., *Proc. Conf. on Atomic Collision Phenomena in Solids, Sussex, 1969*, Amsterdam, North-Holland (1970).
68. Hvelplund, P., *Dan. Vid. Selsk. Mat. Fys. Medd.*, **35**, 10 (1966).
69. White, W. and Mueller, R. M., *Phys. Rev.*, **187**, 499 (1969).
70. Bernhard, F., Müller-Jahreis, U., Rockstroh, G. and Schwabe, S., *phys. stat. sol.*, **35**, 285 (1969).
71. Abel, P., Müller–Jahreis, U., Rockstroh, G. and Schwabe, S., *phys. stat. sol. (a)*, **3**, K173 (1970).
72. Whitton, J. L., *Proc. R. Soc.*, A, **311**, 63 (1969).
73. Samuels, L. E., *Metallographic polishing by mechanical methods*, Melbourne and London, Sir Isaac Pitman and Sons (1967).
74. Tegart, W. J. McG., *The Electrolytic and Chemical Polishing of Metals*; Oxford, Pergamon Press (1959).
75. Quéré, Y. and Couve, H., *J. appl. Phys.*, **39**, 4012 (1968).
76. Whitton, J. L. and Bellavance, G. R., *Rad. Effects*, **9**, 127 (1971).
77. Mayer, J. W., Eriksson, L., Picraux, T. and Davies, J. A., *Can. J. Phys.*, **46**, 663 (1968).
78. Nelson, R. S. and Mazey, D. J., *Can. J. Phys.*, **46**, 689 (1968).
79. Davies, J. A. and Jespersgård, P., *Can. J. Phys.*, **44**, 1631 (1966).
80. Dearnaley, G., Wilkins, M. A. and Goode, P. A., *Second Int. Conf. on Ion Implantation in Semiconductors*, Germany, Springer-Verlag (1971).
81. McCaldin, J. O., *Progr. Solid State Chem.*, **2**, 9 (1965).
82. Andersen, H. H. and Sigmund, P., *Nucl. Instrum. and Meth.*, **38**, 238 (1965).
83. Whitton, J. L. and Bellavance, G. R., presented at *Second Int. Conf. on Ion Implantation in Semiconductors, Germisch-Partenkirchen, Germany, 1971*.
84. Reddi, V. G. K. and Sansbury, J. D., *Appl. Phys. Letts.*, **20**, No. 1, 30 (1972).
85. Dearnaley, G., *Reports on Progress in Physics*, **32**, Pt. 2, 405 (1969).
86. Mayadas, A. F., *J. Electrochem. Soc.*, **116**, No. 12, 1742 (1969).
87. Markus, A. M. and Fainshtein, A. L., *Ukr. Fiz. Zh.* (Russ. Ed.), **15**, No. 10, 1738 (1970).
88. Koch, F. A. and Vook, R. W., *J. appl. Phys.*, **42**, No. 11, 4510 (1971).
89. Noggle, T. A. (1971), Third Int. Symp. on Research Materials for Nuclear Measurements. Gatlinburg, Tennessee; to be published in *Nucl. Instrum. and Meth.*
90. Whitton, J. L. and Quenneville, A. F. (1972), to be published.

CHAPTER IX
Radiation Damage

R. S. NELSON

9.1 INTRODUCTION

The interaction of an energetic heavy ion with a crystalline solid results in atomic collisions of large energy transfer. The recoiling atoms from these primary encounters initiate a cascade of atomic collisions within the solid which give rise to phenomena such as radiation damage and sputtering. From the foregoing chapters on channeling it is readily apparent that, if we can reduce the probability of small impact parameter collisions at the expense of collisions with large impact parameters, then we will inevitably reduce secondary processes such as radiation damage and sputtering. Clearly, therefore, the amount of radiation damage and the sputtering yield will be reduced when the incident ion beam is aligned close to a major channeling direction of a single crystal target.

Although this chapter will deal primarily with the influence of channeling on radiation damage production, we will include a discussion of its effect on sputtering, as, after all, sputtering is essentially the manifestation of radiation damage at the surface of a solid. Furthermore, it was from sputtering experiments that the first hint of an orientation dependence of atomic collision processes was observed. In 1959, Rol et al.[1] at the FOM laboratories in Amsterdam observed an orientation dependence in the high energy sputtering yield (the number of atoms ejected per incident ion) of a {100} single crystal of copper as it was rotated about a [110] direction in its surface. Their results, shown in Figure 9.1, are reproduced for historical reasons. Between then and the accepted date for the discovery of channeling in 1963,[2] numerous investigations studied the orientation dependence of sputtering; for example, Almèn and Bruce,[3,4] Molchanov et al.,[5] Fluit et al.[6] and Southern et al.[7] These studies led to what was known as the 'transparency model' of single crystal sputtering, which supposes that the atoms of the surface layers could be described as hard spheres, and that the sputtering yield was proportional to the effective area presented by those atoms exposed to the incident ion beam. Quantitative calculations by Odintsov[8] and by Southern et al.[7] based on this model gave reasonable agreement with experiment. It was about this time that Robinson and Oen[9] published their computer

259

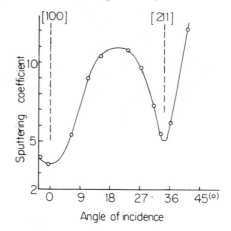

Figure 9.1. The variation in sputtering ratio as a {100} copper single crystal is rotated about a ⟨110⟩ direction in its surface (after Rol *et al.*[1] 1959)

calculations which recognized the fact that the ions trajectories were steered between the open channels in the crystal. The transparency model was then superceded by models based on channeling theories.[10]

In this chapter we will first outline the observed nature of radiation damage in metals and semiconductors to serve as a basis for the subsequent discussion on the influence of channeling. We will then discuss briefly the relevant theoretical material which provides a framework for the interpretation of the experimental results. A variety of experiments will be cited which are taken from many different sources and types of investigation.

9.2 THE OBSERVED NATURE OF RADIATION DAMAGE

9.2.1 Metals

For the purposes of this chapter we will restrict our discussion to the observed nature of radiation damage which results from an irradiation at room temperature. In many cases both the individual interstitials and vacancies are mobile at this temperature and we will be concerned with the physical state of the irradiated solids which is pertinent to this situation. However, in high melting point materials such as tungsten and nickel only the interstitial is thought to be mobile at room temperature. However, for convenience let us first consider the former of those two cases.

The first question to be answered is: what is the configuration of damage in the vicinity of the isolated displacement cascades when the ion dose is sufficiently low that such cascades are well separated? Clearly a substantial degree of mutual recombination will occur within the cascade, thus reducing the total number of defects which are able to diffuse away into the surrounding solid. But perhaps even more important, owing to the dynamic segregation of defects within the collision cascade which occurs predominantly in heavy materials, the vacancy-rich core can rearrange to form a compact defect cluster, so leaving the interstitials and residual vacancies free to migrate away from the damaged regions. The precise morphology of these clusters, i.e. whether they form three-dimensional voids, faulted dislocation loops, unfaulted dislocation loops or stacking fault tetrahedra, depends on a variety of parameters such as stacking fault energy and surface energy. For instance, Sigler *et al.*[11] have made a number of predictions for a variety of pure metals depending on the number of vacancies within the agglomerates. However, owing to the inadequacy of our knowledge of the parameters involved, we must treat quantitative results of such computations with some reservations. In light materials such as aluminium and graphite, in which the heterogeneous nucleation of defect clusters within collision cascades is unlikely, owing to the fact that interstitials and vacancies are created within rather large diffuse cascades, essentially every interstitial and vacancy is created free to migrate through the surrounding lattice. Some experimental evidence for these ideas has been provided by field ion and electron microscopy. Both techniques suggest that in heavy metals such as copper, silver and gold, at temperatures at which defects are mobile, the individual collision cascades leave regions which give contrast consistent with some sort of vacancy agglomerate. Generally this is a small dislocation loop often called a 'black spot defect' as shown in Figures 9.2 and 9.3. However, in light metals bombarded with low doses no such evidence for cluster formation within the collision cascades has been forthcoming.

Next we will discuss the general behaviour as the bombardment dose steadily builds up to a point at which the individual collision cascades overlap. Whilst the individual cascades are well separated a large fraction of the freely migrating defects will be lost to the surface, which in most cases of interest acts as the most dominant sink. However, in heavy metals, as the dose builds up the isolated vacancy agglomerates will build up linearly until overlap and saturation prevails, Figure 9.4. At the same time the interstitial concentration will build up to a sufficiently high level that the homogeneous nucleation of interstitial clusters occurs, most probably in the form of loops as in Figure 9.5. As the dose is increased still further those new vacancies and interstitials which escape recombination will be captured by the existing defect clusters rather than sustain further nucleation. The clusters will therefore grow steadily until they interact and eventually form a complicated dislocation entanglement as illustrated in Figure 9.6. This situation represents the ultimate saturation in the configuration of damage, as further

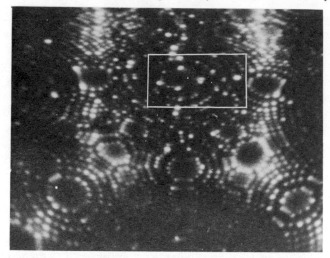

Figure 9.2. Field ion microscope photograph of a dislocation
loop in irradiated iridium. The two white dots define the ends
of the dislocation (after Hudson[12])

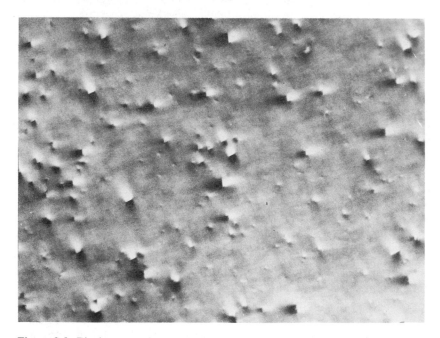

Figure 9.3. Black spot defects in copper irradiated with 30 keV Cu^+ ions, each
spot corresponds approximately to one incident ion. Analysis of the diffraction
contrast suggests that such defects are vacancy-type dislocation loops (after
Wilson[13])

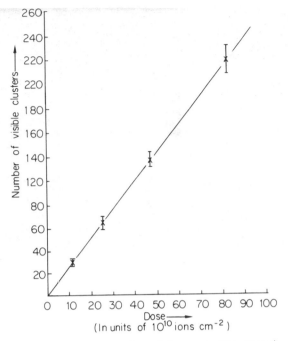

Figure 9.4. The number of clusters in 30 keV Cu$^+$ ion-bombarded copper with diameters greater than 25 Å plotted as a function of dose showing the linear relationship (after Wilson[13])

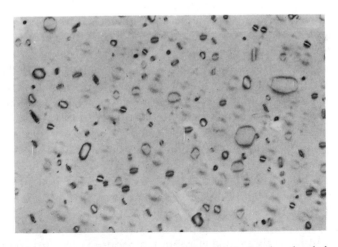

Figure 9.5. Interstitial dislocation loops in ion-bombarded copper.

Figure 9.6. High dose ion-bombarded copper showing the growth of interstitial dislocation loops into a complex dislocation entanglement

irradiation will simply result in the rearrangement of this entanglement by the process of slip and climb. In light metals, such as aluminium, the heterogeneous nucleation of vacancy clusters within cascades is considered unlikely and it is thought that the damage is built up by the simultaneous homogeneous nucleation and growth of both interstitial and vacancy clusters (Beevers and Nelson[14]).

It is of interest to estimate the actual number of atoms which remain displaced from their lattice sites, especially in the vicinity of the surface. If both interstitials and vacancies are mobile, the number of free interstitials and vacancies is essentially zero, as they will either have suffered annihilation at fixed sinks, recombination or agglomeration to form a complex dislocation array. It is well known that dislocations exhibit strong elastic interactions both with each other and with the surface. The surface interaction is manifest in the creation of an image force which acts so as to remove the dislocations from the solid. The net result is that, even under saturation damage conditions, the fraction of atoms which are not situated in equilibrium lattice sites is only between 10^{-4} and 10^{-3}.

In the case of higher melting point metals, in which only the interstitial is thought to be freely mobile at room temperature, the situation is slightly different. Under these conditions, although we expect a rather large fraction of defects to suffer recombination within the lattice, heterogeneously nucleated vacancy clusters and homogeneously nucleated interstitial agglomerates will form on a rather fine scale.

9.2.2 Semiconductors

So far we have considered the nature of radiation damage in metallic solids in which, to all intents and purposes, the solid remains essentially crystalline, even after excessive bombardments under which every atom is displaced many hundreds of times. However, this is not the case for radiation damage in semiconductor materials.

Electron microscopy and diffraction studies of ion-bombarded Ge,[15] Si[16] and GaAs[17] have shown that at high doses radiation damage is manifest in the creation of a noncrystalline, essentially amorphous layer, within which every atom is displaced from its equilibrium lattice site. It is thought that such intense damage results from the accumulation of small disordered regions of approximately 30 Å diameter, which are produced in the vicinity of the most energetic primary recoils as a consequence of displacement cascades. The precise mechanism for their creation and why such a phenomenon does not occur in metals is not yet understood. But the results from recent high dose and dose rate 1 MeV electron irradiations at liquid nitrogen temperatures in the high voltage electron microscope at least support the hypothesis that a necessary requirement for the production of an amorphous surface phase is the existence of displacement cascades containing a high density of displaced atoms.

Clearly not all displaced atoms are contained within the disordered zones, as the interstitials and vacancies produced as a result of low energy primary recoils, together with those produced at the outermost regions of the collision cascades, will be free to migrate randomly through the lattice where they will either recombine or cluster and create dislocation loops or the well known defect centres described, for example, in the book by Corbett.[18]

In the context of the present chapter we will restrict ourselves to the damage which is manifest as the destruction of the crystal structure and commonly called 'amorphicity'. The existence of the amorphous phase is illustrated in Figure 9.7, in which a prethinned (~ 1000 Å thick) specimen of GaAs has been irradiated with 80 keV Ne^+ ions. The left hand area of light contrast (A) was shielded from the ion beam with a grid, whilst the right hand area (B) suffered bombardment with $4{\cdot}0 \times 10^{15}$ ion/cm^2 at 20 °C. Selected area electron diffraction of these regions clearly demonstrates the degeneration from a crystalline to an amorphous state.

During ion bombardment of Si, that part of the surface on which the ions impinge undergoes visible colour changes as the dose builds up.[19] At threshold, faint blue or pink hues are observed, the actual colour depending

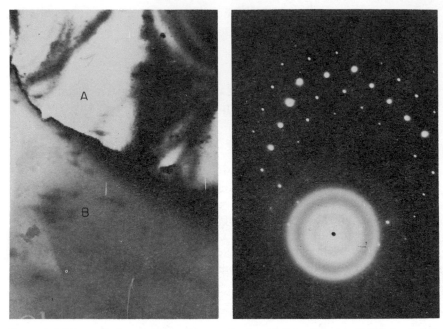

Figure 9.7. Transmission electron micrograph of an Ne$^+$ ion-bombarded GaAs sample at 20°C together with diffraction patterns. The light contrast area has been shielded during irradiation and shows a crystalline diffraction pattern, whereas the dull irradiated area shows a diffuse ring diffraction pattern typical of amorphous material (after Mazey and Nelson[17])

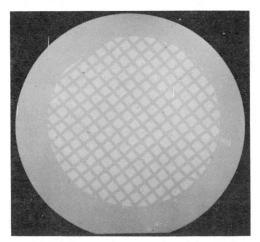

Figure 9.8. Photograph of an ion-bombarded silicon specimen shielded during bombardment with a coarse grid. The 'milky' effect is clearly visible.

on the ion and its energy. However, as the dose increases these eventually saturate into a characteristic 'milky white' appearance. Figure 9.8 shows a photograph of a silicon slice bombarded through a grid so as to provide a series of demarcation lines; the milky appearance is readily seen. It is thought that these visual changes arise as a consequence of the different light scattering properties of amorphous and crystalline silicon. The onset of the milky appearance in Si has been used as a useful guide to the existence of an amorphous surface phase and can be used to ellucidate the effects of channeling on damage production as we will see below.

9.3 THEORETICAL ASPECTS

In this section it is not our intention to present an all-embracing review of channeling theory, but to outline the basic ideas which are relevant to interpretation of experimental results.

Figure 9.9 shows a schematic view of a charged particle beam incident parallel to a major channeling direction. The critical impact parameter ρ_{crit}

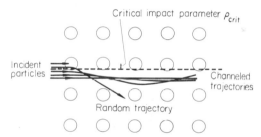

Figure 9.9. A schematic view of a charged-particle beam incident parallel to a major channeling direction. The critical impact parameter which divides the beam into channeled and random trajectories is illustrated

essentially divides the particle trajectories into two components: (i) trajectories with impact parameters greater than ρ_{crit} which are constrained to move in regions of low atomic density—the channeled component, and (ii) trajectories with impact parameters less than ρ_{crit} which have sufficient transverse energy to pass through the channel walls—the random components. In practice it is possible to relate this critical impact parameter to a critical angle ψ_{crit} which separates those trajectories with angles less or greater than ψ_{crit} which enter the channeled or random components respectively. Following Lindhard[20] we can calculate this critical angle from the formula

$$\psi_1 = \left(\frac{2Z_1Z_2e^2}{dE}\right)^{\frac{1}{2}} \quad \text{for } \psi_1 < a_{TF}/d$$

where Z_1 and Z_2 are the atomic numbers of the projectile and target atoms, e is the electronic charge, d is the separation of atoms in the row, E is the projectile energy and a_{TF} is the Thomas–Fermi screening distance. At lower velocities where $\psi_1 > a_{TF}/d$ then Lindhard gives

$$\psi_2 = \left(\sqrt{\frac{3}{2}} \frac{a_{TF}}{d} \psi_1 \right)^{\frac{1}{2}}$$

To a first approximation, therefore, if the angle of incidence to a channel is greater than ψ_1, essentially no particles become channeled. If, however, the incident beam is well collimated and accurately parallel to a channeling direction, then a substantial fraction of trajectories will be constrained to move in regions of low atomic density. As previously mentioned a perfectly channeled beam will therefore suffer somewhat fewer large angle collisions and it might be expected that radiation damage is reduced. The fraction of the incident beam which can enter channels can be estimated simply from the fraction of the space which falls within the area defined by the critical impact parameter to that which falls outside this area. In experiment this is manifest as the magnitude of the minimum yield under best channeling conditions and is therefore given as

$$\chi_{min} = \pi a_{TF}^2 N_0 d \quad \text{for axial channeling}$$

and

$$\chi_{min} = 2a_{TF}/d_p \quad \text{for planar channeling.}$$

where N_0 is the atomic density and d_p is the interplanar spacing.

In practice these simple criteria must be improved to take account of effects such as thermal vibration, surface contamination and dechanneling. Furthermore it makes a significant difference whether the minimum yield corresponds to events at the surface, to events at some specified depth or to the summation of events over the particle's range. Perhaps the best approaches to formulating this problem have been those which make use of computers. An example of such an approach which studies both critical angles and minimum yields as a function of depth, is that by Barrett.[21] In his model a Monte-Carlo computer program follows the trajectories of high energy ions in a lattice and the relevant parameters such as critical angle, *etc.* are determined for a selection of variables such as temperature and energy. However, in the present context, because of the difficulty in making accurate measurements of the effect of channeling on radiation damage, we will restrict our discussions of his work to some of the qualitative predictions.

Perhaps the most significant prediction relevant to radiation damage is that which arises from the bunching of trajectories at certain depths below the surface, commonly called 'flux-peaking' (see Chapter 3-6). The depth distribution of minimum yield is expected to follow a similar oscillatory behaviour. Figure 3.11 shows a specific example for 0·4 MeV protons incident parallel to the {111} planes in aluminium at a temperature of 313 K. Similar results have also been obtained for 60 MeV I ions in gold. Another feature, which was examined by Barrett,[21] was the effect of the divergence of the incident ion beam on the minimum yield, and he found a close interdependence on the effect of temperature.

Some experiments designed to study the effect of channeling on radiation damage production and especially sputtering use rather high ion doses in order to obtain an observable effect. In other words atomic disorder in the form of point defects, dislocation loops, impurities and other complex defect structures, as outlined in Section 9.2 above, will be present in rather large concentrations in the bombarded region. It is well known that such defects lead to significant dechanneling and consequently can influence the effect of channeling on damage production, in as much as the effect will be a function of ion dose. The dechanneling of channeled particles by defects has been considered theoretically in some detail by Quéré[22] and by Morgan and Van Vliet.[23]

9.4 EXPERIMENTAL EVIDENCE FOR THE EFFECT OF CHANNELING

9.4.1 Electron Microscope Studies in Metals

The first electron microscope observation of the effect of channeling on the production of radiation damage was by Noggle and Oen.[24] They studied the formation of the so-called 'black spot defects' produced in {100} single crystal gold foils (~ 2200 Å thick) during irradiation with 51 MeV [127]I ions. As previously discussed black spot defects are the manifestation in the electron microscope of small disordered zones produced by energetic primary recoils. They are observable as small regions of strain which give rise to a dark contrast under bright field conditions (e.g. Figure 9.10 from Noggle and Oen[24]) and are most probably some sort of unresolved vacancy-type dislocation loop. The density of such defects was measured as a function of crystal orientation near to the $\langle 110 \rangle$ axis. The results are shown in Figure 9.11 and clearly indicate a significant reduction in damage as the channeling direction is approached. The statistics indicate a maximum reduction of about 14 times, and the measured half widths agree with the simple theoretical prediction of Lindhard within a factor of two. An advantage of this particular experiment is that gold is likely to have a fairly clean surface and so contamination effects are likely to be minimal. However, perhaps even more important, as the foils are so thin and the dose is low so that the damaged regions do not overlap and effects of dechanneling are

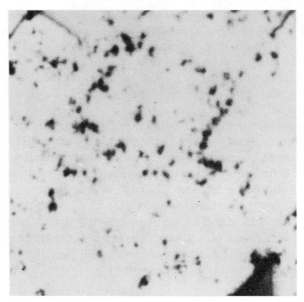

Figure 9.10. Black spot defects produced by 51 MeV ^{127}I
ion bombardment of a gold foil (after Noggle and Oen[24])

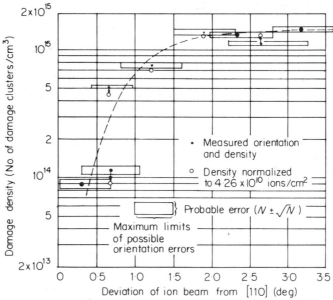

Figure 9.11. Variation in black spot defect density as a function
of angle of incidence about a $\langle 110 \rangle$ direction after Noggle and
Oen[24])

likely to be very small, the measured minimum yield can therefore be compared fairly directly with theory.

A detailed study of the effect of channeling on radiation damage production in copper with low energy Cu^+ ions (up to 100 keV) has been made by Wilson.[25] Black spot defects produced after irradiation at ambient temperature with ion doses between 10^{11} and 10^{12} ions/cm^2 were identified as mainly Frank sessile loops of vacancy character. The density, size and depth distribution of such defects were determined for two orientations, namely a random direction and a $\langle 110 \rangle$ direction. Figure 9.12 shows two micrographs

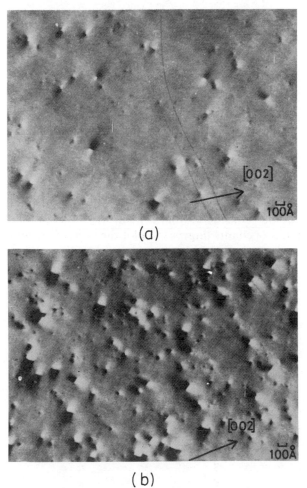

Figure 9.12. Black spot defects produced by 30 keV Cu^+ ions in two copper foils viewed in a [002] reflection with $s = 0$ (after Wilson[25]). (a) 9.8×10^{11} ion/cm^2 incident along a $\langle 110 \rangle$ direction (b) 8.3×10^{11} ion/cm^2 incident along a random direction

which illustrate the result for 30 keV Cu⁺. Quite clearly there is an overall reduction in the density of black spots due to channeling. Normalized size distributions derived from three ⟨110⟩ specimens are shown in Figure 9.13

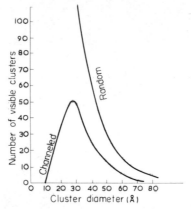

Figure 9.13. Cluster size distribution for copper specimens irradiated with 30 keV Cu⁺ ions incident along both ⟨110⟩ and random directions. The distributions are normalized to a dose of a 1000 ions (after Wilson[25])

together with the mean normalized size distribution from random-orientation irradiations. It was considered that a good way of assessing the reduction under conditions of channeling was to plot the ratio of the number of visible clusters in a given diameter range under channeling and nonchanneling conditions. This ratio is illustrated in Figure 9.14, in which it is readily

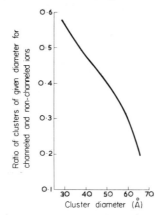

Figure 9.14. The ratio of the number of visible clusters, of a given diameter range, in Cu⁺ bombarded copper under channeling conditions (after Wilson[25])

apparent that not only is the total damage reduced during channeling, but the number of larger diameter clusters is reduced relative to the smaller ones. Perhaps the most interesting result, apart from the overall reduction in cluster density, was the difference in depth of damage between the two irradiation directions. These depths distributions were obtained by stereo-microscopy and are illustrated in Figure 9.15. It can be seen that in the case of ⟨110⟩ irradiations the damage is distributed over a depth of approximately 100 Å with a tendency to show oscillations in density, whereas the random irradiation shows damage mainly confined to within 60 Å from the surface.

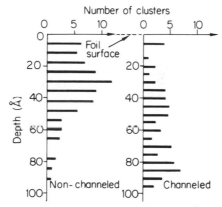

Figure 9.15. Cluster depth distributions for two copper specimens irradiated with 30 keV Cu$^+$ ions incident in a random and a ⟨110⟩ direction respectively (after Wilson[25])

9.4.2 Field-Ion Microscope Studies

Apart from transmission electron microscopy some limited evidence for the effect of channeling on damage production has been obtained using the field-ion microscope, Hudson and Ralph.[26] The metal iridium has a high melting point which makes it suitable for field-ion microscopy and at the same time has an fcc lattice. Using the technique of field evaporation it is possible to strip the specimen atom by atom to depths of several hundreds of ångström units and so unfold radiation damage in the form of both single vacancies and defect clusters. The results indicate that the damage density was considerably less in the case of ⟨110⟩ irradiations than for random bombardments. For example, a specimen of iridium bombarded with 60 keV A$^+$ ions to a dose of 10^{15} ion/cm^2, within 1° of ⟨110⟩, had a reduction in damage density of a factor of five compared with that for a similar irradiation in a random direction. Furthermore the depth of the damage was considerably greater in the channeled case, the deeper component consisted mainly of small dispersed clusters and single vacancies; large clusters and dislocation loops were contained only within the first part of the track.

9.4.3 Semiconductors

The effect of channeling on damage in semiconductors has been studied in a variety of ways. Perhaps the most pictorial is that of Nelson and Mazey.[27] Advantage was taken of the fact that the ion bombardment of Si results in the formation of an amorphous surface phase which has a characteristic milky appearance. If the crystal is then bombarded under conditions for channeling, in which the probability of large-angle elastic collisions with lattice atoms is reduced, the milkiness should appear after a larger ion dose than that corresponding to random trajectories of the incident ions.

A silicon crystal 2·5 cm in diameter and 0·3 mm thick, having either its {111}, {100} or {110} planes parallel to its flat chemically polished surface, was mounted on a rotable target holder which allowed simultaneous rotation about two orthogonal axes (Figure 9.16). The vertical axis allowed

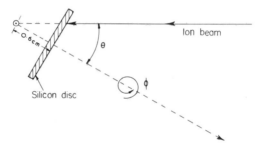

Figure 9.16. Schematic arrangement illustrating the two orthogonal rotations used to study the orientation dependence of radiation damage in silicon (after Nelson and Mazey[27])

variation in the inclination between the crystal and ion beam direction (θ) and the horizontal axis provided continuous rotation about the normal to the crystal surface (ϕ). The face of the crystal was held at a distance of 6 mm from the vertical axis so that, as θ was increased by 1° intervals, the beam would describe a series of concentric circles on its surface. The crystal was set into rotation at a rate of 2 rev/min and bombarded with 80 keV Ne$^+$ ions collimated to 0·08 cm diameter with a divergence of approximately 0·01°. After a dose just sufficient to form the amorphous layer under nonchanneling conditions, as determined by the onset of the milky appearance viewed through the vacuum window, the angle of inclination was reduced by a succession of 1° intervals until the whole sample had been scanned with the ion beam. The irradiation time was carefully adjusted at each angular setting so as to produce a uniform dose over the bombarded surface. Figure 9.17 is a direct photographic reproduction of the crystal surface after bombardment. It is evident, from the regular patterns mapped out on the crystal surfaces, that those parts which do not exhibit a milky appearance

Figure 9.17. Channeling patterns produced on a {111} silicon crystal (after Nelson and Mazey[27])

correspond to angles of incidence close to the major axial and planar channeling directions. It should be noted that the authors considered the apparent milkiness at the centre of the spots to be an optical effect caused by photographic reproduction; it was not actually seen on the Si samples. These results are consistent with a reduction in radiation damage and, hence, the rate of amorphous phase formation, whenever the incident ion was channeled. As the bombardment continued, the milkiness steadily spread into those parts of the patterns which corresponded to channeling, until eventually the whole Si surface had become amorphous. To provide a measure of the reduction in radiation damage under conditions of channeling, the dose required to obliterate the patterns was compared with the dose just necessary to turn the surface milky when the incident ions were not channeled. For the major channeling direction $\langle 110 \rangle$ this occurred after a dose increment of about $8 \times$, and therefore implies a reduction in radiation damage by a factor of eight under conditions of best channeling. Although in this case measurement of the half width of the channeling directions can only provide semiquantitative information of critical channeling angles, general agreement with the expected variation of ψ_2, with the atomic repeat distance of the channel wall was observed. For instance Figure 9.18 shows a logarithmic

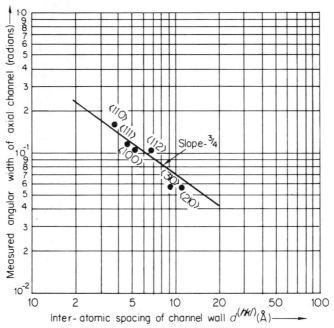

Figure 9.18. Logarithmic plot of the angular channel widths during 80 keV Ne^+ irradiation of silicon. The values are measured from channeling patterns as a function of $d^{(hkl)}$ (after Nelson and Mazey[27])

plot of the measured angular channel width as a function of interatomic spacing $d^{(hkl)}$, together with a line having a slope of $(-\frac{3}{4})$ drawn for comparison. It should be mentioned that during the progressive anodic stripping of such patterns there is a suggestion of an oscillatory behaviour in the degree of milkiness with depth for those parts of the surface which correspond with the good channeling directions. However no quantitative data was obtained for comparison with the predictions of Barrett.[21]

Transmission electron microscopy and electron diffraction has also been used to study the effect of channeling on radiation damage production.

As far as the build up of amorphous material is concerned, the results essentially bear out those quoted above in connection with visual observations. However, the configuration of the residual damage which remains after epitaxial recrystallization of the amorphous layer has also been compared for irradiations under both channeling and random conditions. In general the results are as expected, in as much as under channeled bombardments the dislocation density is reduced and extends to greater depths.

9.4.4 Sputtering Studies

In the Section 9.1 we inferred that it was from sputtering experiments that the first hint of an orientation dependence of atomic collision processes was observed. The sputtering phenomena arises basically from medium-to-large angle atomic collisions in the surface layers of crystals. Clearly, if the specimen is a single crystal and is oriented so that the incident ions are aligned with the major channeling directions in the lattice, the probability of large energy transfers is very much reduced. Consequently the total sputtering yield will vary with orientation. Perhaps the most significant work on this phenomenon was performed at the FOM Laboratories, (see for instance Onderdelinden[10]). In one experiment a {100} Cu crystal was rotated about one of the ⟨110⟩ axes in its surface so that the ion beam explored the ⟨100⟩, ⟨211⟩, and ⟨411⟩ axes. Figure 9.19 shows the total sputtering ratio as a function of the

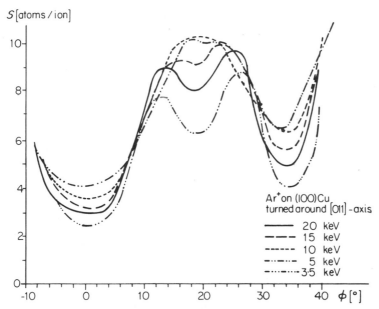

Figure 9.19. Variation in total sputtering ratio as a function of angle of incidence for various energy A$^+$ ions on {100} copper single crystals rotated about the ⟨011⟩ axis (after Onderdelinden[10])

angle of incidence to the surface normal for bombardment with A^+ ions of various energies. It was readily apparent that as the energy was increased the dips which correspond to the major channeling directions became narrower and more pronounced. It was also evident that below about 10 keV the $\langle 111 \rangle$ axial channel was no longer operative. The half widths at half minimum were compared with ψ_2 and found to be in reasonable agreement. Furthermore, the absolute minimum distance of approach to the $\langle 411 \rangle$ channel wall was calculated to be equal to one half the actual channel width at 10 keV, thus providing a simple explanation for the extinction of this channel below approximately 10 keV. It was concluded that only those incident particles which entered the random nonchanneled component could transfer sufficient energy to the surface layers to cause sputtering.

9.4.5 Electrical Resistivity

An interesting observation of the effect of channeling on damage production using the change in electrical resistivity, which occurs when thin foils suffer ion bombardment, as a measure of the radiation damage has been described by Merkle.[28] Thin ($\sim$ 5000 Å) single crystal foils of gold with their $\{100\}$ planes parallel to their flat surface were grown by electron beam evaporation onto rocksalt at 400 °C at a pressure of approximately 10^{-7} Torr. Current and potential leads were incorporated into the shape of the specimen by evaporating through a suitable mask. After dissolution of the rocksalt the specimens were mounted onto a suitable goniometer which could be cooled to liquid helium temperature. The electrical resistance at 4·2 K was then measured after dose increments of 3×10^{12}, 120 keV proton/cm². After every five dose increments the orientation of the crystal with respect to the beam was changed. The resistivity change per unit dose (damage rate) as determined from these values was therefore obtained as a function of crystal orientation, Figure 9.20. It is important to note that the constancy of the

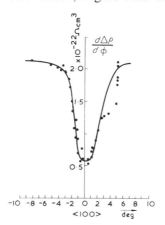

Figure 9.20. The effect of channeling on radiation damage in gold upon rotation through a $\langle 100 \rangle$ direction. The damage is measured by the rate of change of electrical resistivity during 120 keV H$^+$ irradiation at 4·2 K (after Merkle[28])

damage rate over the whole experiment was checked by normalization at one orientation, thus precluding any spurious effects which might occur as a consequence of saturation. The angular width around the $\langle 100 \rangle$ axes was compared with ψ_2 and found to agree quite well. However, it was considered not fruitful to compare the minimum yield with theory as defects which are immobile at 4·2 K would result in dechanneling and consequently raise the minimum yield in the $\langle 100 \rangle$ orientation.

9.5 DECHANNELING AT DEFECTS

As we have previously mentioned, dechanneling which occurs at defects or at radiation damage can influence the overall reduction in radiation damage or sputtering which occurs when the incident ion beam is aligned with a channeling direction. Furthermore dechanneling is also important in the quantitative application of Rutherford scattering techniques to radiation damage or to atom location studies; however we will not pursue this particular point here as it will be dealt with in detail in Chapter 14.

The most comprehensive experimental study of dechanneling at defects has been carried out by the French group under Quéré.[22] Their technique is to pass energetic α-particles from an α-source through thin single crystal foils. The transmitted α-particles are detected simply by recording their tracks in photographic emulsion. An example of the dechanneling effects of dislocations is shown in Figure 9.21 where a heavily cold-worked foil of platinum has been partially recrystallized to produce essentially dislocation free grains. The dark regions correspond to recrystallized material which the α-particles have channeled through the foil, whereas the white regions (where the emulsion has not been sensitized) correspond to those parts of the foil still containing a high dislocation density from the cold-work.

A very direct observation of dechanneling has been made by Schober and Balluffi.[29] In this experiment a thin (~ 650 Å) single crystal gold foil having a $\langle 100 \rangle$ orientation, which was still in epitaxial contact with a silver foil of similar thickness, was irradiated with 40 keV Au^+ ions. The sense of the irradiation was such that the Au^+ ions were directed onto the silver foil parallel to the $\langle 100 \rangle$ axis normal to its surface. The channeled Au^+ therefore passed through the silver foil into the gold foil; the silver foil was sufficiently thick that nonchanneled Au^+ was stopped and did not penetrate into the gold foil below. Subsequent to the irradiation, the silver foil was dissolved and the gold foil examined by transmission electron microscopy.

Figure 9.22 shows a typical result. The majority of the foil contains black spot defects which are presumably caused only by channeled particles and which correspond to the long tail in the black spot defect distribution as seen by Wilson.[25] However adjacent to nearly every stacking fault there is a region clear of defects. This was interpreted as follows. In the sandwich foils

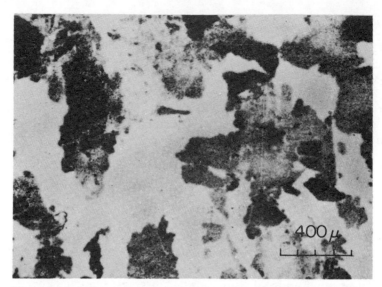

Figure 9.21. Alphagraphy of a partial recrystallized sample of platinum. The white parts correspond to regions still containing a high dislocation density, whereas the darkened regions correspond to those grains which have recrystallized and permit the channeling of α-particles (after Quéré[22])

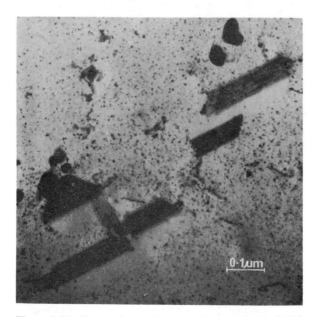

Figure 9.22. Transmission electron micrograph of a {100} gold single crystal layer removed from the epitaxial sandwich after bombardment with 40 keV Au$^+$ ions to a dose of 1.5×10^{14} ion/cm^2. The regions denuded of black spot damage adjacent to the stacking faults are clearly visible (after Schober and Balluffi[29])

some stacking faults were continuous across the silver–gold boundary and the majority of these passed diagonally through the sandwich. It was then supposed that any channeled particle travelling towards a fault within the silver foil became dechanneled and could not continue its passage into the gold foil. The clear regions in the gold foil were therefore shadows of the faults in the silver foil, and were adjacent to the faults in the gold foil purely as a consequence of the fault having been continuous across the boundary.

It should be mentioned that dechanneling at parallel arrays of dislocations, which have grown from a complex dislocation network provided by ion bombardment damage, is thought to be a cause of the regular surface topography which is evolved after the removal of some thousands of ångströms of the surface by sputtering.[30] Regular structures can be seen particularly when flat single crystal surfaces are irradiated along channeling directions, see, for example, Figure 9.23.[31]

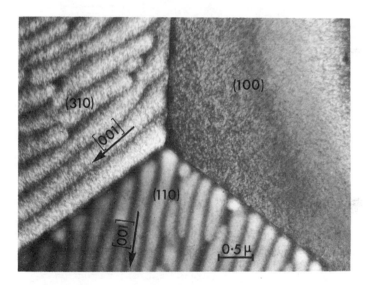

Figure 9.23. Electron micrograph of the surface structures developed on a copper polycrystal during 60 keV Xe$^+$ ion bombardment at normal incidence. The intersection of three differently orientated grains show that whilst sputtering structures have developed on the (310) and (110) grains no structure has developed on the (100) surface (after Mazey *et al.*[31])

9.6 ELECTRON CHANNELING

The accelerating voltages now available in high voltage electron microscopes are sufficient to produce atomic displacements in light and medium atomic weight materials. Makin[32] has demonstrated that interstitial defect clusters are shown to form in copper irradiated at 300 K, the density of clusters as a function of time agreeing well with the number predicted by homogeneous nucleation and growth theory. A layer denuded of clusters is observed around all surfaces and at grain boundaries; the width of this layer also agrees with theory. Figure 9.24 shows a typical example of such damage

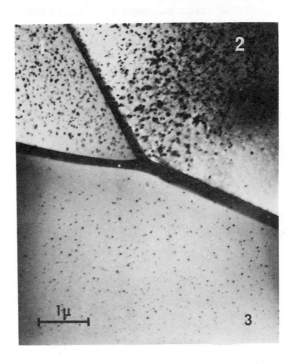

Figure 9.24. Transmission electron micrograph taken in the high voltage electron microscope showing electron damage in a copper foil. The micrograph clearly shows the effect of orientation on the damage produced after an irradiation of 15 min by a 0·4 μA beam of 525 keV electrons (after Makin[32])

produced and observed simultaneously in the high voltage electron microscope. The denuded zone is readily seen at the high index grain boundary but not at the twin boundary. This micrograph also shows a marked effect of orientation of the damage produced after an irradiation of 15 min by a 0·4 μm beam (in a 5 μm spot) of 525 keV electrons which suggests an effect of electron channeling. Makin[32] has in fact studied the effect of electron channeling in some detail. An example is shown in Figure 9.25 where the

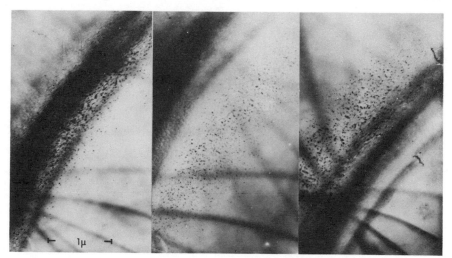

Figure 9.25. The effect of planar electron channeling on either side of a [111] contour. The irradiation was with 600 keV electrons for 15 min using a 1 μA beam current (after Makin[32])

irradiation was carried out in a bent {110} foil containing a {111} contour, for 15 min at 600 keV with a 1 μA (5 μm spot) beam. Clearly there has been heavy damage at the exact $\langle 111 \rangle$ orientation (at the centre of the contour) and at the Bragg positions at the edges of the contour; but no defects have formed in the good transmission regions outside the contour. It is worth pointing out the difference from positive particle effects. In electron channeling the damage rate is a maximum in the exact plane direction and a minimum over two ranges of angles lying on either side of the low index plane direction. Qualitatively these results are exactly as predicted by the two-beam dynamical theory of diffraction,[33] and since we are considering planar channeling the two-beam theory is not a bad approximation. In the two-beam theory the wave vector can be divided into two Bloch wave vectors, the coefficients of which are proportional to $\sin(\pi gr)$ and $\cos(\pi gr)$ for type 1 and 2 waves respectively; g is the reciprocal lattice vector and r is a position vector perpendicular to the beam direction. When the spacing of reflecting planes is d then $g = 1/d$ and hence when $r = 0$ and d, then $\sin(\pi gr) = 0$ and $\cos(\pi gr) = 1$. Hence wave 1 peaks halfway between the lattice planes and wave 2 peaks at the planes, and because of this wave 2 will be more heavily scattered than wave 1. The relative amplitudes of the Bloch waves depend on the sign of the deviation of the planes from the Bragg angle. The amplitudes of the two waves are proportional to $\cos(\beta/2)$ and $\sin(\beta/2)$ respectively, where $\cos \beta = S\xi g/(1 + S^2\xi^2 g)^{\frac{1}{2}}$. ξ is the extinction distance and S is the deviation parameter and is positive when the angle between the beam and the reflecting planes exceeds the Bragg angle and negative when less. Wave 1 is therefore the good channeling wave so that at angles greater than the Bragg angle there is good transmission and a low damage rate, whereas at angles less than the Bragg angle where wave 2 dominates, and is heavily scattered, we find poor transmission and a high damage rate. It is not yet known whether the simple two-beam theory quantitatively accounts for the observed effects, and it is not understood as to what extent the observations can be explained by a classical theory of electron channeling.

References

1. Rol, P. K., Flint, J. M., Vichbock, F. P. and De Jong, M., *Proc. IVth Int. Conf.*, *Ion Phenomena in Gases, Uppsala*, p. 257, Amsterdam, North Holland (1959).
2. Robinson, M. T. and Oen, O. S., *Appl. Phys. Lett.*, **2**, 285 (1963).
3. Almèn, O. and Bruce, G., *Nucl. Instrum. and Meth.*, **11**, 257 (1961).
4. Almèn, O. and Bruce, G., *Nucl. Instrum. and Meth.*, **11**, 279 (1961).
5. Molchanov, V. A., Telkovski, V. G. and Chicherov, V. M., *Sov. Phys. Dokl*, **6**, 222 (1961).
6. Fluit, J. M., Rol, P. K. and Kistemaker, J. *J. appl. Phys.*, **34**, 690 (1963).
7. Southern, A. L., Willis, N. R. and Robinson, M. T., *J. appl. Phys.*, **34**, 153 (1963).
8. Odintsov, D. D., *Sov. Phys-Solid St.*, **5**, 813 (1963).
9. Robinson, M. T. and Oen, O. S., *Phys. Rev.*, **132**, 2385 (1963).
10. Onderdelinden, D., *Can. J. Phys.*, **46**, 729 (1968).
11. Sigler, J. A. and Kuhlman-Wilsdorf, D., *phys. stat. sol.*, **21**, 545 (1967).
12. Hudson, J. A., *Thesis*, University of Cambridge (1969).
13. Wilson, M. M., *Thesis*, University of Oxford (1970).
14. Beevers, C. J. and Nelson, R. S., *Phil. Mag.*, **8**, 1189 (1963).
15. Parsons, J. R., *Phil. Mag.*, **12**, 1159 (1965).
16. Mazey, D. J., Nelson, R. S. and Barnes, R. S., *Phil. Mag.*, **17**, 1145 (1968).
17. Mazey, D. J. and Nelson, R. S., *Rad. Effects*, **1**, 229 (1969).
18. Corbett, J. W., 'Electron Radiation Damage in Semiconductors and Metals', *Solid State Physics*, Suppl. 7, New York, Academic Press (1966).
19. Nelson, R. S. and Mazey, D. J., *Can. J. Phys.*, **46**, 689 (1968).
20. Lindhard, J., *Dan. Vid. Selsk. Mat. Fys. Medd.*, **34**, No. 14 (1965).
21. Barrett, J. H., *Phys. Rev.*, B, **3**, 1527 (1971).
22. Delsarte, G., Jousset, J. C., Mong, J. and Quéré, Y., *Proc. Conf. on Atomic Collision Phenomena in Solids, Sussex*, 1969, Amsterdam, North Holland.
23. Morgan, D. V. and Van Vliet, D., *Proc. Conf. on Atomic Collision Phenomena in Solids, Sussex, 1969*, Amsterdam, North Holland.
24. Noggle, T. S. and Oen, O. S., *Phys. Rev. Lett.*, **16**, 395 (1966).
25. Wilson, M. M., *Phil. Mag.*, **24**, 1023 (1971).
26. Hudson, J. A. and Ralph, B., *Phil. Mag.* **25**, 265 (1973).
27. Nelson, R. S. and Mazey, D. J., *J. Mater. Sci.*, **2**, 211 (1967).
28. Merkle, K. L., *Proc. Conf. on Atomic Collision Phenomena in Solids, Sussex, 1969*, Amsterdam, North Holland.
29. Schober, T. and Balluffii, R. W., *phys. stat. sol.*, **27**, 195 (1968).
30. Hermanne, N. and Art, A., *V Yugoslav Symposium on the Physics of Ionized Gases*, Herieg-Novi, Yugoslavia, July 1970. Ljubljana, Inst. 'J. Stefan'.

31. Mazey, D. J., Nelson, R. S. and Thackery, P. A., *J. Mater. Sci.*, **3**, 26 (1967).
32. Makin, M. J., *Proc. Conf. on Atomic Collision Phenomena in Solids, Sussex, 1969*, Amsterdam, North Holland.
33. Hirsch, P. B., Howie, A., Nicholson, R. B., Pashley, D. W. and Whelan, M. J., *Electron Microscopy of thin crystals*, London, Butterworths (1965).

CHAPTER X
Channeling of β-Particles*

L. T. Chadderton

'*Now behold how nature, favouring our needs and wishes, presents us with two striking conditions no less than different than motion and rest*'

Galileo Galilei

10.1 SOME IMPORTANT CONCEPTS

Even a casual reader of the channeling literature will be aware of the controversy regarding wave/particle duality in channeling phenomena. One reason for the debate is the extremely basic nature of the physics. We are not here concerned with small quantal variations on a classical envelope, variations which might be smeared out by secondary processes (as, for example, with heavy particles[1,2]) but whether one analysis is merely sufficient or whether both are necessary and sufficient. Therefore it is important that certain concepts, some of which may appear trivial at first are made clear from the start.

To begin with it should be emphasized again that the star patterns seen in transmission through single crystals with heavy particles are the complete analogue of Kikuchi patterns.[3] This statement seems to have been accepted in a superficial way but is not fully understood in depth. For every feature of a star pattern there is an analogous feature in a Kikuchi pattern,[4] and in each case that feature is a 'signature' of the physics of the scattering processes taking place in the crystal. As an example we cite intensity changes seen with varying crystal thickness or angle of beam incidence. In both cases these are caused by a Pendellösung behaviour, associated either with orbital wavelengths λ (classical mechanics) or extinction distances ξ_g (quantum mechanics).

However, such patterns are normally observed with thicker crystals, in which there is considerable inelastic scattering, and it is not yet generally realized that there are also perfectly analogous features to be seen with thin crystals both in quasi-channeling patterns[5-7] and spot-diffraction patterns— patterns which show more clearly the striking differences between a classical

---

* (Comments on the scattering of charged particles by single crystals III.)

and quantal behaviour. This prompts us to deplore the fact that, because thick crystal patterns are dominated by the consequences of inelastic scattering processes, those same processes are frequently appealed to in the literature as having a major influence, owing to the smearing effect they produce, on the quantum/classical transition. The truth is that the transition has deeper and more fundamental roots.

Secondly, therefore, and without wishing to become involved in any deep epistemological discussion of quantum mechanics (whether the statistical interpretation[8] is appropriate etc.) it is worth re-emphasizing that the *general* criteria needed for quantum mechanics are well laid down. The necessity for a quantum description derives from experiment—the inability to describe scattering by conventional classical mechanical orbits. Therefore in quantum mechanics we find that events are not discrete and that continuous $|\psi|^2$ probability distributions are common. Conversely, when classical mechanics becomes sufficient, Ehrenfest's theorem:

$$\frac{d}{dt}\langle \mathbf{p}\rangle = -\int \psi^*(\nabla V)\psi \, d\tau = -\langle \nabla V\rangle = \langle \mathbf{F}\rangle \tag{10.1}$$

which is simply Newton's second law, valid for expectation values, applies.

Consider again[1] the simpler example of two-body repulsive scattering at *finite* distances, for the case of a Coulomb interaction. In the classical analysis[9] the intensity profile of the 'shadow' is characterized by an infinitely sharp edge at some critical angle. This singularity is a fundamental failure of the classical treatment. The quantum theoretical solution to the same problem involves the confluent hypergeometric series $F(-i\gamma|1|y)$, itself a regular solution of an equation of the Laplace type, so that:

$$|\psi|^2 = |Ae^{ikz}F\{-i\gamma|1|ik(r-z)\}|^2 \tag{10.2}$$

with $\gamma = Z_1Z_2e^2/\hbar v$ and A an adjustable constant. It does not matter if we introduce electronic screening in the problem. The net result is the same. The singularity is replaced again by an interference function which, similar to equation (10.2), yields a finite system of interference resonances damped out towards higher scattering angles. The point is that the singularity is brought down initially, not by multiple scattering, thermal vibrations, etc., all of which can be treated in quantum mechanics, but by the very quantum mechanics itself. Actually the singularity is a 'rainbow' phenomenon[10] and we have here just one example of a general property of wave mechanics whereby, because of the uncertainty principle, discontinuous classical limits are approached nonuniformly by Fresnel-type transients.

With increasing distance the quantum effects gradually disappear for heavy particles at low energies and for very large distances the asymptotic Rutherford value for the intensity function, corresponding to equation (10.2), is obtained.

It is interesting to note, moreover, that even at infinity, with increasing particle energy, diffraction features reappear in a similar form, owing to effects associated with the Coulomb barrier. Frahn[11,12] shows that the analytical expression in the strong absorption model for the ratio of the differential scattering cross section $\sigma(\theta)$ to the Rutherford cross section $\sigma_R(\theta)$ reduces to the simple formula:

$$\sigma(\theta)/\sigma_R(\theta) = \tfrac{1}{2}[\{\tfrac{1}{2} - C(w)\}^2 + \{\tfrac{1}{2} - S(w)\}^2] \qquad (10.3)$$

where $C(w)$ and $S(w)$ are Fresnel integrals of argument:

$$w = (\Lambda/\pi \sin \theta_c)^{\frac{1}{2}}(\theta - \theta_c) \qquad (10.4)$$

with

$$\theta_c = 2 \tan^{-1}(Z_1 Z_2 \alpha/\Lambda \beta) \qquad (10.5)$$

and Λ the cut-off angular momentum. That equation (10.3) should be of the same form as the intensity distribution of light diffracted by the edge of an opaque screen in the vicinity of the shadow cone is no accident. The Coulomb field distorts the incoming plane wave appreciably. In semiclassical* terms particles scattered through the grazing (critical) angle θ_c then appear to originate from a virtual point source situated at a finite distance from the scattering centre. Fresnel effects are predicted, and observed.[14]

Actually it should be evident that a proper understanding of two-body scattering is basic to an understanding of channeling itself. On the one hand it is clear that the steps from single (two-body) scattering to scattering by pairs of atoms and thence to scattering by an atomic string, are logical and direct. After all, the atomic continuum is the model on which the immensely successful classical theory of channeling[9] is formed. Conversely, when channeling occurs, particles are restricted to loosely defined critical angular limits in the forward direction and the transition from a channeling to a random condition (dechanneling) takes place by close encounter single collisions which give rise to Rutherford scattering. Under these circumstances the continuum potential concept is, of course, violated. Furthermore by virtue of the arguments leading up to equation (10.2) residual quantum effects are predicted to be present even for heavy particles (protons, deuterons) at angles of the order of two or three times the critical angle. Such effects are unlikely to be observed in heavy particle transmission studies since classically channeled and multiply scattered particles dominate a 'complete angular scan'.

Since our earlier paper,[1] in which we described an unsuccessful search for such quantum effects, using X-ray excitation, Grasso *et al.*[15] have made much more careful systematic measurements of angular yields in backscattering of 300 keV protons and deuterons from beams impinging along different axes of silicon. They find several oscillations in the random level lying in the range of a few times the critical angle. These fluctuations, which exhibit good reproducibility, cannot be attributed

---

* The transition from geometrical to physical optics is, of course, exactly paralleled by the transition from quantum to classical mechanics through the semiclassical WKB approximation.[13]

to the lattice geometry because their angular positions shift when the bombarding beam is changed from protons to deuterons of the same energy. The oscillation period evidently scales as $(m_{p+}/m_{d+})^{-\frac{1}{3}}$ in agreement with a quantum theoretical calculation using a screened potential.

With β-particles, however, we shall see that even in transmission and even at high energies the quantum resonances associated with diffraction and the three-dimensionality of the crystal are dominant, and are plainly observable under proper experimental conditions.

But what are proper experimental conditions? Typical arrangements using accelerators are depicted schematically in Figure 10.1. The crystal can be tilted and the transmitted (Figure 10.1(a)) or scattered (Figure 10.1(b))

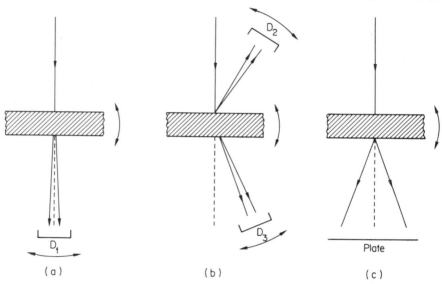

Figure 10.1. Schematic diagram of typical β-particle 'channeling' experiments in which either the crystal or detectors (D) may be tilted. The transmitted (a) and scattered (b) β-particle distributions can be examined, or a net intensity profile obtained on a screen or a photographic plate (c)

particle distributions can be observed, or the crystal can be held fixed and the detector moved*. If the angular acceptance is poor, however, (with or without energy analysis) it cannot be expected that such experiments, with either light or heavy particles, necessarily discriminate between classical and quantal behaviour. Alternatively, if an arrangement of the kind used in conventional or high voltage electron microscopes (HVEM, in the diffraction mode), with a photographic plate as detector (Figure 10.1(c)), is employed,

---

* See Fujimoto *et al.*[16] for a discussion of reciprocity (waves), the analogue of reversibility (particles) (cf the Stokes' law of optics).

then while a good angular representation of the transmitted pattern is obtained the intensity profile can only be regarded as qualitative. Clearly the results of such experiments must be interpreted with extreme care, bearing in mind the efficiency *and* adequacy of the arrangement. As we have seen for heavy particles, apparent arrival at a valid classical mechanical analysis over a certain range of angles does not exclude quantal features elsewhere. Are we measuring intensity profiles or angular widths?

A somewhat related factor is the angular divergence of the incident beam. In a heavy particle channeling experiment it is obvious that poor collimation ($\Delta\theta \gtrsim \psi_c$) will render the observations useless, on account of the increased fraction of incident particles deposited immediately into the 'random' beam. Because of this it has often been assumed that the same applies to β-particles. In fact deliberate convergent and divergent beam techniques are frequently employed as analytical tools for probing the solid state (e.g. Kikuchi and Kossel–Mollenstedt patterns) and the features observed are clearly quantal in nature, despite poor collimation. Sometimes large angular divergences are also called upon in arguments using the Born–Heisenberg uncertainty principle, the implication being that it is therefore possible to locate the β-particle in a channel and thence use orbital theories. The uncertainty principle, of course, is much more fundamental than that; and if invoked is a *de facto* requirement for quantum mechanics in the basic scattering process, exclusive of experimental primary beam collimation.

Finally we emphasize the nonmagic nature of the 1 MeV energy level. As we have seen[2] transmission experiments with electrons are easily performed up to just such energies in the new generation HVEMs and quantum features *dominate* the patterns. They are not merely present as 'wiggles' on a classical envelope. At above 1 MeV, however, the approach to classical behaviour accelerates. We therefore choose to concentrate on energies at and above 1 MeV for convenience, though it should be perfectly clear that the quantum/classical transition is both gradual and progressive. The choice of this energy is therefore arbitrary.

10.2 THE STRING

The step from scattering by an atomic pair to scattering by a string of atoms is an obvious and natural one, since it implicitly deals with effects seen when crystals are oriented with major axes parallel to the incident beam. Scattering by a string alone, however, does not even contain all of the classical mechanical physics of channeling, since restriction to forward angles and (heavy positive ion) Pendellösung effects are due to subsequent collisions with neighbouring strings. Furthermore in the string, or continuum potential model, it is normal *to neglect completely* effects associated with transverse periodicity.

10.2.1 Scattering in the Transverse Plane

In the classical theory of heavy particle channeling[9] it is convenient and useful to separate the motion of the particle into longitudinal and transverse components. In a scattering event with a single string (or plane) the total transverse energy is conserved, so that in the transverse plane the collision is thematically similar to a two-body collision, with no immediate necessity to consider deflections (or critical angles) associated with the longitudinal motion. Thus a positron sees a repulsive scattering potential identical with that seen by a proton, and an electron sees an inverted attractive potential centred on atomic sites. Out of such considerations, coupled with large multiple scattering at rapidly rising repulsive potentials in the immediate vicinity of nuclear sites operating on both particles, arise ideas that electrons should be 'less classical' than positrons. Similarly, as we shall see, while estimates of the number of quantum levels in the appropriate potential trough are informative, there are other quantum effects owing to both longitudinal *and* transverse periodicity which cannot *per definitionem* be contained in the continuum string model. For example, positron experiments of the kind shown in Figure 10.1(b) yielded results which were at first interpreted only in terms of a classical intensity envelope, comprising unresolvable fine structure owing to bound (single channel) states, in much the same way that fine structure appears in a quantum mechanical treatment of simple harmonic oscillation. Nevertheless a consideration of β-particle–string scattering is a logical next step. Furthermore it should not seem odd that we begin by discussing low energy ion–atom scattering; the potential has repulsive and attractive components and there is a very close formal identity with electron–string scattering in the tranverse plane.

Ford and Wheeler[17] single out the Lennard–Jones 6–12 potential for ion–atom collisions:

$$V(r) = 4\varepsilon\{(\sigma/r)^{12} - (\sigma/r)^6\} \tag{10.6}$$

and define dimensionless quantities:

reduced separation:	$r^* = r/\sigma$
reduced impact parameter:	$b^* = b/\sigma$
reduced collision energy:	$(g^*) = \frac{1}{2}\mu g^2/\varepsilon$
reduced angular momentum:	g^*b^*

The reduced effective potential then becomes:

$$\phi_{\text{eff}}^* = (4/r^{*12}) - (4/r^{*6}) + (g^*b^*/r^*)^2. \tag{10.7}$$

A Ford–Wheeler topological illustration of classical scattering by this potential is shown in Figure 10.2, in the plane of reduced collision energy $(g^*)^2$ and reduced angular momentum (g^*b^*). For present purposes we

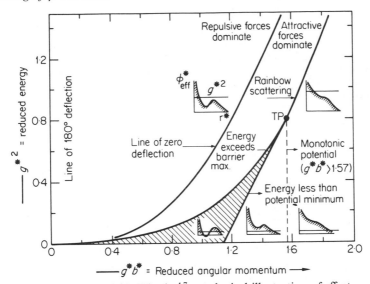

Figure 10.2. A Ford–Wheeler[17] topological illustration of effects which can occur during low-energy-ion–atom scattering, depending on energy and angular momentum. In the crosshatched area of the plane two kinds of classically permitted motion coexist, giving way to a triple point (TP) behaviour in the vicinity of the energy barrier maximum. Electron–string scattering in the transverse plane is very similar (after Ford and Wheeler, 1959)

draw attention to the simultaneous existence of rainbow and glory scattering, and orbiting (spiral scattering) within the plane. Consider a fixed reduced energy and the impact parameter varying from infinity to zero (moving from right to left along a horizontal line). The deflection is at first caused by a weak attraction, which becomes stronger, and then (depending on the energy) approaches infinity (orbiting) or reaches a minimum (rainbow). Thereafter the repulsive forces take over and the deflection angle moves through zero from negative to positive values. As the impact parameter approaches zero the deflection approaches π—the scattering always exhibits a forward glory. It should be especially noted that there is a permitted classical motion within the potential trough, accessible via quantal tunnelling through the small potential hill. Thus there is a small region of the topological $(g^*)^2/(g^*b^*)$ plane (cross hatched) for which *two* kinds of classically allowed motion exist. More importantly quantal scattering features, associated with singular behaviour of the classical deflection function, are predicted.

It is commonly supposed that a large number of waves always necessarily goes hand in hand with classical mechanics. Colliding low energy atoms and ions, for example, are sufficiently massive that the number of partial waves influenced in the scattering process is very great. This might lead one to assume that the scattering could be treated purely classically, and that no significant quantum features would remain to be observed. However, when special features such as rainbow scattering are present, the differential cross section approaches the classical cross section very slowly with increasing angular momentum quantum number l, and even for a low resolution experiment marked quantum effects remain when several hundred partial waves contribute significantly to the scattering.[17]

Within classical mechanics the electron–string transverse scattering phenomenon is extremely similar to that for ion–atom collisions and the relevant topological illustration closely corresponds to that in Figure 10.2. On the other hand there is now no repulsive term in the real potential, only in the effective potential $W(r, L)$ comprising the attractive continuum string $U(r)$ and the rotational energy $L^2/2mr^2$, which results from the angular momentum L of the electron parallel to the string:

$$W(r, L) = U(r) + L^2/2mr^2. \tag{10.8}$$

Accordingly values of tranverse energy E and angular momenta L exist for which electrons are bound to classical orbits around the string since they are confined to a potential trough (trapping zone) defined by equation (10.8) (Figure 10.3(a)).

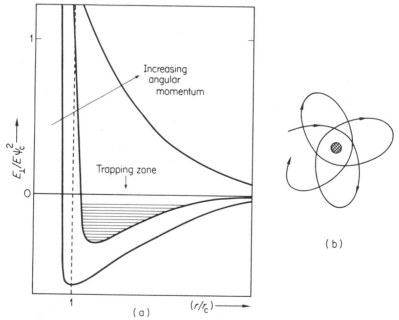

Figure 10.3. For median values of angular momenta electrons can be classically bound in their transverse motion to a trapping zone (a), so that ideally they execute the rosette orbits (b) of Kumm and colleagues[18]

10.2.2 Longitudinal Motion

When the forward classical motion of the electron is taken into account it is the interplay between the terms $U(r)$ and $L^2/2mr^2$ in equation (10.8) which determines the behaviour. This has been pointed out clearly by Kumm and colleagues.[18]

If the angular momentum L is small the potential $W(r, L)$ (equation (10.8)) becomes repulsive for distances smaller than the critical distance r_c, within which the continuum potential breaks down. In this case the electron sees the nuclear sites and classically suffers severe scattering. It joins the 'random' beam.

For somewhat higher angular momenta ($W(r, L)$ repulsive for $r > r_c$) a trapping of electrons with low $E_\perp$ in the potential valley takes place. The most simple forward motion of the particle is then seen as a helical orbit, the electron spiralling about the string. In the transverse plane, owing to deviations of the attractive force from a strict r^{-2} dependence there is a precession of the orbit, yielding 'rosette' figures (Figure 10.3(b)) which can be calculated from simple computational models.[18] The reader should compare such patterns with the Lissajous-like figures for heavy atom channeling predicted by Robinson and Oen.[19]

Finally, for very high L values the net potential $W(r, L)$ is everywhere repulsive, bound orbits cannot exist even for small $E_\perp$, and all electrons are scattered away from the string.

The case for the electron and the string is closely paralleled for the positron and attractive potential of an axial channel. The latter is trivially similar to proton channeling and deserves no elaboration.

10.2.3 Quantum String and Plane Effects

Certain quantum effects for string scattering in the general case, but with particular reference to heavy particles and the longitudinal periodicity of a string, as opposed to the continuum potential, have been discussed in detail by Lervig *et al.*[20] It was shown that at *no* energy are heavy particles expected to penetrate perceptibly into classically forbidden regions. For β-particles, on the other hand, we expect exactly the opposite.

In considering electron–string scattering in the transverse plane we noted above that glory, rainbow and spiral* scattering processes, along with interference phenomena, owing to the nonmonotonic dependence of the deflection function on impact parameter, were to be expected. When forward motion of the electron is taken into account it should not, of course, be

---

* For clarity it should be remarked that this spiral scattering is not the same as the spiral motion of classical rosette orbits, although it is closely phenomenologically related. The spiral scattering of quantum mechanics is associated with a local potential hill. There is here an unfortunate ambiguity of terminology.

supposed that these particular (two-body) quantal features would survive. On the other hand there is a category of quantum effect which is a *consequence* of the forward motion and *can* survive.

We note first that, quantum mechanically, it is possible for electrons to tunnel from one channel to the next. However, this naturally leads to a discussion of transverse periodicity which is excluded from consideration in a single string or a single channel model. Secondly, and in contrast, strong electron scattering from nuclear sites occurs for small and large angular momenta L. Classically this occurs when the continuum string potential breaks down $(E_\perp > E\psi_c^2)$ and there is no conservation of transverse energy $E_\perp$. Quantum mechanically the very appearance of longitudinal periodicity demands wave interference. For an initial wave moving at an angle δ to the axis the scattered waves are in phase for angles ϕ_n such that:

$$d(\cos \delta - \cos \phi_n) = n\lambda. \qquad (10.9)$$

Then, if good 'channeling' conditions prevail $(\delta = 0°)$, we have:

$$\cos \phi_n = \{1 - n\lambda/d\}. \qquad (10.10)$$

Concentric intense circles of increasing order n are therefore predicted by equation (10.10) and can be observed experimentally in a modified form in electron scattering from crystals.[6] In every sense these circles are directly equivalent to the axial blocking shoulders of classical mechanics, which they indeed become when the quantum/classical transition is complete although, at this stage, that change is perhaps better imagined for positrons rather than electrons. Additionally it is clear that interference phenomena of the kind described by equation (10.10) may not be observed if incoherent scattering is a factor to be reckoned with, as it is with increasing energy. Practically, however, the resonances are easily seen with electrons at 1 MeV.

There is also a zeroth $(n = 0)$ order constructive interference a spot—in the direction parallel to the axis. In kinematic diffraction, for a string of length t, it is trivial to show that the total angular width $2\psi_0$ (minimum to minimum) of this central spot is determined by:

$$\psi_0 = \sqrt{2\lambda}/t. \qquad (10.11)$$

Finally, the rosette figures $(E < 0)$ of spiral electron orbits are also the classical limit of an inherent quantum mechanical feature. With either electrons (about a string) or positrons (in a 'proper' channel), providing that $E < 0$ we have a particle in a potential well. In the semiclassical WKB approximation the one-dimensional quantization rule, which may be regarded as a condition for standing waves, follows from the action or phase integral:

$$J = \oint p \, dq = (n + \tfrac{1}{2})h. \qquad (10.12)$$

This equation, determining the number of quantum levels n, only differs from the Bohr–Sommerfeld quantization rule of the old quantum theory by the presence of half integral quantum numbers. Moreover the physical meaning of equation (10.12), that the area of phase space between one bound state and the next equals h, simply carries over to systems whose number of dimensions is η, if one takes h^η as the unit of extension in phase. Using the standard atomic potential[9] for a string:

$$U_1(r) = \frac{Z_1 Z_2 e^2}{d} \ln\left\{\left(\frac{Ca}{r}\right)^2 + 1\right\} \tag{10.13}$$

the quantization rule then yields as approximate condition[21] for the number of bound electron orbits about an axis:

$$n_a \simeq \left(\frac{4a_0}{d}\right)\gamma Z_2^{\frac{1}{3}} \tag{10.14}$$

where γ is the ratio of the relativistic to the rest electron mass and d the atomic spacing. Similar conditions for planes (spacing d_p) and positrons are shown in Table 10.1. Conventionally the classical limit of this particular quantal feature is to be associated with the limit $h \to 0$, whence we may write:

$$n_{ap}^{-+} \gg 1 \tag{10.15}$$

so that strings are 'more classical' than planes (Figure 10.4), positrons 'more classical' than electrons (Table 10.1).

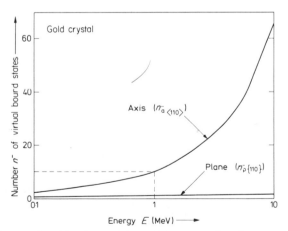

Figure 10.4. Numbers of bound states n^- for electrons at $\langle 110 \rangle$ axes and $\{110\}$ planes in gold, as a function of energy. Above about 1 MeV the axis becomes rapidly classical, whilst for planes the small number of bound states renders a classical analysis doubtful

Table 10.1 Semiclassical estimates of numbers of bound states for relativistic electrons and positrons: single string or single channel.

	Electron	Positron
Axis	$n_a^- \simeq (4a/d)\gamma Z_2^{\frac{1}{3}}$	$n_a^+ \simeq (1/\pi Nd^2 a_0)\gamma Z_2$
Plane	$n_p^- \simeq (0.4)\gamma^{\frac{1}{2}}$	$n_p^+ \simeq (Nd_p^3)^{\frac{1}{2}}\gamma^{\frac{1}{2}}Z_2^{\frac{1}{3}}$

Several authors have drawn attention to an early prediction of a classical channeling effect by Stark[22], a prediction lost in subsequent preoccupation with de Broglie matter waves. It is less well known that a similar process for electrons was invoked by Emslie[23] in an effort to explain the radii of intense circles seen in reflection Kikuchi patterns. According to Emslie 'the mechanism whereby an electron wave packet becomes localized may be as follows. The original plane wave electron loses a little energy to an atom and becomes an approximately spherical wave radiating from that atom. If it leaves the atom in a direction nearly parallel to the axis it may continue in that direction, *entrapped in a kind of a potential tube*.* It is scattered finally by the potential variations along the length of the tube.'

Shortly thereafter Shinohara[24] showed that details of the circles were best described in terms of secondary envelope phenomena. Emslie's 'channeling' was again lost in the main studies directed at unravelling quantal intensity features in low energy electron scattering from crystal surfaces.

10.3 THE CRYSTAL

Without reservation we affirm categorically that a better model than that of the 'string' is the 'crystal'. The ideal solutions should then emerge from asking what happens when a wave packet, representing the incident particle, falls at some arbitrary angle on a fully three-dimensionally periodic potential field, representing the crystal, inclusive of the influences of relativity, particle spin, inelastic scattering, etc. Such a problem, however, is extremely difficult. The usual solutions are obtained from much simpler models in which spin is neglected, relativity only involves an increase in mass, and the incident particle is represented simply by a plane wave. In one guise or another the formalism appropriate to diffraction of high energy electrons by single crystals then naturally appears. Furthermore the effects predicted by such an analysis show that quantal features related to the three dimensional periodicity in the crystal occur over a huge range of scattering angles, even for very high energy electrons. Only over a very limited angular range in the forward direction do classical aspects slowly appear.

Regrettably there are minor difficulties associated with confusing terminology, of which we shall mention two. The first of these relates to the Bormann effect.[25]

* My emphasis.

First observed for X-rays the Bormann effect for electrons* is associated with an anomalously large scattering of one of the two Bloch waves which, in the simplest case, may be assumed to propagate through a crystal.[27] If the first Bloch wave is highly scattered the second is, by comparison, well transmitted, so that in electron microscopy and diffraction many effects occur which are due to this 'anomalous transmission'. Since classical channeling is also an anomalous transmission phenomenon there is therefore a clear verisimilitude for the two processes. Moreover, although in two-beam electron diffraction the anomalous transmission is not in the forward direction as it is with channeling, it does switch in a very important way to the forward direction at higher energies as higher order Bloch waves are excited. Nevertheless the conventional two-beam Bormann effect should not as it stands be *equated* directly with channeling.

The second difficulty in terminology is closely related to the former, and perhaps illustrates the point better. Since at higher energies some of the Bloch waves are strongly 'absorbed' whilst others are readily transmitted, observable phenomena in the high voltage electron microscope, for example at bend and thickness contours on crystal images, are conveniently understood in terms of the periodic waves which either do or do not have nodes at lattice sites. On the basis of these concepts workers in the field of HVEM now frequently refer to 'Bloch wave channeling' without necessarily wishing to imply arrival at a classical particle behaviour customarily associated with mass independent critical angles ψ_c. The same can also be said of the pseudo-Kikuchi patterns seen in scanning electron microscopy, also sometimes called channeling patterns. We prefer to use the word channeling for a demonstrably classical particle behaviour, saving 'Bormann' and 'Kikuchi' for the appropriate wave mechanical effects.

Since necessarily we shall be referring to electron microscopy as well as electron diffraction the point should be made at the outset that patterns seen on crystals with the machine in the microscopy mode are in *no way* equivalent to star patterns seen for heavy particles. The former are observed at the exit face of the crystal and are normally associated with local curvature of the crystal (Figure 10.5). The latter are observed at infinity.

There is no space here to discuss fully all parts of a natural development of electron diffraction important to an understanding of the transition to classical mechanics. Where necessary, the reader should refer to the original articles, though we particularly wish to cite the book by Hirsch and colleagues,[28] and the work of De Wames, Hall and Lehmann.[29-32] Basically it is the relativisitic increase in mass[33] with its requirement for many Bloch waves, *followed* by the increased probability of inelastic scattering, which at

---

* See Tikhonova[26] for a derivation of the Bormann effect using a method comparable to the WKB method, and a discussion of the zeroth approximation in $\hbar$.

Figure 10.5. Extinction contours on a [011] aluminium single crystal as observed by 100 keV electron microscopy are due to local curvature of the sample. They are in no way equivalent to classical star patterns. Magnification × 11000 (Micrograph courtesy of E. Johnson)

extremely high energies sets the trend for classical mechanics, As always we are concerned with what may be observed—specifically in experiments of the kind shown in Figure 10.1(b) and (c). The order of the following is then most natural.

10.3.1 Thin Crystals

(1) *The Phase Grating Approximation*

The phase grating approximation (PGA) is an approximation first developed by Cowley and Moodie[34] which applies particularly to very thin crystals. In addition, however, it provides a thread (frequently concealed) which runs throughout all electron diffraction theory. For that reason and because we also want to know what the quantum/classical transition looks like in very thin crystals, we first discuss the PGA.

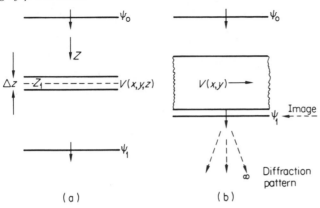

Figure 10.6. Basis for the phase grating approximation (PGA): (a) thin potential slab; (b) thin crystal, producing a diffraction pattern at infinity and an image at the exit surface

A plane electron wave, energy E, falls on a potential slab $V(xyz)$ of thickness Δz located at z (Figure 10.6(a)). For a de Broglie wavelength λ the phase change is

$$\Delta\phi = [\{1 + V(xyz_1)/E\}^{\frac{1}{2}} - 1]2\pi\Delta z/\lambda. \tag{10.16}$$

If $V(xyz_1) \ll E$:

$$\Delta\phi = \frac{\pi\Delta z}{\lambda} \frac{V(xyz_1)}{E}. \tag{10.17}$$

In this way the effect of the phase change on the wave function can be shown to be

$$\psi_1 = q(xyz_1)\psi_0 = \exp\{i\sigma V(xyz)\Delta z\}\psi_0 \tag{10.18}$$

where $\sigma = 2\pi m\lambda/h^2$.

In the limit where the number of such potential slices tends to infinity the 'transmission function' becomes:

$$q(xy) = \exp\{i\sigma V(xy)\} \tag{10.19}$$

where

$$V(xy) = \int_{-\infty}^{\infty} V(xyz_1)\,dz.$$

We make first contact here, therefore, with the basic continuum feature of classical channeling, namely a smearing of the potential over the z-coordinate.

For a weak potential ($\sigma V(xy) \ll 1$)

$$q(xy) = 1 + i\sigma V(xy) \tag{10.20}$$

and the intensity is given by

$$|\psi_1|^2 = q^2(xy)|\psi_0|^2. \tag{10.21}$$

Now we consider the simple case of systematic reflections only with a periodic planar lattice potential:

$$V = V_0 \cos(2\pi x/a) \tag{10.22}$$

The image observed at the exit face of the crystal—under microscopy conditions—then consists of a set of fringes with the periodicity of the lattice planes (Figure 10.6(b)):

$$|\psi_1|^2 = |\psi_0|^2\{1 + \sigma^2 V_0^2 \cos^2(2\pi x/a)\} \tag{10.23}$$

whilst the amplitude of the electron distribution observed at infinity—under diffraction conditions—is the Fourier transform $Q(u_1 v)$ of $q(xy)$:

$$Q(u, v) = \delta(u, v) + i\sigma E(u, v). \tag{10.24}$$

The intensity distribution of this diffraction pattern (Figure 10.6(b)) is

$$I(u, v) = |Q(u, v)|^2 = \delta(u, v) + |\sigma E(u, v)|^2 \tag{10.25}$$

the $\delta(u, v)$ being the sharp transmitted beam and the $|\sigma E(u, v)|^2$ being diffraction spots. The PGA therefore, although simple, contains two of the basic elements of electron (elastic scattering) diffraction, namely the projection approximation—z-smearing of the lattice potential—and a line of sharp clean resonances observable at infinity.

(2) *Kinematic Diffraction*

In kinematic theory the crystal is still thin but has measurable thickness and only one diffracted beam which is weak in relation to the transmitted beam is considered. Constructive interference occurs at the Bragg angle θ_B:

$$\lambda = 2d_{hkl} \sin \theta_B \tag{10.26}$$

Amplitudes scattered by atoms are added taking due account of phase (phase/amplitude diagrams). However, whilst the geometry of the diffraction process is predicted correctly (location of Bragg spots) the appropriate intensity terms are erroneous, on account of the large scattering of electron waves compared with X-rays, for which the theory is more successful. The Ewald construction in reciprocal space, consisting of a sphere of reflection, is useful for determining the direction of propagation of the diffracted wave and for visualizing the diffraction process.

Kinematic electron diffraction theory is most helpful for elastic scattering, but has a limited use in explaining some inelastic effects (e.g. Kikuchi lines) at low energies.

(3) Dynamical Diffraction

For somewhat thicker crystals (~ 2000 Å) the dynamical theory of diffraction is necessary in order to account, in the two-beam simplification, for an interplay between the incident and diffracted beams which varies with depth.[28] At the Bragg reflection position (equation (10.26)) standing waves of the form $\exp(2\pi i\mathbf{k}\cdot\mathbf{r}) + \exp(2\pi i(\mathbf{k} + \mathbf{g})\cdot\mathbf{r})$ are set up, where $\mathbf{g}$ is the reciprocal lattice vector, so that the electron current flows in sheets parallel to the atomic planes.

The total electron wave function obeys Schrödinger's equation:

$$\nabla^2\psi + (8\pi^2 me/h^2)(E + V(\mathbf{r}))\psi = 0 \qquad (10.27)$$

with the periodic planar lattice potential given by

$$V(\mathbf{r}) = \sum_g V_g \exp(2\pi i\mathbf{g}\cdot\mathbf{r}) = \left(\frac{h^2}{2me}\right)\sum_g U_g \exp(2\pi i\mathbf{g}\cdot\mathbf{r}). \qquad (10.28)$$

The more general (*n*-beam) solution of equation (10.27) which describes the electron wavefunction inside the crystal then consists of a linear combination of *n* Bloch waves:

$$\psi(\mathbf{r}) = \sum_{j=1}^{n} \psi^{(j)}b^{(j)}(\mathbf{k}^{(j)}, \mathbf{r}) \qquad (10.29)$$

labelled $b^{(j)}(\mathbf{k}^{(j)}\mathbf{r})$, whose wave vectors $\mathbf{k}^{(j)}$ differ only in their component normal to the crystal surface. The excitation amplitudes ψ^j of the Bloch waves are determined by wavefunction matching at the entry surface of the crystal and by normalization

$$\left(\sum_g^n |C_g^{(j)}(\mathbf{k}^{(j)})|^2 = 1\right)$$

so that

$$\psi^{(j)} = C_0^{(j)}(\mathbf{k}^{(j)}). \qquad (10.30)$$

In spite of much contrary comment the two-beam dynamical theory is immensely successful, not only in that it contains the same geometrical features as the PGA and kinematic diffraction, but in its accurate prediction of intensities for perfect and defective crystals in the 100 keV electron microscope. At infinity the intensity I_g of the diffracted beam relative to the incident beam I_0 is

$$\frac{I_g}{I_0} \simeq \frac{\sin^2\{t(1 + y^2)^{\frac{1}{2}}/\xi_g\}}{(1 + y^2)} \qquad (10.31)$$

where t is the crystal thickness, $\xi_g (= \hbar v/2|V_g|)$ the extinction distance and y a simple deviation parameter.[35] The dynamical theory of diffraction can handle inelastic scattering.

10.3.2 Thick Crystals

In fact it is growth of the *inelastically* scattered component of an incident electron beam which signifies that a crystal is becoming thick. In the beginning the sharp diffraction resonances described by the simple theories become broader and more diffuse as the crystal is made thicker. Beyond a certain point, however, corresponding to a fanning-out of the incident radiation owing to inelastic scattering processes, background Kikuchi effects begin to appear in experiments of the kind shown in Figure 10.1(c) and eventually dominate the transmission pattern (Figure 10.7a).

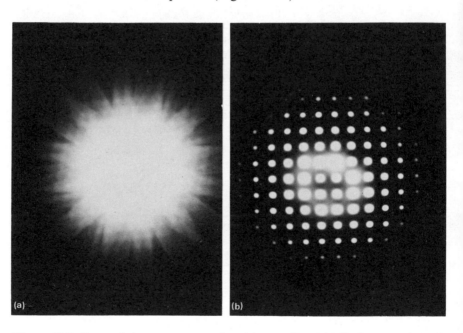

Figure 10.7. Transmission patterns produced by sending 1 MeV electrons through thick crystals are dominated by strong background Kikuchi phenomena ((a) $\langle 110 \rangle$ silicon). With thin crystals, if the incident beam divergence is adequate, Bragg spots are obtained ((b) $\langle 100 \rangle$ gold)

There are no great difficulties in calculating Kikuchi intensities by dynamical theory, but it *is* a long and tedious business. Excellent treatments have been given by Kainuma[36] and by Gjønnes.[37] The geometrical aspects of Kikuchi patterns, on the other hand, are readily understood. One simply imagines emission of a spherical electron wave from a lattice site on a truly three-dimensionally periodic array—the initial inelastic event—and asks what effect subsequent *elastic* scattering will have. The following, all characterized in reciprocal space by Bragg angles (equation (10.26)) are basic:

(i) *Kikuchi line pairs*. Parallel, bright and dark, caused by elastic scattering by planes inclined to the direction of the incident beam.

(ii) *Kikuchi bands*. Symmetrical, excess or deficient depending on thickness and angle, caused by planes parallel to the direction of the incident beam.

(iii) *Primary Kikuchi envelopes*. Parabolas caused by interference between intersecting Kikuchi lines of the same sign.

(iv) *Secondary Kikuchi envelopes*. Circles caused by interference between asymptotically intersecting parabolas.

A more detailed discussion of Kikuchi phenomena can be found in a previous review.[6]

Two aspects of Kikuchi patterns deserve special emphasis. First we remark that the concentric circles which are the secondary Kikuchi envelopes have radii determined by equation (10.10). This shows a clear link with the string effects discussed in Section (10.2.3). Close examination of the circles, however, particularly the inner ones, shows them to be polygonally constructed[24] and to derive from sets of lattice *planes*. Therefore, although the sizes and positions of the circles are determined by equation (10.10), clearly the intensities can only be calculated using dynamic theory with the full lattice periodicity.* Secondly there is a region *within* the first circle which is generally not considered in discussions of Kikuchi patterns. At low energies ($\sim 100\,\text{keV}$) this consists of extremely complicated interferences, comprising multiply intersecting Kikuchi bands and centred on the intersection of the major axis with the photographic plate (Figure 10.1(c), see also Figures 13 and 14, reference 6). We refer to this as the Kikuchi zone.

That all the effects we have described are governed at low energies by Bragg angles indicates conclusively that, in spite of inelastic scattering, coherence of the electron wavefunctions is maintained. The principal consequence of the scattering process with depth is a progressive divergence of the beam, so that rocking curves are produced. At these small electronic masses ($\gamma \simeq 2$) energy losses (to plasmons, interband transitions etc.), are extremely small.

Extension of the above arguments to positrons involves a trivial potential inversion.

10.3.3 High Energy Diffraction by Real Crystals

As the electron energy is increased and relativistic influences become important the nature of the scattering processes changes. The radius of the Ewald sphere increases owing to the decreasing de Broglie wavelength and,

---

* The circles also delineate the different ordered Laue zones[2] of reinforcement.

for a given elastic reflection, the electron scattering factor increases because of the relativistic increase of mass. Therefore, when the crystal is oriented for strong diffraction from one set of planes, raising the electron energy increases the simultaneous excitation of high order systematic reflections, and many-beam interactions between the reflections become important. At a given energy these n-beam interactions are stronger when the atomic number Z_2 of the crystal is high, or when the planar spacing is large—effects which maybe visualized in terms of the distance of reciprocal lattice points from the flattened Ewald sphere.

These changes are of extreme importance for high voltage electron microscopy,[38-41] particularly when absorption (inelastic scattering) is taken into account. The even-numbered Bloch waves of equation (10.29) generally have absorption minima at the odd-numbered systematic reflections (1, 3, etc.) and absorption maxima for the even numbered (0, 2, etc.). The opposite is true for the odd-numbered waves. Normally at 100 keV and with good so-called 'channeling' conditions (the symmetry position) Bloch wave 2 is strongly excited (and absorbed) so that there is poor transmission, whilst at 1 MeV (3-beam interactions) we have a strong Bloch wave 3, and good transmission. For crystals of different atomic number and for a given set of planes the 'critical energy', at which the excitation amplitudes of waves 2 and 3 are interchanged,[38] varies. With uranium, for example, it can be as low as 100 keV. Above the critical energy image patterns of the kind shown in Figure 10.5 can then be quite different, and bend contours corresponding to low index planes develop a central peak of high transmission. Such image changes are accompanied by a profound alteration of the structure of Kikuchi bands.[39]

It turns out, too, that high energy electron irradiation damage in single crystals is well explained on the same basis. From equation (10.15) the fractional intensity excited in each Bloch wave at the entry surface of the crystal is:

$$I^{(j)} = |C_0^{(j)}(\mathbf{k}^{(j)})|^2. \tag{10.32}$$

A phenomenological representation of absorption makes the lattice potential complex, which amounts to letting $\mathbf{k}^{(j)} = \mathbf{k}^{(j)} + i\mathbf{q}^{(j)}$. The intensity of wave j after progressing to a depth z is then attenuated by a term $\exp(-4\pi q^{(j)}z)$, whence the total Bloch wave intensity at the exit surface of crystal of thickness is:

$$I(t) = I_0 \sum_{j=1}^{n} (C_0^{(j)})^2 \exp(-4\pi q^{(j)}z). \tag{10.33}$$

Bloch wave radiation damage coefficients $\bar{v}^{(j)}$ are simply proportional to their respective absorption coefficients:

$$\bar{v}^{(j)} = mg^{(j)}. \tag{10.34}$$

Thomas[42] obtains excellent correlation between this approach and experiment.

As the de Broglie wavelength becomes small and many waves are excited the Bragg angles also decrease, with the result that in reciprocal space *all* of the angular features in diffraction patterns shrink inwards towards the central transmitted beam. At the same time, as γm_0 grows, the proportion of the incident beam which is inelastically scattered appreciably increases. Strong Kikuchi patterns are also naturally generated, therefore, through increasing γ as well as through increasing thickness t.

What happens to the inelastically scattered component should not be allowed to obscure the fact that elastic scattering can still occur and be seen for thin samples. In the 1, 3, and, presumably, in the forthcoming 10 MeV electron microscope, diffraction conditions correspond to the impact of an extremely highly collimated monoenergetic electron beam onto a crystal which is frequently oriented along a major axis or plane. This is a good channeling experiment of the kind shown in Figure 10.1(c). Widths of Bragg spots in the elastically scattered component are determined *only* by the experimental beam divergence $\Delta\theta_i$ and spots can be seen easily at high energies (see Figure 10.7(b)). In this case elastic and inelastic electron components should not be directly equated with the channeled and random components respectively for heavy particles. For the moment the distinction between the two former is drawn sharply, whilst in classical mechanics and for thin crystals (or small depths) it is recalled that a channeled distribution runs somewhat smoothly into a random one via the quasi-channeled component.

10.3.4 Multiple Beams

The problem of understanding the quantum–classical transition therefore boils down to understanding multiple beam diffraction. More than this, it also boils down to understanding multiple beam diffraction for inelastic *and* elastic scattering, as we shall illustrate.

The two kinds of classical channeling experiment which will concern us are those in Figures 10.1(b) and (c) and of these the kind in Figure 10.1(b) is by far the most frequently encountered arrangement. We have seen that an experiment in which the crystal is tilted and the yield in a detector observed at some high angle, is quite simply an experiment in which the particle probability density at the nuclear sites is monitored. Tilting the crystal, on the other hand, is perfectly identical with tilting the incident beam, as may be realized from arguments based on either reciprocity or reversibility. If instead of using the word tilting we use the word rocking, it can immediately be seen that the channeling experiment of Figure 10.1(a) is the same as examination of intensity profiles in Kikuchi patterns alone and, depending in which direction we rock, either axial or planar effects, or both, are being explored. In saying that we are concerned with inelastic scattering, therefore, we do not wish at this stage to imply that we are interested in energy loss— only that primary inelastic scattering events are responsible for giving the angular spread we require. It follows that, as relativity produces higher electronic masses, changes in the Kikuchi pattern are the key to any transfer to Newtonian mechanics in channeling experiments of the type shown in Figure 10.1(b). By identical arguments it will be obvious that experiments of the type shown in Figure 10.1(c) produce, for electrons, an amalgam of Kikuchi/channeling experiments (Figure 10.1(b)) and, providing that crystal thickness and incident angular and energy spreads permit it, elastic processes giving rise to Bragg-type peaks.

There are very good reasons for the current paucity of analytical multiple beam diffraction theories. Firstly, the complexity and size of the matrices involved are such that numerical solutions are conveniently obtained by computation. Secondly both channeling and high voltage electron microscopy have only recently emerged as important experimental tools. One formulation of the problem is due to Kagan and Kononets,[43] hereafter referred to as KK.

For the 'particle'/crystal combination the Hamiltonian describing the system is

$$\hat{H} = \hat{H}_1 + \hat{H}_2 \tag{10.35}$$

in which $\hat{H}_1$ refers to the particle and $\hat{H}_2$ to the crystal. The total density matrix $\hat{\rho}$ for the entire system has the usual form (in units of $\hbar = 1$):

$$i\frac{\partial \hat{\rho}}{\partial t} = [\hat{H}, \hat{\rho}] \tag{10.36}$$

with $\hat{\rho}$ represented by the product

$$\hat{\rho} = \hat{\rho}_1 \hat{\rho}_2 \tag{10.37}$$

where $\hat{\rho}_1$ is the density matrix in the function space of the particle and $\hat{\rho}_2$ is the density matrix of the crystal. The problem is then tackled by seeking solutions for the density matrix in the systematic reflection (planar chaneling?) condition, with the potential periodic in one dimension only. Not surprisingly this simplification reduces the complexity of the problem enormously, but in principle axial channeling can be dealt with by judicious use of symmetry properties and of the theory of groups.[30] In considering boundary conditions the longitudinal motion of the particle is assumed to be classical so that instead of the variable of depth z in the crystal the variable of time t is used; interaction between particle and crystal being 'switched-on' at $t = 0$. Values for the particle density $n(x, t)$, where x is the transverse coordinate, can then be calculated.

The KK treatment is important for the following reasons:

(i) It specifically deals with the classical mechanical (proton) limit for planar channeling in silicon and germanium. The results can therefore be considered to be appropriate for relativistic positrons in the extreme high energy.

(ii) The critical angle emerges naturally as a width in a complete angular scan across the planes. The arbitrary nature of the critical angle, however, is clarified, corresponding to the condition which we consistently refer to as quasi-channeling. Moreover, even for incidence parallel to the planes the particle density at the lattice sites (classically zero), while relatively small, remains finite.

(iii) Residual quantal effects on the shoulders of the distribution are also clearly predicted (see Section (i), above).

(iv) The total scattered angular distribution (cf Figure 10.1(b)) can be associated with that limit in which the parallel different order resonances of a Kikuchi band from a single set of planes become a star pattern streak owing to planar blocking.

(v) Arrival at this limit is properly related to transverse *energy bands* (numbering ~ 100) whose population density is a function of angular deviation.

(vi) The total forward scattered distribution (cf Figure 10.1(c)) can be associated with (a) blocked (above-the-barrier) plus (b) channeled (below-the-barrier) contributions, whose integrated population is a function of angular deviations.*

Whilst satisfying in terms of the end result the KK analysis does not treat the gradual progress from quantum to classical mechanics, and the density matrix to some extent hides the interesting fundamental changes which the Bloch waves undergo. Thus it should be emphasized yet again that the geometrical details and intensity variations of Kikuchi patterns are explained only through Bragg peaks; a fact which tends to be obscured if the word 'inelastic' is employed too frequently (and all too readily equated with 'incoherent'). Furthermore we are specifically concerned with the quantum–classical transition as it may be observed experimentally. For these and other reasons the high energy formalism of Berry,[45] although preceded by KK, turns out to be more instructive.†

We are under no illusions as to the difficulties in reviewing the Berry theory and in picking out the parts which are relevant here. The mathematics are lengthy and the notation differs from that which we have used to describe multiple-beam diffraction so far. Therefore we proceed in a simple stepwise fashion.

(1) *Basics of the Model*

A plane electron wave with vector $\mathbf{k}_0$ (length $k = 2\pi/\lambda$) falls on a crystal of thickness t (Figure 10.8(a)). Since spin is neglected the wavefunction is a

---

* This corresponds to experiments on silicon performed with normal[5] or super collimated[45] beams, in which total transmitted distributions are observed on a fluorescent screen.
† The reader will find in this single paper a unique contribution to our understanding of duality for this problem crystallized, perhaps, in the question: 'What happened to the waves?'

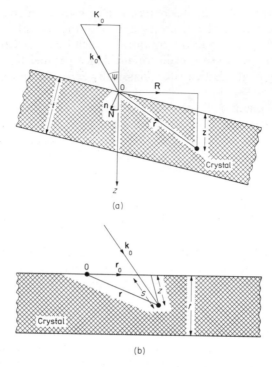

Figure 10.8. (a) Geometry and coordinate system for analysis of multiple beam β-particle diffraction (b) Corresponding geometry and coordinates for the limit of the phase grating approximation (after M. V. Berry,[45] 1971)

solution of the Schrödinger equation:

$$\{\nabla^2 + k^2 - U(\mathbf{r})\}\psi(\mathbf{r}) = 0 \tag{10.38}$$

for which there are the corresponding relativistic expression:

$$k = \frac{1}{\hbar}(2m_0 E + E^2/c^2)^{\frac{1}{2}} \tag{10.39}$$

$$U(\mathbf{r}) = 2(m_0 + E/c^2)\{V(\mathbf{r})/\hbar^2\}. \tag{10.40}$$

For a reduced wave function $\rho(\mathbf{r})$:

$$\psi(\mathbf{r}) = \exp(i\mathbf{k}_0 \cdot \mathbf{r})\rho(\mathbf{r}) \tag{10.41}$$

a general solution (Fresnel approximation) of the wave equation (10.38) can be obtained if the condition

$$(kt\theta^4/8) \ll 1 \tag{10.42}$$

is met, where θ is a typical maximum angle of diffraction. Under the still more restrictive condition

$$(kt\theta^2/2) \ll 1 \tag{10.43}$$

the solution for $\rho(\mathbf{r})$ (Figure 10.8(b)) becomes

$$\rho(\mathbf{r}) = \exp \left(-\frac{i}{2k} \int_0^s U(\mathbf{r}') \, ds' \right) \tag{10.44}$$

which *is* the phase grating approximation (equation (10.19)) valid for thin crystals, involving the projection of the potential along the direction of the incident beam.

From the more general expression corresponding to equation (10.42) it transpires that the region which contributes to the wave field in the crystal at $\mathbf{r}$ is a parabolic cone with apex at $\mathbf{r}$ and axis along the incident beam direction $\mathbf{k}_0$. However, for electrons, the wave propagation proceeds according to the phase grating approximation over regions near the apex of the cone which are larger than a lattice spacing, so that the z-smearing of potential implicit in (10.44) is perfectly adequate for the general real crystal case.

This contact of the phase grating approximation with high energy diffraction and with a continuum potential is extremely important. It provides a common thread between the classical and quantum theories.

For systematic reflections $\mathbf{G}$ the lattice potential

$$\bar{U}(\mathbf{R}) = \sum_{\mathbf{G}} U_{\mathbf{G}} \exp(i\mathbf{G} \cdot \mathbf{R}) \tag{10.45}$$

corresponding to equation (10.28), is

$$\bar{U}(x) = \sum_n u(x - na). \tag{10.46}$$

In other words we construct a repetitive continuum planar potential from

$$u(x) = \frac{2\pi a}{\Omega} \int_{|x|}^{\infty} r\phi(r) \, dr \tag{10.47}$$

where Ω is the volume of the unit cell and $\phi(r)$ a screened atomic potential:

$$\phi(r) = \frac{-Z_2 e^2}{r} \chi(r). \tag{10.48}$$

Defining again a new wavefunction $\tau(\mathbf{R}, z)$ (coordinate system of Figure 10.8(b)):

$$\rho(\mathbf{r}) = \exp(-i\mathbf{K}_0 \cdot \mathbf{R}) \tau(\mathbf{R}, z) \tag{10.49}$$

which can be decomposed into eigenfunctions as in equation (10.29):

$$\tau(\mathbf{R}, z) = \sum_j C_j \tau_j(\mathbf{R}) \exp(-is_j z/2k) \exp(iK_0^2 z/2k) \qquad (10.50)$$

we arrive at a normal two-dimensional Bloch wave problem. In the (simplified) real lattice space the Bloch waves $\tau_j(\mathbf{R})$, wave vector $\mathbf{K}_0$, are solutions of:

$$\{\nabla_{\mathbf{R}}^2 + s_j - \bar{U}(\mathbf{R})\}\tau_j(\mathbf{R}) = 0 \qquad (10.51)$$

for which eigenvalues s_j are to be discovered. In reciprocal space the amplitude diffracted into a reflection at $\mathbf{G}$, under condition (10.42), is

$$A_{\mathbf{G}}(z) = \frac{\exp(iK_0^2/2k)}{S} \sum_j \left\{ \left[\int_{\text{cell}} \tau_j^*(\mathbf{R}) \exp(i\mathbf{K}_0 . \mathbf{R}) \, d\mathbf{R} \right] \right.$$

$$\left. \times \left\{ \int_{\text{cell}} \tau_j(\mathbf{R}) \exp(-i(\mathbf{K}_0 + \mathbf{G}) . \mathbf{R}) \, d\mathbf{R} \right\} \exp\left(\frac{-is_j z}{2k} \right) \right. \qquad (10.52)$$

where S is the area of a cell face in the plane of $\mathbf{R}$.

We remark on the strong dependence of C_j, s_j and τ_j on the relativistic mass of the electron (equation (10.40)). This, then, is the basic reason for requiring an increasing number of Bloch waves.

(2) *Transverse Energy Bands*

For systematic reflections with either positive or negative β-particles (Figure 10.9) equation (10.51) becomes one-dimensional:

$$\left\{ \frac{d^2}{dx^2} + s - U(x) \right\} \tau(x, s) = 0 \qquad (10.53)$$

where $\tau_1(x, s)$ is an eigenfunction in the zeroth potential cell ($|x| < a/2$). The general solution is a linear combination of negative and positive travelling waves:

$$\tau_j(x) = A_j \tau_1(x, s_0) + B_j \tau_2(x, s_j) \qquad (10.54)$$

where

$$\tau_2(x, s) = \tau_1(-x, s). \qquad (10.55)$$

Assuming a plane wave incident on the zeroth potential hill and directed along x-positive there will be reflected and transmitted amplitudes $R(s)$ and $T(s)$ respectively:

$$T = |T| \exp(i\mu)$$
$$R = -i \exp(i\mu)|(1 - |T|^2)^{\frac{1}{2}} \qquad (10.56)$$

where $\mu(s)$ is the phase of T. The allowed energy bands are then obtainable by

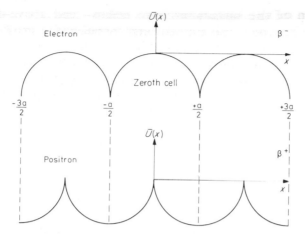

Figure 10.9. Transverse potentials for electrons and positrons in the case of systematic reflections. Strong nuclear scattering potentials at the cusps are not indicated. (after M. V. Berry,[43] 1971)

graphical solution of the dispersion relation:

$$|T| \cos K_0 a = \cos(\beta a + \mu) \tag{10.57}$$

where

$$\beta = \{s - \bar{U}(-a/2)\}^{\frac{1}{2}}.$$

Two extremes, because they illustrate the importance of the strength of the potential, are important. Firstly, for a high and strong potential, $|T|$ is zero. Then the allowed energy bands are of very narrow width with eigenvalues given by

$$\beta(s_j)a + \mu(s_j) = (j + \tfrac{1}{2})\pi. \tag{10.58}$$

Secondly, when the particles are so unaffected by the lattice planes that they may be considered free in their transverse motion, $|T|$ is unity. The allowed energy bands are broad and

$$\beta(s_j^{\pm})a + \mu(s_j^{\pm}) = (\pm K_0 a + 2\pi j) \tag{10.59}$$

predicts the central eigenvalues.

These arguments parallel those we have used previously.[1] The classical aspects of 'channeled' and 'random' trajectories depend on the quantal aspects of bound and free states.

(3) *Semiclassical* (WKB) *Illustration of the Convergence to Classical Mechanics*

The most striking common feature of the KK, Berry and our analysis[1] is the separation of the solutions into below- and above-the-barrier components. Of course for weak potentials and therefore for small numbers of Bloch waves (and reflections) no such partition can be made. But with huge relativistic increases in electronic mass it can be argued that the two components, corresponding to $|T|$ unity and $|T|$ zero, are joined together by only a very narrow range of energy values near the tops of the barriers ($s = 0$). Then the above-the-barrier solution is a travelling wave:

$$\tau_1(x, s) = \frac{\exp\{i\phi(0, x)\}}{\{s - \overline{U}(x)\}^{\frac{1}{4}}} \tag{10.60}$$

where

$$\phi(a, b) = \int_a^b \{s - \overline{U}(x)\}^{\frac{1}{2}}\, \mathrm{d}x \tag{10.61}$$

is the WKB classical phase integral (compare equation (10.12)) and correspondingly, if $x_0^{\pm}(s)$ are the classical turning points for negative s, the below-the-barrier solution is:

$$\tau_1(x, s) = \frac{\{\tfrac{3}{2}\phi(x, x_0^-)\}^{\frac{1}{6}}}{\{s - \overline{U}(x)\}^{\frac{1}{4}}}\, \mathrm{Ai}[-\{\tfrac{3}{2}\phi(x, x_0^-)\}^{\frac{2}{3}}] \tag{10.62}$$

where $\mathrm{Ai}(\omega)$ is the Airy function.

This solution should be compared with that given by Lervig *et al.*[20] (equation (3.a)), for semiclassical scattering by a single atom. The implications of the identity in physics of the two expressions, in view of our previous comments,[1,2] are then self evident.

From equation (10.60) and the asymptotic form of the normalized eigenfunctions τ_j, solutions of (10.61) can be constructed so that, if $\mathbf{K}_G = \mathbf{K}_0 + \mathbf{G}$, the one-dimensional (planar channeling) expression for the diffracted amplitude (equation (10.62)) becomes

$$A_G(z) = \frac{\exp(iK_0^2 z/2k)}{a} \sum_j \left\{ \int_{-a/2}^{+a/2} \tau_j^*(x) \exp(iK_0 x)\, \mathrm{d}x \right\}$$

$$\times \left\{ \int_{-a/2}^{+a/2} \tau_j(x) \exp(-iK_G x)\, \mathrm{d}x \right\} \exp(is_j z/2k). \tag{10.63}$$

The Bloch waves contributing to $A_G(z)$ in this approximation have eigenvalues s_j which fall within a band of width $|u(-\tfrac{a}{2})|$ with an upper edge at $s_j^{\max} = K^2$ (Figure 10.10). It then follows that for diffraction directions corresponding to vectors $\mathbf{G}$ being negative—pointing in the opposite direction to K_0—there is no support to $A_G(z)$ from the free (travelling wave) states. Quite logically the amplitudes of 'reflected' beams are then sensitively dependent on whether $|K_0|$ is greater or less than a well defined value:

$$|K_{0_c}| = \{|u(-\tfrac{a}{2})|\}^{\frac{1}{2}}. \tag{10.64}$$

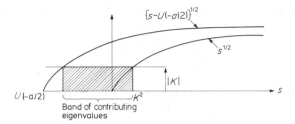

Figure 10.10. The dispersion relation
restricts the contributing eigenvalues in
multiple beam diffraction to a finite band
(crosshatched) as the graphical solution
illustrates. In practice the band of Bloch
waves contributing significantly to the
amplitudes of diffracted beams has diffuse
edges (after M. V. Berry,[45] 1971)

This defines a critical angle of incidence

$$\psi_c = \{|u(-\tfrac{a}{2})|\}^{\frac{1}{2}}/k \tag{10.65}$$

if $\psi > \psi_c$ Bragg 'reflection' $(\mathbf{K_G} = -\mathbf{K_0})$ does not occur.

This solution should be compared with the Lindhard critical angle ψ, derived from the single continuum potential model.[9] Bearing in mind our use of the reduced potential (equation (10.40)), we are evidently exploring the same physics. However, it should be remarked that ψ_c is always greater than ψ, since the latter approximates progressive angular breakdown of the continuum potential.

The maximum number of visible diffraction resonances in the directions for which $\mathbf{K_G}$ is parallel or antiparallel to $\mathbf{K_0}$, n^+ and n^- respectively, depends also on the angle of incidence ψ (Table 10.2). For normal incidence where channeling conditions are optimized $(\psi = 0)$ the *total* number of diffraction resonances is

$$N = (2n^+ + 1) = (a/\pi)(|u(-\tfrac{a}{2})|)^{\frac{1}{2}} + 1 \tag{10.66}$$

Table 10.2 Maximum numbers of diffraction resonances in β-particle scattering for planar potentials as a function of angle of incidence $\psi(= K_0/k_0)$

n^+	n^-	ψ
$\dfrac{a}{2\pi}\left[\left\{K_0^2 + \left|u\left(-\dfrac{a}{2}\right)\right|\right\}^{\frac{1}{2}} - K_0\right]$	$-\dfrac{a}{2\pi}\left[\left\{\left|u\left(-\dfrac{a}{2}\right)\right|\right\}^{\frac{1}{2}} + K_0\right]$	$< \psi_c$
$\dfrac{a}{2\pi}\left[\left\{K_0^2 + \left|u\left(-\dfrac{a}{2}\right)\right|\right\}^{\frac{1}{2}} - K_0\right]$	$-\dfrac{a}{2\pi}\left[K_0 - \left\{K_0^2 - \left|u\left(-\dfrac{a}{2}\right)\right|\right\}^{\frac{1}{2}}\right]$	$> \psi_c$

and the *total* number of Bloch waves contributing to $A_0(z)$ (equation (10.63)) is

$$N_B(0) = \frac{\int_{-a/2}^{+a/2} \{-u(x)\}^{\frac{1}{2}} \, dx}{2\pi} \qquad (10.67)$$

Equation (10.66) and (10.67) and those in Table 10.2, clearly bring out the importance of the strength $\{|u(-\frac{a}{2})|\}$ and range a of the planar potential.

It remains only to note that convergence towards classical mechanics must be associated with a large number of bound states, since specific channel localization of the particle, whether electron or positron, is required. The number of contributing Bloch waves must be very large, i.e. $N_B(0) \gg 1$.

This solution should be compared with equation (10.15). It is trivial to show that, for planes, the two are identical.

The fundamental general trend towards classicality is now clear. Separation of solutions of the wave equation into transversely free and bound components exactly mirrors a splitting into classical random and channeled trajectories which is imposed by the available range of impact parameters at the free surface of the crystal. In some respects the small angle elastic (Bragg spot) and high angle inelastic (Kikuchi) contributions to a low energy transmission pattern (Figure 10.1(c)) can now be understood by comparison with corresponding phenomena in the classical limit. Bragg spots arise from scattering processes for which there is no energy loss. Channeled (below-the-barrier) trajectories are dictated by conservation of energy in the transverse plane and ideally there are no nuclear energy losses. Conversely background Kikuchi bands arise from large angle scattering processes for which, initially, there are energy losses, however small, and independent of subsequent diffraction. Blocked (above-the-barrier) trajectories are subject to strong energy loss processes, independent of any subsequent rechanneling.

The parts played by the Bragg angle θ_B and the critical angle ψ_c are also put into sharper focus through the analysis leading up to equation (10.65). Thus in two-beam dynamical diffraction theory θ_B divides reciprocal space into angular bounds within which either Bloch wave 1 or 2 is most excited, an effect which is clearly evident in Kikuchi band profiles (Figure 10.11). Coupled with anomalous absorption and transmission this prescribed division gives rise to the Bormann effect. In classical mechanics ψ_c similarly separates angular regions of high and low transmission. The route from θ_B to ψ_c, however, is the route of many beams. The excitation of each Bloch wave must be accounted for,* followed by the probability that it be absorbed. The product of integrals in equation (10.63) is the first step on that route. Anomalous transmission and absorption of waves in the two-beam Bormann effect can then be compared with the minimal and maximal scattering processes to which the channeling and blocking orbits (positive particles) of classical mechanics are respectively subjected.

Finally we recall how, for protons, the above-the-barrier scattered component degenerated into an angular scan of the image of a single atom, characterized by Fresnel-type fringes beyond the critical angle (equation (10.2)). What, then, are we to expect for the below-the-barrier contribution to a transmission pattern? Berry notes that, if proper approximations are used for the eigenstates, valid for all eigenvalues s,

---

* For normal incidence *all* odd bound solutions are not excited—hence the factor of $\frac{1}{2}$ in equation (10.67).

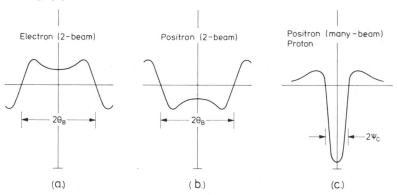

Figure 10.11. Schematic illustration of Kikuchi band profiles as they might be obtained with a complete angular scan in an experiment of the kind shown in Figure 1(b) with detector D_3 fixed and the crystal rotated. The electron profile is 'normal' for thin crystals or high angles. For thick crystals and small angles the profile converts to one similar to that shown for positrons owing to inelastic scattering. The many-beam positron or proton profile is the 'blocking' curve—the ultimate classical limit of a position Kikuchi band

then the edges of the band of contributing Bloch waves (Figure 10.10) are not sharp. In fact the diffracted beams outside the strictly proscribed range do not have zero strength, and the decay in amplitude is 'described by the same Fresnel integrals as those governing the transition from light to shadow behind a sharp edge'.

(4) *Arrival at Classical Mechanics*

The slow progression towards a classical effect, described and illustrated in the previous section, is patently fundamental in nature. The critical angle appears naturally as a limiting boundary for the many diffraction resonances excited within a narrowly defined forward direction. In practice, however, actual arrival at classical mechanics is as much determined by experimental conditions, the criterion for wave mechanics being observation by experiment, of features which require it.

Berry writes equation (10.63) in the simple asymptotic form:

$$A_G(z) = \sum_i a_i(K_G, z) \exp\{ib_i(K_G, z)\} \tag{10.68}$$

where the running index i determines the different contributions from topologically different classical paths,[45] a_i is a path 'density' and b_i is a classical action function. Then the intensity:

$$|A_G(z)|^2 = \sum_i \sum_j a_i a_j \exp\{i(b_i - b_j)\} \tag{10.69}$$

is obviously an interference expression. For very high β-particle energies, on the other hand, the phases b_i become very large. Under these conditions an inadequately collimated primary beam or a redistribution of directions and energies owing to inelastic scattering events in the crystal imposes a finite limitation to the depth z' over which coherence of the wave function can be assumed to prevail. In the language of matrix mathematics it is the strength of those elements for scattering into *transverse* energies which determines the magnitude of z'. In either event for $t > z'$, only an average value is required:

$$\overline{|A_G(t)|^2} = \sum_i a_i^2(K_G, t). \qquad (10.70)$$

Finally, if the angular detection efficiency of the experimental assembly (Figure 10.1(a)) is so poor that the individual resonances (10.70) are not distinguished, the intensity profile between $-\psi_c$ and $+\psi_c$ can be satisfactorily explained on the basis of Newtonian mechanics and orbits in planar channels.

10.3.5 Axes

Progress from dynamical β-particle diffraction to classical mechanics with increasing energy is leisurely. It is not characterized for a particular set of planes by a particular energy, as is the switch from Bloch wave 2 to Bloch wave 3, because criteria demanding that the numbers of Bloch waves or bound states be large must only be assessed in the context of all known experimental conditions (divergence, angular resolution, monochromaticity of incident beam, crystal thickness etc.). From the work of Berry and De Wames and Hall[31] on transmitted and emitted β-particle distributions respectively, however, the energy at which restriction of scattered waves either to or beyond a crudely defined critical angle occurs is obviously an important milestone.

Two problems remain. The first of these concerns inelastic scattering. In Section 10.3.4 all of the examined multiple beam effects concerned elastic scattering. The additional problem of dealing thoroughly with inelastic scattering is overwhelming—in addition to scattering vectors **G** we need also inelastic scattering vectors **S**, which means more waves, and which therefore brings classical mechanics closer still. Fortunately the geometrical effects of inelastic scattering are rather well visualized. For the case of systematic reflections the individual spots $A_G(z)$ blend indistinguishably with their neighbours so that equation (10.70) represents an intense streak set perpendicular to the intersection of the planes with the photographic plate and limited, more or less, by $\pm\psi_c$.

The second problem—the axis—is more fundamental. Again it is obvious that for axes there are more reflections than for planes and more Bloch waves are involved. By the same token we should accordingly be closer to classical mechanics, although this time the number of waves reflects the basically stronger potentials produced when planes intersect. Therefore, even if an analysis for axes like that for planes is prohibitively complex it is useful at least to examine the underlying physics in a simple way, emphasizing the geometry.

(1) *Simple Criteria*

The two-beam dynamical theory is remarkable for its success. Not only does it work perfectly well for low energy β-particles, but the Bloch wave behaviour and the anomalous phenomena associated with the Bormann effect are clear pointers to relativistic multi-beam effects. Whilst it might at first appear paradoxical we appeal, now, to two-beam theory for an understanding of the axis. Purely for simplicity we consider first positrons.

Equation (10.31) expresses quite simply the intensity of a single diffracted beam. By the arbitrary but reasonable definition of a diffraction 'resonance width' at $y = 1$ we can write (see reference 6):

$$(\theta_B - \theta') = \Delta\theta_B = \frac{|V_g|}{(\sin 2\theta_B)E} \tag{10.71}$$

for small angles, where θ' is the angle at which y is unity and V_g is the first Fourier coefficient of a planar potential (10.28). The advantage of expressing $\Delta\theta_B$ in this way is that it can then be compared with θ_B, using equation (10.26):

$$\alpha_g = \left|\frac{\Delta\theta_B}{\theta_B}\right| = \left|\frac{4\gamma V_g m_0}{\hbar^2 g^2}\right| \tag{10.72}$$

so that α_g takes on the significance of a correspondence parameter, similar to the κ used by Bohr[46] for screened atomic collisions. Of course, α_g, even though it contains γ, the potential, $\mathbf{g}$ and $\hbar$ in the correct way, is different from the classical phase space estimates of Table 10.1 and equation (10.67); but then we want it to be. For the moment we wish to retain the elements of lattice periodicity inherent in $\mathbf{g}$ and V_g. In fact a large α_g, which can be achieved by making γ large, implies an overlapping of the zeroth and that single diffracted beam so that, using the uncertainty principle, we are as justified in using the condition

$$\alpha_g \gg 1 \tag{10.73}$$

as any other for the onset of classicality. It is perhaps more interesting that the different α_g, α_{2g}, α_{3g} etc., for a line of systematic reflections in multi-beam diffraction, become smaller for higher scattering vectors $n\mathbf{g}$. This shows rather nicely how the classical behaviour spreads outwards, with increasing γ, from the centre of a transmission pattern, and fits well with what we should expect from our understanding of Bloch waves and bound and free states. Assisted by inelastic transitions Kikuchi bands slowly double and treble their widths (measured in Bragg angles), whilst the transmitted beam becomes indistinguishable from the first diffraction spot, which becomes indistinguishable from the second, and so on.

For the axis a similar transition takes place. The difference lies in the fact that, because several sets of planes are now involved, the transition appears to proceed somewhat faster. Fujimoto and colleagues[16] construct an axial criterion from the geometrical mean of α_g and $\alpha_{g'}$, which corresponds to two sets of intersecting planes of spacing d_g and $d_{g'}$, respectively. For a screened Coulomb potential:

$$V_g = Ze^2 N/\pi g^2 \qquad (10.74)$$

where N is the atomic density. Since

$$N(d_g d_{g'} d) = 1 \qquad (10.75)$$

and

$$(d_g d_{g'}) = (gg')^{-1} \qquad (10.76)$$

we can write

$$\alpha_a^+ = \sqrt{\alpha_g \alpha_{g'}} = (4/\pi N d^2 a_0)\gamma Z_2 \qquad (10.77)$$

This should be compared with the expression for n_a^+ (Table 10.1), estimated semiclassically.

(2) *Laue and Kikuchi Zones*

No electron microscopist who has also performed channeling experiments could fail to make a comparison of the intensity changes which occur at the zeroth order Laue zone, on tilting the crystal, with corresponding effects in, for example, transmitted proton star patterns. For an electron beam diffracted parallel to an axis the zeroth-order Laue zone is an intense disc of diffraction spots falling off into a dark background of Kikuchi lines and bands. On tilting the crystal the zone becomes a skewed annulus whose radius grows with angle of tilt and which later carries the intensity from the Kikuchi pattern with it. Apart from the spots exactly the same happens with a proton star pattern.

In kinematical theory the diffuse angular limits placed on the zeroth-order Laue zone (cf equation (10.11)) are purely those imposed by the crystal thickness. In diffraction effects at an axis our experience with planes tells us that the limitation at high energies will be more fundamental—owing to many-beam dynamical diffraction theory itself. In fact the critical angle for axes, related as it is to the axial potential hill, is the angle which replaces ψ_0 in equation (10.11), a many-beam extinction distance (classical channeling wavelength) playing the part of the thickness t.

By analogy with planes, therefore, the influence on pattern geometry at axes of increasing the energy of the β-particle becomes clear. As before, the entire diffraction pattern shrinks towards the central transmitted beam, Kikuchi effects associated with the axis become prominent (particularly secondary Kikuchi envelopes) and diffraction spots grow very diffuse. The zeroth-order Laue zone becomes our circular Kikuchi zone which, at sufficiently high energies, is restricted to an intense disc with a roughly defined angular radius ψ_c.

A Fresnel-like diffraction pattern at the edges of the disc is as natural a consequence of the slow transfer to classical mechanics as in the converse case, where Fresnel fringes were seen on the shoulders of a blocking shadow (équation (10.2)). This is because in both cases we are looking at the edge of the same shadow, where the continuum potential breaks down. On the one side we have the above-the-barrier and on the other below-the-barrier states, whilst the top of the barrier itself is necessarily approached in either direction by oscillatory Fresnel transients.

It is interesting to note that there is here that same physics which is present in Babinet's principle. The near-field vector amplitude produced at a point by an opaque disc ('blocking') is complementary to that produced by a circular hole of the same size in an opaque sheet ('channeling').

10.4 EXPERIMENT

In our opinion it is a mistake to attach too much significance in experiment to evaluations of the critical angle only. The lesson of residual quantal effects at blocking shoulders for protons and deuterons illustrates the fact that the critical angle is only indicative of certain aspects of classical mechanics. This is even more obvious when it comes to β-particles. Nevertheless experiments with positrons and electrons frequently concentrate on measurements of the critical angle alone. It will be one of our points that this does not necessarily mean that quantal influences are no longer detectable.

It should also be emphasized again that for β-particles we expect transmission patterns (Figure 10.1(c)) to be of the general type first described by Kikuchi[47] and, that for beam orientations parallel or nearly so with an axis, these will have all of the characteristics discussed in Section 10.3.2. With increasing energy the angular features will shrink towards the centre of a pattern, Bragg spots will blur and disappear, Kikuchi phenomena will become stronger and the central Kikuchi zone will first become an intense disc with bounds described by the axial critical angle. It is expected that the last visible vestige of quantum mechanics will be the disappearance of higher order resonances about low order (weak planar potential) Kikuchi bands, seen at very high scattering angles, so that ultimately only diffuse shoulders remain. This consumption of large angle quantal features by small angle classical phenomena is one indication of how relativistic effects become important for high energies.

Finally we record the fact that many experiments with β-particles tend to be of the kind shown in Figure 10.1(b). This is to be regretted since not only is attention focused strongly on the ubiquitous critical angle but important information in the channeled/diffracted (forward scattered) component is lost.

The following discussion is not intended to be either exhaustive or complete. Several interesting experiments with β-particles are described in the proceedings of recent conferences.[48-51]

10.4.1 Electrons

(1) Transmission

To begin with we show in Figure 10.12(a) the systematic reflections seen for orientation of a highly collimated 1 MeV beam of electrons (HVEM) parallel to the $\{4\bar{2}0\}$ planes in silicon. The discrete Bragg spots line up perpendicular to the intersection of the plane with the plate and the excess Kikuchi band in the background is clearly to be seen. For comparison Figure 10.12(b) shows the corresponding effect with 400 keV protons and $\{100\}$ planes in gold.[5] Whilst every feature in Figure 10.12(a) is patently quantal in nature, ruled by θ_B, everything in Figure 10.12(b) is classical. The streak lying perpendicular to the planes terminates abruptly at $\pm\psi_c$, and the crystal is too

(a)

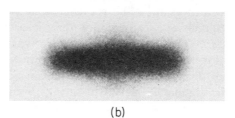

(b)

Figure 10.12. (a) Systematic reflections in transmission of 1 MeV electrons parallel to $\{420\}$ planes in silicon. Discrete Bragg spots line up perpendicular to the intersection of the planes with the plate, whilst the parallel excess Kikuchi band is to be seen plainly in the background. (b) 400 keV quasichanneling proton pattern obtained by transmission parallel to $\{100\}$ planes in gold, the classical equivalent of (a) (contrast reversed in photography)

thin to show any signs of the defect blocking which would correspond to a Kikuchi band. However, suggestions of intensity streamers, perpendicular to the streak at its ends and midpoint are due to multiple scattering of protons parallel to the planes. They reflect the ease with which a proton deep in the channel or at the top of the assumed ideal barrier, positions of stable and unstable equilibrium respectively, can in this aspect of its lateral motion be moved freely by inelastic electronic and nuclear collisions.

Figure 10.13(a), on the other hand, shows the circumstance for 1 MeV electrons transmitted down the $\langle 111 \rangle$ axis of silicon. The number of reflections excited is markedly increased but, apart from the anomalously high intensities of the innermost Bragg spots, the differences between Figures 10.12(a) and 10.13(a) are only geometrical. Then, when the crystal is slightly tilted (Figure 10.13(b)), much of the ordered array of Bragg spots gives way to strong inelastic scattering, made manifest in the enhanced diffuse transmission at small angles and the strong Kikuchi effects, including primary envelopes, at high angles. The same occurs for beams exactly parallel to poor high-indexed axes, when secondary envelopes appear (Figure 10.13(c)). We maintain that the details of all these geometrical effects for 1 MeV electrons through silicon are only to be understood in terms of arguments based on waves, of the kind presented in earlier sections. Furthermore the crystal thickness (0.5–1.0 μm) and the angle of tilt, through their influence on multiple inelastic scattering processes, are evidently playing the critical part we anticipate.

More interesting results, however, have been obtained by Fujimoto and colleagues.[16] In these experiments the transmission patterns obtained with crystals of silicon, copper and gold are examined for different axes, for different electron energies up to 1 MeV and for different crystal thicknesses. The kinds of pattern observed fall into two classes—flower and geometric— the principle difference being that the geometric pattern is of the type normally encountered at low energies while the higher energy flower pattern, which is thickness *independent*, is characterized by axial critical angles. Furthermore the importance of the strength of the potential, through the order of the axis examined and the atomic number of the crystal, is demonstrated clearly. We conclude from these results that, for major axes at 1 MeV, there is a tendency towards classicality. Yet it should be emphasized that determining a limitation of the flower pattern by a critical angle is only a part of the story. Before concluding that everything within the Kikuchi zone is to be described classically it is necessary to show that no quantal features remain on a detailed intensity scan (Figure 10.1(a)) across the axis. Thus we note that the patterns for beam incidence parallel to axes do indeed show quantal fine-structure[16] and that the intensity 'envelope' through an axis may itself require a quantum mechanical explanation. The greater part of each pattern, namely the part beyond the axial critical angle, is, of course,

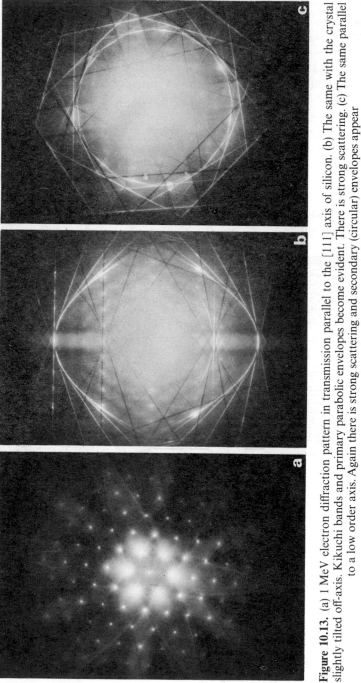

Figure 10.13. (a) 1 MeV electron diffraction pattern in transmission parallel to the [111] axis of silicon. (b) The same with the crystal slightly tilted off-axis. Kikuchi bands and primary parabolic envelopes become evident. There is strong scattering. (c) The same parallel to a low order axis. Again there is strong scattering and secondary (circular) envelopes appear

dominated by Bragg spots, Kikuchi envelopes and modestly broadened Kikuchi bands.

A second interesting feature is the conversion of the intense central disc to a deep trough with increasing crystal thickness, and the spreading of defect Kikuchi bands out into those regions which are normally excess (Figure 10.11(a), large angles; Figure 10.11(b), small angles). This phenomenon has been reported by several authors in recent years. Quantum mechanically the solution lies in introducing absorption into the theory of Kikuchi patterns.[52-53] In fact a recent very elegant treatment of the case for thin crystals[54] shows both experimentally and theoretically that the process is initiated by phonon and electron core excitation. This gives rise, in the first instance, to two defect discs joined at a circumferential point which coincides with the central spot. Further dispersion of the incident beam with increasing crystal thickness produces a spreading of the defect zone out along the band.

Classically this behaviour is, after a qualitative fashion, well understood. The strong axial electron channeling in the thin crystal is rapidly attenuated with thickness by scattering events in the vicinity of atomic nuclei, so that the particles are ejected into above-the-barrier states. The deflection is stronger for axes than for planes, just as the initial channeling is stronger, and we have a circumstance exactly opposite to that for positive particles, where the channel is not studded with scattering centres. For planes, where they enter an axis, the scattering is also strong but becomes less so as the angular distance from the axis increases. We therefore have rather an interesting, though natural, difference between negative and positive particles. For the former channeling is unstable, and after a certain thickness may only be seen in planes and at high angles, the central part of the pattern being dominated by above-the-barrier (blocking) states. For the latter, channeling is strong and only the outer part of the pattern is dominated by blocking. We recall, however, that for protons in thick crystals there is a rapid diffusion of axially channeled particles into neighbouring planes. Proton bands in a star pattern therefore change their intensity from excess to deficient as we pass from smaller to higher scattering angles.

Kikuchi patterns observed in the HVEM leave the question as to what happens to the dominant below-the-barrier (elastic) resonances unanswered. There appears to be no doubt that they are subjected to strong inelastic scattering, smear out, subscribe to 'channeling' in the Kikuchi zone and are finally pushed into above-the-barrier states. Certainly the centre of a Kikuchi zone from a thick crystal becomes rapidly devoid of Bragg spots.

The only other extensive piece of work on high energy electron transmission seems to be that by Kumm and colleagues.[18] Using 20 MeV electrons collimated to a divergence of less than 0·05° and passed through an 8 μm single crystal of magnesium oxide they find a similar behaviour for the [100] axis. There is a strong increase of scattering as the axis is approached, but a measurable increase of transmission in a small angular zone about the axis itself. The angular width of the peak agrees well with classical estimates, though defect Kikuchi bands radiating from the axis have widths determined by Bragg angles.

Qualitative results of a similar kind for bremsstrahlung production in silicon have also been reported.[55] Generation of coherent radiation by channeled charged particles is a topic in itself.[56]

(2) *Reflection and Emission*

Indirect investigations of anomalous electron penetration have been made by scattering experiments at high angles (Figure 10.1(b)), and by emission experiments. Either the β-particles themselves are detected or the yield of some secondary process, such as K X-ray generation is monitored.[57,58] Energies range from one or two hundred keV[59,60] up to 1 MeV.[61,62]

Without exception strong peaks are found for electrons for both major axes and planes and the qualitative predictions of both the wave and particle theories are fulfilled. When it comes to the question of whether the classical theory provides an adequate quantitative explanation, however, there are problems.

First it is to be acknowledged that the critical angles, on which the experiments concentrate their measurements, are of the right order and scale in the right way as functions of crystallographic direction and atomic number. However, the best experiments[62] sometimes show structure on axial peak edges which make an unambiguous determination of the angular width difficult. The structure is ascribed to planar effects. If that is so then the effect is basically quantal, since on both the models which we have outlined the planes scatter according to wave mechanics up to exceedingly high energies. Secondly, the experimental conditions—in terms of angular beam divergence, detection angles, crystal thickness, etc.—are such that a smearing-out of the scattered particle distributions would make discrimination between a basic quantal and classical behaviour difficult, if not impossible. Finally we emphasize yet again that arrival at critical angles is only the first step in a slow progression to orbital mechanics. The work of De Wames and Hall[32] shows quite conclusively that even after the disappearance of resonance (Bragg peak) structure in Rutherford scattering experiments the remaining angular pattern of intensity retains a significant dependence on Planck's constant and is accordingly still quantum mechanical.

10.4.2 Positrons

Fewer experiments have been performed with positrons but results are becoming available. Amongst these we cite K X-ray excitation in rubidium bromide by positrons with energies up to 480 keV[63] and transmission experiments through silicon and anthracene at 247 keV.[64] In each case a qualitative behaviour similar to that observed for protons is observed, but experimental limitations make a comparison with either theory both difficult and dangerous.

At higher energies all of the experiments[65,66] described are investigations of Rutherford scattering of positrons emitted from ^{58}Co and accelerated to a mean energy of 1·2 MeV. The later work on silicon[66] rather than gold, is distinguished by the fact that a clear trend from more to less immediately observable quantal behaviour is apparent in changing from lower to higher order crystallographic planes or to axes. For planes a careful comparison with many-beam diffraction calculations is made with excellent agreement. For axes the half width at half minimum fits well with relativistic classical estimates of the critical angle. At the shoulders the statistics are insufficient to detect the quantum features which are lodged there, though it should be noted that the minimum (small angle) yield χ_{min} is anomalously high.

10.4.3 A Final Note

The aim of experiment is to investigate and one purpose of theory is to explain. There is no question, therefore, that classical and quantum mechanics are in any competition to describe anomalous effects seen with β-particles. It is perfectly clear, for example, that the low energy experiments of Ok[64] are well explained by classical theories. By the same token the results of electron microscope experiments at these and higher energies literally force us into wave mechanics, complementarity and the dual nature of matter.

In the vicinity of 1 MeV, for strong axes, for large atomic numbers and for rather small scattering angles there is a clear trend towards classical mechanics. Critical angles emerge as a controlling feature and resonances are hard to detect. Again, if no attention is paid to detailed variations of intensity, classical mechanics is sufficient for an understanding. Indeed the physics of channeling is helpful in getting a feeling for many high voltage electron microscope phenomena. However, if electron and positron transmission patterns are not viewed through half-closed eyes, and the detailed effects of scattering into large angles, as well as small, are investigated, we are once more faced with the necessity for wave mechanics. Kikuchi lines, bands and envelopes, as well as Bragg spots, for which there is not even a superficial Newtonian mechanical description, are manifestly of quantal origin.

10.5 CONCLUSION

The scattering of a collimated beam of β-particles by aligned single crystals can be discussed on the basis of two models. In one a single string or plane of atoms provides the starting point. In the other—the crystal—at least one element of transverse lattice periodicity is retained. For both models forward motion can generally be considered classical. In the transverse motion the crystal model predicts resonance features which the single string or plane cannot. In spite of a slow relativistic trend towards classicality at quite small scattering angles the higher angle resonance features persist even in simple transmission patterns and more sophisticated reflection or transmission experiments may well require quantum theory throughout.

The critical angle is a signal point which indicates initiation of the quantum–classical transition. Predictions of the two models are only identical when confinement of the β-particle to a wavepacket in the transverse plane is complete. This occurs for large numbers of Bloch waves, synonymous with large numbers of bound states. A separation into above and below-the barrier states parallels random and channeled orbits, transgression of the barrier top itself, corresponding to continuum potential breakdown, being achieved through the medium of Fresnel transients.

For the two extremes—namely two-beam dynamical theory and full classical mechanics—there is a fundamental difference in nature between the two Bloch waves, and the channeled and blocked orbits respectively. One of the Bloch waves, and one set of orbits are subject to strong scattering processes. In the dynamical theory of diffraction this gives rise to the Bormann effect. In the high energy ultimate limit of classical mechanics, however, the influence of the scattering processes depends on the sign of the β-particle. With positrons we have an action akin to that of protons, whereby blocked orbits are strongly scattered and highly penetrating channeled orbits have long lifetimes. Conversely, for electrons, the channeled orbits are dispersed rapidly so that, for thicker crystals, only those orbits conventionally referred to as blocked can be sustained.

References

1. Newton, C. S. and Chadderton, L. T., *Rad. Effects*, **10**, 33 (1971).
2. Johnson, E. and Chadderton, L. T., *Risø Report No.* 241 (1971).
3. Chadderton, L. T., *Phys. Lett.*, **23**, 303 (1966).
4. Chadderton, L. T., *Phil. Mag.*, **18**, 1017 (1968).
5. Chadderton, L. T. and Eisen, F. H., *Phil. Mag.*, **20**, 195 (1969).
6. Chadderton, L. T., *J. appl. Cryst.*, **3**, 429 (1970).
7. Van Vliet, D., *Rad. Effects*, **10**, 137 (1971).
8. Ballantine, L. E., *Rev. Mod. Phys.*, **42**, 358 (1970).
9. Lindhard, J., *Dan. Vid. Selsk., Mat. Fys. Medd.*, **34**, 1 (1965).
10. Newton, R. G., *Scattering Theory of Waves and Particles*, New York, McGraw-Hill (1966).
11. Frahn, W. E., *Phys. Rev. Lett.*, **26**, 568 (1971).
12. Frahn, W. E., *Nucl. Phys.*, **75**, 577 (1966).
13. Merzbacher, E., *Quantum Mechanics*, New York, Wiley (1961).
14. Baker, S. D. and McIntyre, J. A., *Phys. Rev.*, **161**, 1200 (1967).
15. Grasso, F., Lo Savio, M. and Rimini, E., *Rad. Effects*, **12**, 149 (1972).
16. Fujimoto, F., Takagi, S., Komaki, K., Koike, H. and Uchida, Y., *Rad. Effects*, **12**, 153 (1972).
17. Ford, K. W. and Wheeler, J. A., *Annals of Physics*, **7**, 287 (1959).
18. Kumm, H., Bell, F., Sizmann, R., Kreiner, H. J. and Harder, D., *Rad. Effects*, **12**, 53 (1972).

19. M. T. Robinson and Oen, O. S., *Phys. Rev.*, **132**, 2385 (1963).
20. Lervig, P., Lindhard, J. and Nielsen, V., *Nucl. Phys.*, A, **96**, 481 (1967).
21. Lindhard, J., abstract in *Atomic Collision Phenomena in Solids*, Amsterdam, North Holland (1970).
22. Stark, J., *Phys. Zeitschrift*, **13**, 973 (1912).
23. Emslie, A. G., *Phys. Rev.*, **45**, 43 (1934).
24. Shinohara, K., *Phys. Rev.*, **47**, 730 (1935).
25. Bormann, G., *Z. Phys.*, **42**, 157 (1941).
26. Tikhonova, E. A., *phys. stat. sol.*, (b), **49**, 783 (1972).
27. Hashimoto, H., Howie, A. and Whelan, M. J., *Proc. R. Soc.*, A, **269**, 80 (1962).
28. Hirsch, P. B., Howie, A., Nicholson, R. B., Pashley, D. W. and Whelan, M. J., *Electron Microscopy of Thin Crystals*, London, Butterworths (1965).
29. DeWames, R. E., Hall, W. F. and Lehman, G. W., *Phys. Rev.*, **148**, 181 (1966).
30. DeWames, R. E., Hall, W. F. and Lehman, G. W., *Phys. Rev.*, **174**, 392 (1968).
31. DeWames, R. E. and Hall, W. F., *Acta. Cryst.*, A, **24**, 206 (1968).
32. DeWames, R. E. and Hall, W. F., *Rad. Effects*, **5**, 197 (1970).
33. Fukuhora, A., *J. Phys. Soc.*, *Japan*, **18**, 496 (1963).
34. Cowley, J. M. and Moodie, A. F., *Acta. Cryst.*, **10**, 609 (1957).
35. Zachariasen, W. H., *Theory of X-ray Diffraction in Crystals*, New York, Wiley (1945).
36. Kainuma, Y., *Acta. Cryst.*, **8**, 247 (1955).
37. Gjønnes, J., *Acta. Cryst.*, **20**, 240 (1966).
38. Thomas, L. E. and Humphreys, C. J., *phys. stat. sol.*, **3**, 599 (1970).
39. Humphreys, C. J., Thomas, L. E., Lally, J. S. and Fisher, R. M., *Phil. Mag.*, **23**, 87 (1971).
40. Humphreys, C. J. and Fisher, R. M., *Acta. Cryst.*, A, **27**, 42 (1971).
41. Lally, J. S., Humphreys, C. J., Metherell, A. J. F. and Fisher, R. M., *Phil. Mag.*, **25**, 321 (1972).
42. Thomas, L. E., *Rad. Effects*, **5**, 183 (1970).
43. Kagan, Yu. V. and Kononets, Yu. V., *Sov. Phy.–J.E.T.P.*, **31**, 124 (1970).
44. Armstrong, D. D., Gibson, W. M. and Wegner, H. E., *Rad. Effects*, **11**, 123 (1971).
45. Berry, M. V., *J. Phys. C.: Solid St. Phys.*, **4**, 697 (1971).
46. Bohr, N., *Dan. Vid. Selsk.*, *Mat. Fys. Medd.*, **18**, 8 (1948).
47. Kikuchi, S., *Sci. Pap. Inst. Phys. Chem. Res. Tokyo*, **18**, 223 (1928).
48. 'Atomic Collision and Penetration Studies', Chalk River, Canada, *Can. J. Phys.*, **46**, 449 (1968).
49. Proc. Conf. on *Atomic Collision Phenomena in Solids*, Sussex, Ed. D. W. Palmer, M. W. Thompson and P. D. Townsend, Amsterdam, North Holland (1970).
50. Proc. Int. Conf. on *Solid State Physics Research with Accelerators*, Brookhaven National Laboratory, Ed. A. N. Goland, BNL 50083 (C-52) (1967).
51. Proc. Conf. on *Atomic Collisions in Solids IV*, Ed. S. Andersen, K. Bjorkquist, B. Domeij and N. G. E. Johansson, London, Gordon and Breach (1972).
52. Kainuma, Y., *J. Phys. Soc. Japan*, **8**, 685 (1953).
53. Hall, C., *Phil. Mag.*, **22**, 63 (1970).
54. Okamoto, K., Ichinokawa, T. and Ohtsuki, Y.-H., *J. Phys. Soc. Japan*, **30**, 1690 (1971).
55. Walker, R. L., Berman, B. L., Der, R. C., Kavanagh, T. M. and Khan, J. M., *Phys. Rev. Lett.*, **25**, 5 (1970).
56. Belyakov, V. A., *ZhETF Pis. Red.*, **13**, 254 (1971).
57. Gracher, B. D., Komar, A. P., Korobochko, Yu. S. and Mineev, V. I., *Sov. Phys.– Solid St.*, **10**, 1894 (1969).

58. Bronder, Th. and Jakschik, J., *Phys. Lett.*, A, **29**, 580 (1969).
59. Uggerhøj, E., *Phys. Lett.*, **22**, 382 (1966).
60. Uggerhøj, E. and Andersen, J. U., *Can. J. Phys.*, **40**, 543 (1968).
61. Howie, A., Spring, M. S. and Tomlinson, P. N., in ref. 49.
62. Uggerhøj, E. and Frandsen, F., *Phys. Rev.*, B, **2**, 582 (1970).
63. Behrisch, R., Bell, F. and Sizmann, R., *phys. stat. sol.*, **33**, 375 (1969).
64. Armagan Ok, H. H., *Z. Physik*, **240**, 314 (1970).
65. Andersen, J. U., Augustyniak, W. M. and Uggerhøj, E., *Phys. Rev.*, B, **3**, 703 (1971).
66. Pedersen, M. J., Andersen, J. U. and Augustyniak, W. M., *Rad. Effects*, **12**, 47 (1972).

CHAPTER XI

Scattering (Blocking) Patterns Applied to Crystallography

C. S. BARRETT

11.1 INTRODUCTION

A collimated beam of ions impinging on a crystal is scattered forward, side-ward, and to the rear to produce, on films or fluorescent screens, patterns which contain much crystallographic information. These patterns are frequently called blocking patterns because they appear as if they represent the blocking of the emergence, in certain directions in the crystal, of ions scattered by the atoms on lattice points. Extensive studies by physicists have shown, however, that the basic mechanism is better characterized as a deflection of the scattered ions in collisions with strings or planes of atoms in the crystal, and not merely an abrupt stoppage such as is implied by the term 'blocking pattern'.

Scattering patterns on a flat film consist of intensity-deficient straight lines located along the extension of the crystallographic planes, and spots always located at line intersections, i.e. at places on the film where rows of atoms in the crystal, if extended, would intersect the film. To produce the pattern reproduced in the enlarged print of Figure 11.1, the initial beam of protons from a 150 keV Cockroft–Walton accelerator was directed through a 0·2 mm diameter pinhole collimator; it passed through the hole in the film and backscattered from a cobalt crystal about 10 mm beyond the film. The orientation of the film parallel to the (0001) plane of the crystal preserved the sixfold symmetry of [0001], and the deviation ($\sim 2°$) of the initial beam from this axis intensified the scattering by reducing the channeling. Since the apparatus for producing such patterns is not very complex or expensive and since much crystallographic information is obviously contained in the patterns, a small group of us undertook an exploration of the information that could be extracted from them and of its potential applications. Other chapters in this book cover a number of these matters such as the nature and crystallographic position (interstitial or substitutional) of impurity atoms, and research dealing with imperfections and radiation damage. The present chapter is concerned with symmetry, indexing, orientation, phase identification, the

331

Figure 11.1. Print of a blocking pattern for cobalt with 100
keV proton beam incident nearly parallel to [0001][2]

relation of the pattern to crystal structure, and micrography. It should merely
be stated here that damage from the incident beam of ions does in fact
gradually degrade the patterns if the beam continues to hit one spot on a
crystal, but fortunately our experience with many different crystals has been
that very little damage results from the amount of radiation required to
produce a pattern. Thus, substantially damage-free patterns can be had or,
if desired, one can study the progress of damage with much longer exposures
to the beam. Good patterns, of course, require the absence of abraded or
contaminated surface layers; it is also important to prevent a local charging
of the surface by conducting the current through the crystal or by neutralizing
it on the surface.

11.2 ORIENTATION OR INDEXING

Symmetrical patterns, such as that shown in Figure 11.1, immediately reveal symmetrical orientations, and comparison with published charts quickly identifies all planes and spots. For body-centred, face-centred, and diamond cubic crystals, charts have been published that indicate planar and directional indices and show both the geometry of the patterns and the relative promin- ence of the lines (the dip in intensity below the surrounding background).[1] Indexed patterns for hexagonal crystals have also been published,[2] but a chart would be geometrically accurate only for crystals of a single axial ratio for these and any other noncubic crystal. In fact, from the geometrical rela- tions in a pattern the axial ratio or ratios (but not absolute unit cell dimensions) can be obtained quite simply, and nonsymmetric orientations can be de- termined. The presence or absence of a centre of symmetry in a crystal cannot be recognized by pattern geometry (as is also the case with Laue photographs made with X-rays), but experience shows that it is possible to see differences in the relative line intensities in the patterns of different crystals having the same macroscopic symmetry but different atom distributions. For example the patterns of copper and silicon differ even though both have face-centred cubic space-lattices and both belong to the same Laue symmetry group ($m3m$).[1,3] A method that has been used for precision alignment of thin crystals is to note where a transmitted beam that has channeled through a crystal strikes a screen, compared with the place where the incident beam is imaged when the crystal is removed.[4] Accuracy with this procedure reached 0·02° when 11 MeV protons were used.

11.3 RELATED TYPES OF PATTERNS

Blocking patterns have somewhat the appearance of Kikuchi transmission electron diffraction patterns, or the more recent selected-area electron- channeling patterns (SACP) made in a modified type of scanning electron microscope.[5,6] The latter are produced by the diffraction of a beam of elec- trons striking a point on the specimen from directions that are continuously varied throughout a small angle range, with results closely related to those obtained with a converging bundle of electrons by Kossel and Mölenstadt,[7] and by Goodman and Lehmpfuhl.[8] Kikuchi pattern maps and channeling pattern maps, by which crystal orientations can be quickly recognized, have been published. The accuracy in orientation determination with the SACP technique is lower than that obtainable with blocking patterns because of distortions in lens and display equipment; uses and errors are summarized by Booker.[9] Kossel patterns, on the other hand, provide precision orienta- tions. They also provide precision data on lattice constants[10] and can be made by instruments used for scanning electron microscopy, if designed or modified suitably.

11.4 OBSERVATION OF LINE INTENSITIES

It is not difficult to arrange the various lines of a pattern in a sequence from most prominent to least, i.e. with respect to what might be termed their integrated intensity. A densitometer tracing across the lines may be used, but about equal success can be had with the unaided eye. A gauge prepared by a film recording of slightly diffuse shadows of a set of wires of different diameters is helpful.[2] Attention is directed to segments of the lines between intersections and at roughly equal distances from the centre of the pattern, where background intensities are approximately equal.

The lines are always wider than can be accounted for by the range of angles in the initial beam or by the diameter of the irradiated spot; line breadths decrease with increasing accelerating voltage. Proton patterns are conveniently sharp for crystallographic work at voltages in the 100 keV range, as may be seen from Figure 11.1. A wide band of proton energies is involved in such a pattern, but profile studies of lines show that most of the detail is contributed by the higher-energy components; in fact protons of energy below perhaps 50 keV may be absorbed in the anti-abrasion coatings of some types of film, and backscattered protons are also absorbed in the crystal— the lower their energy and the deeper their origin below the surface, the more are absorbed.

By inspecting the relative intensities of lines in a pattern and relating them to the crystal structure it is possible to obtain crystallographic information, despite the fact that each pattern is made by ions of a wide range of energies and each pattern lacks any information from which the absolute values of interplanar spacings can be deduced. The sequence of integrated intensity-dips, I_{obs}, relative to that for the strongest line of a pattern, may be estimated by eye in several steps or by approximate numerical values obtained using densitometer or shadow gauge; this sequence remains substantially unchanged when the orientation of a crystal is changed, or when the energy of the initial beam is altered through a wide range.

11.5 LINE INTENSITIES AND CRYSTAL STRUCTURE

The intensity problem involved here is the complex one of evaluating the intensity of a beam scattered in the direction OA, Figure 11.2, when scattered ions arrive at O from all directions and all energies as a result of a beam of ions striking the crystal and scattering to O by all kinds of scattering processes, both single and multiple, at various depths below the surface. The calculations are very involved but have been carried out with high speed com-

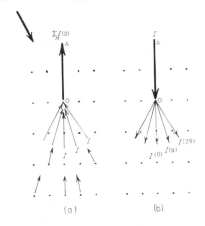

Figure 11.2. (a) Scattered beams arbitrarily assumed to have intensity I, reaching point O, are again scattered in the direction OA with resultant intensity $I(A)$ (b) Beam in direction AO reaching O with intensity I excites diffracted beams $I(O)$, $I(\mathbf{g})$, $I(2\mathbf{g})\ldots$, whose sum $\sum\limits_{\mathbf{g}} I(\mathbf{g})$ is equal to $I(A)$ by the reversibility theorem

puters for individual cases, using the classical approach of computing particle trajectories individually. By tracing out enough trajectories, an approximation of the blocking pattern has been reached in a few instances, and the profiles of some typical lines and line intersections have been mapped out. The high computing costs of this procedure render it useless for most applications of ion scattering patterns to crystallographic problems.

Although the classical approach when employed with certain simplifying assumptions has been successful in accounting for the major features of blocking patterns, and is the normal one for high speed ions,[11] we have chosen to treat the patterns in an approximate way as a diffraction problem of an unusual type, recognizing that treatments based on diffraction involve reflections so broad that the different orders of diffraction overlap, and also that inelastic scattering events are more prevalent than are encountered with X-rays and electrons.

We find that these characteristics of the diffraction approach have not precluded its use in practice as a way to arrive at useful semiquantitative results. We have employed a procedure that is such a crude approximation to the true situation that its degree of success is surprising, but is very welcome, since the computations involved are easy—they involve programs that are familiar to all crystallographers and can be done in seconds on a high speed computer.

The problem can be thought of as somewhat analogous to the calculation of a Kikuchi pattern made by high voltage electrons. Thomas and Humphreys[12] have treated electron Kikuchi patterns on the basis of rocking curves: a Kikuchi band profile is viewed as the sum of rocking curves. A rocking curve represents the variation of intensity with angle in a diffracted beam when the incident beam is fixed at or near a reflecting position, or the variation in intensity of a diffracted beam when the incident beam is varied in angle around the direction for maximum diffracted intensity (often called the 'Bragg angle'). Von Laue's reversibility (reciprocity) theorem is employed, which holds not only for elastic scattering in Kossel and Kikuchi patterns, but also for inelastically scattered electrons[13]—provided that the energy losses are small compared with the incident energy. The theorem has also been shown to hold for proton scattering.[14,15] It can be stated as follows: the intensity of a beam arriving at position A from a source at B is equal to the intensity at B from the same source at A. If one wants to compute the intensity leaving a crystal in a certain direction so as to arrive at a point A, Figure 11.2(a), located, say, on a deficiency line in a proton blocking pattern, the problem can conveniently be solved by considering the reverse problem, i.e. by assuming that a beam of intensity I strikes the crystal from a source at A, Figure 11.2(b), and computing what diffracted beams would arise from it in the crystal. If these beams are of intensity $I(O)$, $I(\mathbf{g})$, $I(2\mathbf{g})\ldots$, where $I(\mathbf{g})$ is the intensity of the diffracted beam for which the diffraction vector is $\mathbf{g}$, then the entire group of diffracted beams excited by the beam from point A is summed to give an intensity $I(A)$ given by

$$I(A) = \sum_{\mathbf{g}} I(\mathbf{g}) \qquad (11.1)$$

According to the principle of reversibility, the intensity $I(A)$ is then equal to the intensity of the diffracted beam that arrives at point A outside the crystal, if the scattered beams within the crystal that converge upon point O are each of intensity I.

If point A lies on a line of the blocking pattern it lies on the extension of a plane (hkl) of the crystal and will receive the zero order diffracted beam $I(O)$ from this plane in the crystal. The higher orders leave this plane at increasing angles with increasing order number. Because of the very short wavelengths involved with fast ions, e.g. protons in the 100 keV range with wavelengths of order 0·001 Å, many orders will diffract simultaneously from the (hkl) plane—even more than are seen in electron diffraction patterns for the systematic cases (100 keV electrons have 0·037 Å wavelengths). The situation approximates diffraction from a two-dimensional grating lying in a plane normal to OA, Figure 11.2. A true calculation would involve a dynamic theory computation for n beams with n a large number, and with absorption. However, it is known that in high voltage electron microscopy the weaker, high-indices reflections have relatively weak many-beam effects compared with the two strongest reflections;[16] in fact some calculations have indicated that even two-beam dynamic theory for plane waves predicts a deficiency band for positrons rather than an excess band,[17] and even kinematic theory has been used in weak-beam high-resolution electron microscopy of defects.[18,19] The rocking curves used in the computation of Kikuchi profiles by Thomas and Humphreys[12] are sensitive to extinction and absorption parameters which vary with wavelength. These parameters also vary with the wavelengths of ions; hence the wide range of wavelengths involved in producing a scattering pattern, and the unknown distribution of intensity over these wavelengths, together with the need to integrate effects throughout the depth of ion penetration, prevent rigorous computation of Kikuchi profiles or integrated intensity deficiencies of lines in the patterns.

In view of the barriers that would be encountered in any attempt at a rigorous solution of intensities in blocking patterns and the cumbersome nature of even a close approximation to one, the author undertook to see what results could be achieved with a computation of an extremely simple type based on kinematic rather than dynamic theory, and with factors of secondary importance being neglected.

Inelastic scattering is assumed to provide the background that varies only slowly over the pattern. The dip in intensity at a line of the pattern is ascribed to elastic scattering, i.e. diffraction, which in effect removes intensity from the line and spreads it nearby in the form of broad, overlapping beams of many orders of diffraction. The reversibility principle is invoked to reduce the problem to the one of Figure 11.2(b), in which the intensities of individual orders are computed, then summed to give the integrated intensity of the bundle as a whole, which is then taken as the integrated intensity dip of the line.

The method[3,20] consists in first computing the structure factor $F(hkl)$ for the plane of Miller indices (hkl) and of periodic spacing d_{hkl} using the familiar relation of X-ray crystallography:

$$F(hkl) = \sum_j f_j \exp\{2\pi i(hu_j + kv_j + lw_j)\} \exp(-B/4d_{hkl}^2) \qquad (11.2)$$

where the summation extends over the atoms in the unit cell of the crystal, and f_j is the atom scattering factor of the jth atom, which has coordinates $u_j v_j w_j$. B is a temperature factor analogous to the factor used in the corresponding X-ray computation and unless chosen otherwise is assumed to be approximately the same as would be used for X-rays; for simplicity it is assumed to be isotropic. The factor f_j is taken as listed in tables for electron diffraction (preferably those computed without exchange). The integrated intensity deficiency in the line, I_{calc}, is then taken as proportional to the sum of the F^2 values of the orders,

$$I_{calc} = \sum_n F_n^2 \qquad (11.3)$$

where the sum extends to all orders n that make any significant contribution to the total. The values of I_{calc} for all lines of a pattern are expressed as a per cent of I_{calc} for the strongest line of the pattern.

Let it be clear that a kinematic approach of this kind deviates from the actual situation: the beams that make a blocking pattern have a broad spectrum of wavelengths and come from a wide range of depths, regardless of whether or not the incident beam is monoenergetic. All wavelengths of importance in the pattern making are so small that a diffraction treatment should be of the n-beam dynamic type with n a very large number (even though the diffraction is, in general, of the systematic, symmetrical type); absorption of both incoming and outgoing beams are important processes and are wavelength dependent, as is the extinction distance. The elongation of points in the reciprocal lattice varies inversely as the extinction distance (unless the scattering is from a very thin film, in which case it is related inversely to the film thickness), consequently the reciprocal lattice-point profiles are an integrated result of wavelength-dependent components. And finally the energy distribution in the spectrum of wavelengths is in general unknown, usually limited on the long wavelength side by the energies required to penetrate the anti-abrasion coating on the photographic film (commonly of order 50 keV), with high energy particles causing most of the dip in photographic density in proton blocking directions.[1,21]

Because of these deviations from actuality we must ask the minimum of the method, and ask merely that it arrange the relative intensities of lines of a pattern in the order actually observed, with discrimination of a few steps of intensity. If we ask no more than this we find that many of the factors mentioned above need not concern us, for experiment has shown that when the energy of the initial beam is changed by as much as a factor of three (~ 50 to 150 keV) and perhaps much more, the sequence of intensities of lines in a pattern remains the same; likewise, when different metals of the same crystal structure are examined (e.g. Al, Cu, Au: fcc) the sequential order of line intensities is the same. We also find that the angle of incidence of the primary beam on the crystal surface is not a critical matter. And we find that the method is applicable to crystals with structures more complex than the simple ones that have one atom on each point of a space lattice, whereas investigators who have made particle-trajectory computations frequently ignore planes of atoms more remote than those of the nearest wall of a channel.

11.6 OBSERVED AND CALCULATED INTENSITIES

When the lines of a pattern such as Figure 11.1 are listed in their order of relative blocking intensity and compared with I_{calc}, the sequence from strongest to weakest as observed matches the sequence of I_{calc} almost as well as intensities can be judged by eye, as will be seen from Table 11.1.[1-3]

Table 11.1 Planes listed in order of decreasing values of their integrated intensity dips I_{obs} for comparison with I_{calc}[3]

Crystal	HKL	I_{obs}	I_{calc}	Crystal	HKL	I_{obs}	I_{calc}
Copper	111	100	100	White tin	200	100	100
(using	200	63	75	(using	220	50	48
$B = 1.0$)	220	25	28	$B = 1.0$)	101	27	20
	311	14	17		112	W	21
	331	W	7		420	W	14
	420	W	6				
	422	W	4	Silicon	220	100	100
	511	W	3	(using	111	94	76
	531	W	1	$B = 0.4$)	400	34	34
					422	20	18
Tungsten	110	100	100		311	19	14
(using	200	44	44		331	11	6
$B = 0.2$)	211	24	26		620	W	7
	310	8	13		511	W	3
	222	W	9		642	W	4
	321	W	7		531	W	2
	411	W	4				
	210	W	4	ζ-CuGe	0002	VS	100
	332	W	3	(hcp: using	10$\bar{1}$1	S	57
				$B = 1.0$)	10$\bar{1}$0	S	31
					11$\bar{2}$0	S	30
					11$\bar{2}$2	M	18
					10$\bar{1}$3	M	16
					20$\bar{2}$1	M	12
					10$\bar{1}$2	M	11
					11$\bar{2}$4	W	6
					20$\bar{2}$3	W	6
					10$\bar{1}$4	W	3

Numerical agreement between I_{calc} and I_{obs} is not to be expected, since the I_{obs} scale is rather arbitrary, and more than five to seven steps in observed intensities cannot usually be distinguished reliably either by eye or by densitometer.

For the simplest structures, an equal success in relative ordering of the lines can be reached by merely assuming that the atom population density on the different planes produces a proportional effect on the lines, as in Lindhard's theory, but I_{calc} computed as mentioned above applies not only to the simplest crystals but also to complex crystals, where parallel planes of given indices are stacked in a complex pattern of repetition and of relative atom densities.

The degree of success with the crystals mentioned in Table 11.1 was also obtained with several other crystals of cubic symmetry,[1] and also several of

Table 11.2 Relative intensity dips for two hexagonal crystals[2,20]

CdSe			CdS		
HKL	I_{obs}	I_{calc} $(B = 0.275)$ $(u = 0.38)$	*HKL*	I_{obs}	I_{calc} $(B = 0.8)$ $(u = \frac{3}{8})$
0001	100	100	0001	100	100
$10\bar{1}1$	90	82	$11\bar{2}0$	70	81
$11\bar{2}0$	80	73	$10\bar{1}0$	60	71
$10\bar{1}0$	40	33	$10\bar{1}3$	40	38
$11\bar{2}2$	30	31	$11\bar{2}2$	40	28
$10\bar{1}3$	20	19	$10\bar{1}1$	30	30
$10\bar{1}2$	10	16	$10\bar{1}2$	20	16
$20\bar{2}1$	<10	16	$20\bar{2}3$	20	17
$11\bar{2}4$	<10	16	$10\bar{1}5$	10	12
$20\bar{2}3$	<10	9	$20\bar{2}1$	<10	9
$10\bar{1}5$	<10	8	$21\bar{3}3$	<10	9
$21\bar{3}1$	<10	8	$20\bar{2}5$	<10	7
$30\bar{3}2$	<10	7	$30\bar{3}2$	<10	7
$20\bar{2}5$	<10	6	$12\bar{3}0$	nil	6
			$11\bar{2}1$	nil	5
			$11\bar{2}6$	nil	5
			$21\bar{3}1$	nil	5
			$11\bar{2}4$	nil	4
			$12\bar{3}2$	nil	3
			$10\bar{1}4$	nil	2
			$11\bar{2}3$	nil	2
			$10\bar{1}6$	nil	2
			$30\bar{3}1$	nil	0.3

hexagonal symmetry[2,3,20] including not only the hexagonal close-packed phase ζ-CuGe of Table 11.1 but also the hexagonal compounds of Table 11.2. Whether or not serious exceptions would be found with some other crystals cannot be stated at this time, but our experience with over a dozen crystals suggested that it might be possible to determine an atom position parameter by comparing the correlations obtained when different values of the parameter were assumed.

The results obtained when parameter determinations were attempted on single-parameter crystals were as follows.[22] The poorest correlation between I_{calc} and I_{obs} was obtained with magnetite which has a cubic unit cell containing eight formula units of Fe_3O_4. The sequence of relative intensities of I_{calc} did not fully match the sequence of I_{obs}, nevertheless there were eight pairs of I_{calc} values that suggested the parameter u was about 0.25. For example the approximate equality $I_{211} \approx I_{311}$ implies $u = 0.25 \pm 0.02$, the errors being estimated from the relative slopes of the curves of intensity against u for the planes. The value determined by X-rays is $u = 0.2548$.

Patterns of bismuth were compared with curves of intensity against the parameter z. Eight intensity ratios among the weaker lines were useful in that each ratio implied a well defined value of z and each of these values fell between 0.22 and 0.24. An average of the eight gave 0.233 ± 0.002, remarkably close to the X-ray value 0.23389.

A simple graphical method was developed for these studies (see Figure 11.3), consisting of a plot of $\ln I_{calc}$ against z for each line, and a logarithmic scale of I_{obs} which could be moved parallel to the $\ln I_{calc}$ axis to different z positions until a given observed ratio $I_{HKL}/I_{H'K'L'}$ matched the corresponding calculated ratio.[22] This procedure as well as the computations of I_{calc} could be done on a computer, but the graphical solution helps one to choose the more reliable ratios—the *weaker* lines are preferable, and discrepant values such as those indicated by broken lines in Figure 11.3 are discarded in favour of alternative ones that are consistent with the rest when two possible values of z exist for a given intensity ratio.

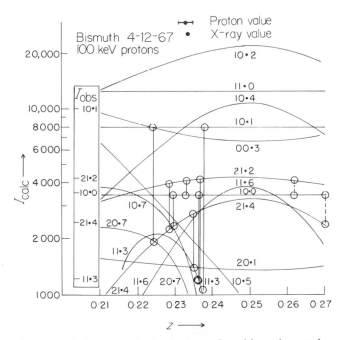

Figure 11.3. Curves of $\ln I_{calc}$ against z for a bismuth crystal, with vertical bars indicating ratios of I_{obs} for pairs of lines. The average of the z values indicated by the pairs is indicated by the horizontal bar at the top and is compared with the X-ray value, plotted as a dot[22]

With the same technique, 16 pairs of lines in patterns of CdS yielded a parameter value 0.381 ± 0.004; the X-ray value is 0.375. Similarly 14 pairs in patterns of CdSe yielded 0.375 ± 0.003; the X-ray value has been given in the literature as 0.375 and later as approximately 0.39. These tests of the method have been too few to permit an evaluation of general reliability. It is only safe to assume that exceptions will be found and that X-rays provide a more powerful and reliable method of parameter determination, capable of precision determinations of large numbers of parameters in a crystal.

11.7 PHASE IDENTIFICATION

Problems requiring the identification of phases can be solved with the patterns by making use of (i) pattern geometry or symmetry (ii) axial ratios derived

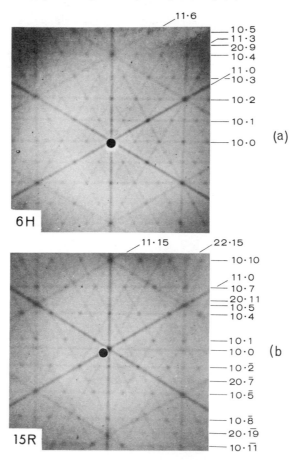

Figure 11.4. Blocking patterns of SiC crystals, with film parallel to the basal plane (0001). (a) Hexagonal structure type 6H. (b) Rhombohedral structure type 15R[23]

from the patterns and (iii) the relative intensities of lines or spots at line intersections. The identification of a crystal deposit on the surface of a solid (together with its orientation) and phases existing within the first few thousand ångströms of the surface, that can be reached with an impinging beam of particles, may turn out eventually to be the major field of application of these patterns.

As an example we can cite the determination of the structure types that were found in synthetic samples of SiC, a compound that is known to exist in over 50 different structure types. Identification was done from patterns, including those of Figure 11.4, by a four-step procedure.[23,24] (1) The symmetry established whether the irradiated spot was hexagonal or rhombohedral; (2) indices were assigned; (3) axial ratios were determined (but the indices and axial ratios alone left ambiguities unresolved); (4) I_{calc} values were compared with I_{obs} to eliminate ambiguities. Figure 11.5 shows I_{calc}

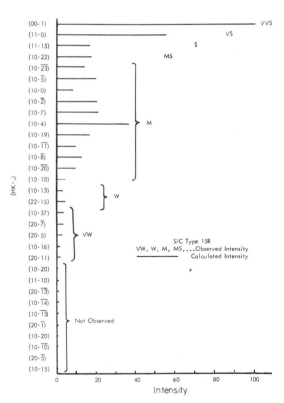

Figure 11.5. Line intensities in 100 keV blocking patterns of a SiC crystal with indices and I_{calc} for structure type 15R[24,35]

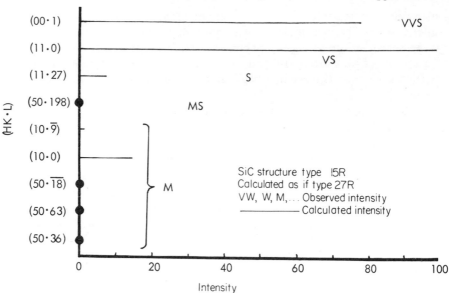

Figure 11.6. Line intensities for the same crystal of Figure 11.5, but with indices and unsatisfactory I_{calc} values as if it were of structure type 15R[24,35]

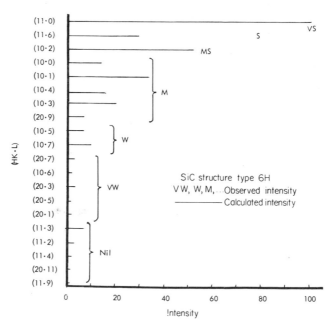

Figure 11.7. Line intensities for a SiC with indices and I_{calc} values for structure type 6H[24,35]

for indexed lines of one of the SiC crystals computed as if it is of the structure type 15R (with 15 molecules of SiC in a rhombohedral unit cell); the observed relative intensities from very strong (VVS) to weak (W) or nil are indicated. From the correlation shown it was concluded that phase type 15R accounted for the relative intensities. An alternative solution to the geometry of the pattern, 27R, required higher indices and yielded relative computed intensities that correlated very poorly with observation, as seen from those plotted in Figure 11.6. Figure 11.7 shows the results obtained with another sample, for which the correlation was taken as evidence that the structure was the hexagonal type 6H (with six molecules per unit cell).

11.8 THERMAL VIBRATIONS

The effect of the temperature factor in equation 11.2 was seen in a series of patterns in which the atomic numbers of the samples were nearly identical and the crystal structure type was the same, but the mean square amplitude of vibration of the atoms, u_{rms}^2, differed from sample to sample:[25]

	Metal:	Rhodium	Palladium	Silver
Atomic number:		45	46	47
u_{rms}^2:		0·0036	0·0055	0·0082 Å^2

It was found that the lines with high indices became relatively weaker as u_{rms}^2 increases in this series,[25] as is accounted for by the larger exponent in the temperature factor of equation (11.2), since the thermal parameter B when assumed to be isotropic is given by $B = 8\pi u_{rms}^2$.[26] A similar result was obtained in comparing platinum and gold patterns, with u_{rms}^2 equal to 0·0038 and 0·0071 Å^2 respectively, and again with chromium and β-CuZn ($u_{rms}^2 = $ 0·0042 and 0·0098 Å^2 respectively). For example, the $\{531\}$ and $\{221\}$ lines were seen in platinum patterns but not in gold patterns, and in rhodium but not in silver.

The effect of the temperature factor on the relative intensities of spots where lines intersect is a problem in n-beam dynamic theory for which n is even a larger number than for lines, and relative intensities are more difficult to judge on a film. Appearances often suggest that spot intensities can be viewed as the sum of the intensities of the lines intersecting at each of the spots, and in fact this rule seems to be followed by the majority of spots in patterns from Cu, Bi, Si, CdSe and CdS.[26] That it is not a general rule, however, is proved by the pattern of rhombohedral SiC reproduced in Figure 11.4 and this fact together with the difficulty of judging relative spot intensities corroborates the view that lines rather than spots should be chosen as the basis of applications involving intensities. For those spots that do appear to be an additive effect of intersecting lines, it has been noted that the effects on the spots of thermal agitation are qualitatively accounted for by the temperature factor contained in the I_{calc} formula for each of the lines that cross at the spot.[26]

11.9 IONOGRAPHY

'Ionography' is a name that might be applied to the imaging of crystalline solids by the use of ions transmitted through thin specimens. If positive ions, say protons, from a source are accelerated by a potential of 10^5–10^6 volts to strike a thin specimen, then those that are transmitted through it and are registered on a photographic emulsion that is placed in contact with the back of the specimen will produce an image of the specimen.[27] When the specimen has a thickness of a fraction of a micrometre or so, chosen so that only a few of the ions that enter the specimen along nonchanneling directions reach the film, the photographic image is made almost entirely by the ions that are channeled. There are degrees of channeling—that is, different crystallographic planes and directions allow different fractions of the incident beam to reach the film, just as there are different intensities of lines and line interactions in blocking patterns. Thus differently oriented grains or subgrains will be registered with different photographic densities. By using fine-grained film in intimate contact with the specimen, subsequent enlargement should yield micrographs with resolution comparable to that obtained in microradiography or in X-ray topography with Berg–Barrett or Lang X-ray topographs.[28] Imperfections and inclusions will be imaged when their dimensions are large enough and their scattering power is sufficiently different from that of the matrix surrounding them, since they cause dechanneling. An analogue is the topography of single crystals by the anomalous transmission of X-rays (i.e. by using the Bormann effect), a technique well established for imaging dislocations and other defects that cast shadows in such a beam. For protons of a given energy the desirable specimen thickness for ions of a given initial energy can be judged from published data on proton ranges. In this method a wide area can be irradiated at once from a relatively distant source (as in Berg–Barrett topographs), and the voltage can be adjusted so as to yield optimum contrast in the image.

A second method of producing ionographs should be mentioned, based on *backscattered* particles, a method that is applicable to a bulk sample whatever its thickness.[27] A grain into which the ions channel with slight attenuation is a grain that backscatters only a small amount; an imperfection that causes dechanneling is one that will show increased backscattering. Therefore if a well collimated or focused beam scans the surface of a specimen, and if backscattered ions are caught in a detector and displayed as a television-type image, the image will be a micrograph of the sample and should be useful in revealing grains, subgrains, twins, inclusions, precipitates and imperfections. Alternatively, the specimen current rather than the backscattered current can be used to create the image. The analogue of this type of ionography is the imaging of polycrystalline samples by electron channeling in a scanning electron microscope; it has recently been shown that differently oriented grains and subgrains can be seen in selected area scanning patterns because of the differing electron channeling in the different grains.[29]

At the time of writing the author has not had an opportunity to test either of these methods, nor is he aware of anyone else having discussed either of them, which is a strange situation, especially in view of the simplicity of the first method.

11.10 ALPHAGRAPHY AND FISSIOGRAPHY

French investigators have recently developed microscopic methods based on the channeling of alpha particles and fission fragments through thin specimens.[30] For a source of alpha particles they deposit on one side of the specimen a layer of americium-235, which provides particles of energies from 2 MeV to 4·5 MeV for the technique they term 'alphagraphy'. For 'fissiography' they employ the fission fragments from a vapour deposit of uranium-235 irradiated in a flux of thermal neutrons. They insert an amorphous filter between the source and the specimen to adjust the particle energies so that only the channeled particles emerge to the recording foil on the back of the specimen (a cellulose nitrate foil for α-particles and a mica foil for fission fragments). After etching the recording foil there are black tracks in the foil that are easily seen under the microscope; enlargement of about 100 diameters is used, although considerable graininess is seen in the recorded patterns at this magnification.

These methods have been used to register the images of grain boundaries,[31] twin boundaries,[31,32,33] stacking faults,[31] dislocations,[32,34] deformation bands, interstitial atoms, irradiation defects and to study the phenomena of recrystallization and precipitation.[30]

References

1. Barrett, C. S., Mueller, R. M. and White, W., *J. appl. Phys.*, **39**, 4695 (1968).
2. Barrett, C. S., Mueller, R. M. and White, W., *Trans. Met. Soc. AIME*, **245**, 427 (1969).
3. Barrett, C. S., *Adv. in X-ray Analysis*, **12**, 72 (1969).
4. Dearnaley, G. and Wilkins, M. A., *J. Sci. Instrum.*, **44**, 880 (1967).
5. Van Essen, C. G. and Schulson, E. M., *J. Mater. Sci.*, **4**, 336 (1969).
6. Van Essen, C. G., Schulson, E. M. and Donaghay, R. H., *Nature*, **225**, 847 (1970).
7. Kossel, W. and Möllenstadt, G., *Ann. der Phys.*, **5**, 113 (1939).
8. Goodman, P. and Lehmpfuhl, G., *Acta Cryst.*, **22**, 14 (1967).
9. Booker, G. R., in *Scanning Electron Microscopy/1971*, Proc. 4th Annual Scanning Electron Microscope Symposium, I.I.T. Research Institute, Chicago, 1971, p. 467.

10. Barrett, C. S. and Massalski, T. B., *Structure of Metals*, 3rd Ed. McGraw-Hill, New York, p. 96 (1966).
11. Chadderton, L. T., *Phil. Mag.*, **18**, 1017 (1968).
12. Thomas, L. E. and Humphreys, C. J., *phys. stat. sol.* (a), **3**, 599 (1970).
13. Pogany, A. P. and Turner, P. S., *Acta Cryst.*, A, **24**, 103 (1968).
14. Bøgh, E. and Whitton, J. L., *Phys. Rev. Lett.*, **19**, 553 (1967).
15. Anderson, J. U. and Uggerhøj, E., *Can. J. Phys.*, **46**, 517 (1967).
16. Humphreys, C. J. and Lally, J. S., *J. appl. Phys.*, **41**, 232 (1970).
17. Whelan, M. J., in *Atomic Collision Phenomena in Solids*, Ed. D. W. Palmer, M. W. Thompson and P. D. Townsend, Amsterdam, North Holland p. 3 (1970).
18. Cockayne, D. J. H., Ray, I. L. F. and Whelan, M. J., *Phil. Mag.*, **20**, 1265 (1969); *Proc. 7th Int. Conf. Electron Microscopy*, Grenoble, **2**, 321 (1970).
19. Cockayne, D. J. H., Jenkins, M. L., Ray, I. L. F. and Whelan, M. J., in *Electron Microscopy and Analysis*, London, Inst. of Physics, p. 108 (1971).
20. Barrett, C. S., *Trans. TMS-AIME*, **45**, 429 (1969).
21. Tulinov, A. F., *Sov. Phys. Usp.*, **8**, 864 (1966).
22. Barrett, C. S., *Met. Trans.*, **1**, 171 (1970).
23. Barrett, C. S., Barrett, M. A., Mueller, R. M. and White, W., *J. appl. Phys.*, **41**, 2727 (1970).
24. Barrett, C. S., *Met. Trans.*, **1**, 1601 (1970).
25. Mueller, R. M. and White, W., *Phys. Lett.*, A, **31**, 431 (1970).
26. Barrett, C. S., *Adv. in X-ray Analysis*, **14**, 1 (1971).
27. Barrett, C. S., unpublished work.
28. Barrett, C. S. and Massalski, T. B., *Structure of Metals*, 3rd Ed., New York, McGraw-Hill, p. 421 (1966).
29. Joy, D. C. and Newbury, D. E., in *Scanning Electron Microscopy, 1971* (Part I), Proc. Fourth Annual Scanning Microscope Symposium I.I.T. Research Institute, Chicago, Ill. 1971, pp. 113–120, and private communication.
30. Delsarte, D., Jousset, J. C., Mory, J. and Quéré, Y., in *Atomic Collision Phenomena in Solids*, Ed. D. W. Palmer, M. W. Thompson and P. D. Townsend, Amsterdam, North Holland, pp. 501–512 (1970). A bibliography of alphagraphy and fissiography is included.
31. Quéré, Y., Resneau, J. C. and Mory, J., *C.R. Acad. Sci.*, **262**, 1528 (1966).
32. Mory, J. and Delsarte, G., *Rad. Effects*, **1**, 1 (1969); Mory, J., *Rev. Phys. Appl.*, **3**, 387 (1969).
33. Quéré, Y., *J. Phys.*, **29**, 215 (1968).
34. Quéré, Y., and Couvé, H., *J. appl. Phys.*, **39**, 4012 (1968).
35. Barrett, C. S., *Naturwissenschaften*, **57**, 287 (1970).

CHAPTER XII

Blocking Measurements of Short Nuclear Decay Times

W. M. GIBSON and M. MARUYAMA

12.1 INTRODUCTION

Since the blocking of charged particles in single crystals was reported[1,2,3] in 1965, it has been applied in various fields as is described in the preceding chapter by C. S. Barrett. One of the most unusual and attractive applications of this effect is in the measurement of the decay time of short-lived compound nuclear states. It is our intent in this chapter to discuss the current status of this application with emphasis on the physical basis of the technique and on what we regard as the most important experimental and analytical problems, especially those still awaiting satisfactory solution. It is beyond the scope of this chapter to consider details of the nuclear physics involved although we will try to indicate the general nature of the nuclear information which is being sought and which one might aspire to obtain using this technique.

The possibility that short nuclear lifetimes might be determined by measurement of particle blocking distributions was suggested by Gemmell and Holland[2] and by Tulinov[5] in some of the first publications announcing the existence of particle blocking effects. It was several years, however, before practical realization of this suggestion was achieved by groups in Canada,[6] Japan,[7] United States,[8] Denmark, the Soviet Union[9,10] and England.[11]

It is interesting that the technique had been anticipated by Treacy in 1960[12] in an attempt to detect a time delay effect in the emission distribution of charged nuclear reaction products from a single crystal. This attempt even predated the discovery of directional effects of charged particle motion in crystals and was based on a model of nearest neighbour lattice scattering. The experimental arrangement used by Treacy bears a striking resemblance to those which will be described in the next section, and he almost certainly would have seen strong directional effects (possible even lifetime effects) had his crystals been of sufficient quality. The first successful observations of the effect of nuclear lifetime on the particle blocking distribution were those of Maruyama *et al.*[7] in a study of the inelastic scattering of protons in germanium crystals and those of Gibson and Nielsen[8] in measurements of proton-induced fission in uranium oxide crystals. At the same time, discussions of the analysis of blocking patterns for extraction of nuclear lifetimes were begun by Komaki and Fujimoto,[13] Gibson and Nielsen,[8] Massa[14] and Sona.[15]

The blocking lifetime method is now being applied in numerous laboratories throughout the world, and those measurements known to the authors will be reviewed in the next section. It can be said that the method has now passed beyond the demonstration of feasibility and is rapidly becoming appreciated as a useful tool in nuclear physics. It is interesting to compare the blocking lifetime method with other time measurement methods used in nuclear physics. Figure 12.1 shows time ranges covered by the various possible techniques. The blocking lifetime method offers a unique opportunity

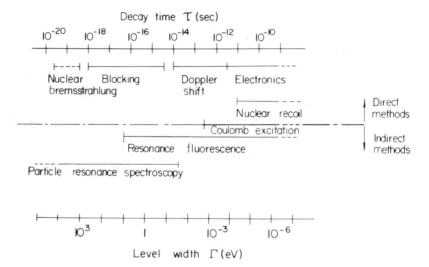

Figure 12.1. Schematic representation of time windows accessible by various direct (upper part of figure) and indirect (lower part) nuclear lifetime measurement techniques. It should be recognized that, in addition to the time windows indicated, each technique has specific problems and limitations which must be considered in its application to any particular case

to measure directly lifetimes in the range of 10^{-14} to 10^{-18} s under favourable conditions.* Although the same range can be covered by resonance fluorescence and particle spectroscopy through application of the uncertainty principle, it is beyond the range of any other direct method. The blocking lifetime method is analogous to the ordinary nuclear recoil method which is

---

* It is interesting to note that the blocking technique is one of the few which allows measurement of truly *nuclear* lifetimes since the level widths in the applicable energy range are controlled with very few exceptions by nuclear emission and absorption. Most other methods of lifetime (or transition rate) determination (Doppler shift, delayed coincidence, Coulomb excitation and resonance fluorescence) deal with levels governed entirely by *electromagnetic* transitions.

applicable to the time range 10^{-7} to 10^{-11} s and is concerned with flight paths around 1 mm. This is also a nuclear recoil method with much shorter flight paths of the order of 10^{-8} to 10^{-7} mm (0·1 to 1 Å).

12.2 BASIC PRINCIPLES

When positively charged particles are scattered or emitted from the nuclei at normal lattice sites in a single crystal, they are restricted from leaving the crystal along low index crystalline axes or planes because of deflection by the positive potential owing to other atoms along those directions. Emergent particle angular distributions show sharp minima along such directions. This is the usual blocking effect.[4] If an excited compound nuclear state is produced by a nuclear reaction at a lattice site by an incident particle beam of several MeV energy, the excited nucleus recoils out of the lattice site with a velocity of the order of 10^8 cm s^{-1}. If the nucleus has a mean lifetime of 10^{-17} s, it will travel on the average 0·1 Å before it emits secondary charged particle decay products. Since the decay function is exponential, some recoils go even further before decaying. If the recoil is perpendicular to a low index plane or axis, then those particles emitted at distances larger than the thermal vibrational amplitude of lattice atoms are not as effectively blocked from motion in the planar or axial direction as they would be if there were no recoil. This results in modification of the blocking pattern in various ways. It can make the minimum yield higher, the dip width narrower and the area or volume of the dip smaller. In some cases, it can even produce a shift or asymmetry of the blocking pattern. From the magnitude of these changes, it is possible, in principle, to determine the mean recoil distance and, from a knowledge of the recoil velocity, the lifetime. Figure 12.2 shows schematically a normal blocking pattern and the change produced by lifetime effects for the usual case in which the crystal is sufficiently thick that asymmetry effects will be averaged out.

Except for special cases where a resonant or strongly energy dependent reaction is being used, it is desirable to use very thin crystals (less than 10 μm). This is in order to confine the range of excitation energy of the excited nucleus, to limit the depth of emission of the charged reaction products in order to avoid excessive multiple scattering or to restrict the energy spread of emitted particles in order to allow definition of the reaction to a particular state of the product nucleus. Obtaining, handling and preserving through the bombardment such thin crystals of the high crystalline perfection and precise and stable orientation necessary, represents a difficult and important experimental challenge. This question will be discussed briefly below.

Figure 12.3(a) shows a typical experimental arrangement for a thick crystal measurement, in this case the reaction under study is heavy-ion-induced fission.[16] Figure 12.3(b) shows a typical experimental setup for

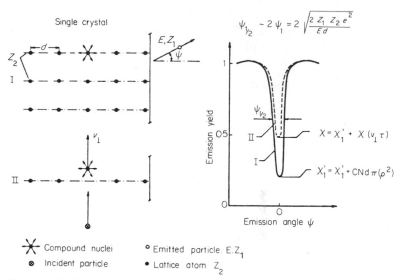

Figure 12.2. Schematic representation of the blocking lifetime measurement technique. The modifications in the emergent particle blocking pattern are characteristic of an emission depth sufficiently large that statistical equilibrium is attained (see text)

measurement of the nuclear reaction time of proton inelastic scattering in a thin crystal.[17] In general, of course, measurement at 90° to the incident beam is most sensitive since the recoil momentum vector is perpendicular to the axis or plane being studied. Measurements at more forward or more backward angles (especially for thick crystal experiments) are also possible and sometimes even necessary if the nuclear reaction being used has a yield minimum at 90°. In fact, control of the emission angle relative to the beam offers a method to shift by as much as a factor of ten the range of lifetimes over which the technique can be applied and can also give information on the nature of the decay function. Lifetimes longer than 10^{-15} s may require that the recoil momentum vector be at relatively small angles to the emission direction.

The detector used can take several forms depending on the charged particles being measured and on the radiation background present. Ideally, it is desirable to measure the entire emission angular distribution in the region of the crystal axis or plane of interest. Plastic or glass track detectors[18] have been used for lifetime studies involving detection of fission fragments[6,8,9,16] or helium ions[19,20] produced in proton or neutron-induced nuclear reactions. After chemical development, individual particle tracks become visible under high magnification and the number of these as a function of position on the detector (each position corresponding to a particular emission direction) are determined. In charged-particle-induced reaction studies, the

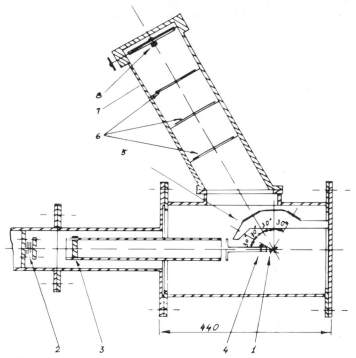

Figure 12.3. (a) Experimental setup for measurement of charged-particle-induced fission lifetimes (after Karamyan *et al.* reference 16). 1, Single crystal; 2, decelerating foils; 3, 4, collimators; 5, 6, glass plates for track registration; 7, tubing; 8, movable semiconductor detector

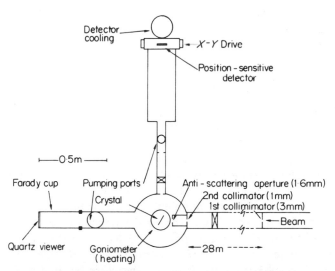

Figure 12.3. (b) Experimental setup for measurement of compound nuclear lifetimes by inelastic scattering in thin crystals (after Gibson *et al.* reference 17)

reaction products of interest are accompanied by a large background of particles elastically scattered from the crystal. Suitable choice and careful development of the detector film are necessary to allow only the heavier particle tracks to be seen and not the tracks from the scattered beam. In many cases, it is also necessary to characterize the energy of the emergent particles. For such measurements position-sensitive semiconductor detectors capable of determining the position of impact of the particle in one or two dimensions as well as its energy have been used.[7,8,11] As the spatial and energy resolution of two-dimensional, energy-sensitive, position-sensitive detectors continues to improve,[21,22] it is expected that they will play an increasingly important role in such measurements.

An X-ray diffraction measurement or particle channeling measurement can be used to orient the crystal relative to the incident beam and detectors. It is often desirable however to monitor the orientation and quality of the crystal during the experiment. When thin crystals are used, this can be done in a convenient qualitative fashion by placing a quartz plate, or phosphor screen in the transmitted particle beam and choosing a crystal orientation such that a blocking pattern falls within the viewing area. In both thick and thin crystal experiments, it is also possible to place a position-sensitive particle detector to monitor continuously the blocking pattern for elastically scattered particles during the measurement. Such a system is shown in Figure 12.4 as used by Sharma et al.[19] in a determination of the lifetime of a ^{32}S compound nucleus as measured by a ^{31}P (p, α) ^{28}Si reaction.

In practice it is not sufficient to measure an emergent particle blocking pattern and compare it with a theoretical distribution in order to obtain a lifetime. Other effects associated with scattering in the crystal or at the crystal surface generally allow only an upper lifetime limit to be set by such a measurement.[6] This is mainly due to the small reaction yield and limited detector solid angle which requires extensive irradiation which in turn produces radiation damage or surface contamination during the measurement. It is necessary therefore to obtain a reference blocking pattern which does not contain the lifetime effect. This should be obtained under the same conditions and preferably at the same time as the measurement. One way to do this is shown in Figure 12.4 where the crystal orientation is chosen such that the blocking pattern for emergence along a $\langle 111 \rangle$ axis at 170° from the incident beam direction is measured simultaneously with a blocking pattern for emergence along a $\langle 111 \rangle$ axis at 81° to the beam. The recoil is nearly parallel to axis A so this should not contain an appreciable lifetime effect and can be used as a control for axis B. In most charged-particle-induced reactions elastic scattering of the incident beam arises from Rutherford scattering and corresponds to an extremely short reaction time ($< 10^{-20}$ s). The blocking pattern produced by these particles can therefore be used as a convenient reference.[7,11]

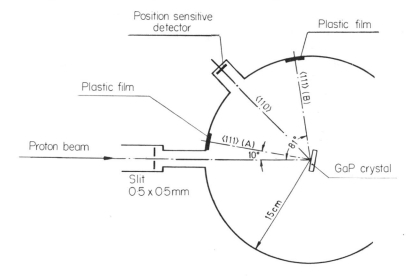

Figure 12.4. Experimental setup for measurement of compound nuclear lifetimes by charged-particle-induced reactions in thick crystals. The semiconductor position-sensitive detector is used in this arrangement to monitor the blocking distribution for elastically scattered beam particles (after Sharma, Andersen and Nielsen, reference 19)

12.3 TYPICAL RESULTS

12.3.1 Fission

Nuclear fission represents a process which is a natural object of such lifetime studies. This is because of its collective nature, its general interest and importance, the energetic, charged, easily detectable particles associated with it and because of the traditionally important role that lifetime measurements have played in fission studies. Indeed, the short time range made accessible by this technique offers the possibility of gaining information about the shape of the fission barrier and also allows flexibility and control over the choice and excitation of the fissioning compound system.

As noted in the previous section, fission studies were among the first to utilize the blocking technique. All of the fission studies to date have used detectors of the same type as those employed in the first measurements of Brown *et al.*[6] Lexan,[6] cellulose acetate[8] or glass[9] track detectors have been used to measure the fission fragment angular distributions. Brown *et al.*[6] and Gibson and Nielsen[8] studied fission of excited neptunium compound nuclei produced by bombardment of uranium dioxide crystals with protons. The latter workers found that perturbation of the blocking pattern was produced for 10 MeV protons when compared with bombardments at higher (12 MeV) and lower (9 MeV) energies. A comparison of the blocking dips for 10 and 12 MeV protons is shown in Figure 12.5. The change in the minimum yield was utilized to give a total lifetime for the 10 MeV bombardments of $1.4 \pm 0.6 \times 10^{-16}$ s. This result was consistent with the lifetime calculated from the change in width of the blocking dip and with the change in the

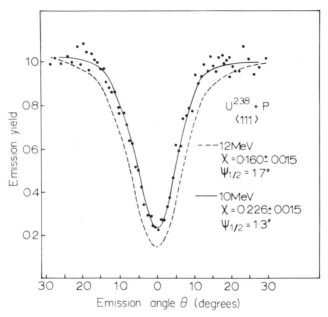

Figure 12.5. Measured angular distributions, normalized at large emission angles, of fission fragments emitted parallel to a $\langle 111 \rangle$ axis from a UO_2 crystal bombarded with 10 and 12 MeV protons. The axial direction was $40° \pm 2°$ from the incident beam direction (after Gibson and Nielsen, reference 8)

blocking dip for particle emission along a $\langle 110 \rangle$ axis at $60°$ to the particle beam compared with emission along a $\langle 110 \rangle$ axis at $175°$ to the particle beam direction. Figure 12.6 illustrates the reaction sequence involved in this measurement and serves to illustrate some of the complications associated with fission lifetime studies as well as to show why a lifetime effect could be observed only over a narrow range of incident proton energies. At all proton energies sufficient to penetrate the uranium Coulomb barrier (~ 7 MeV), the compound nucleus $^{239}Np^*$ has very high excitation energy and decays (principally by fission and neutron emission) in a time too short to be measured by this technique. Up to almost 10 MeV proton energy, this is the only source of fission products. At about 10 MeV the fission fragments from

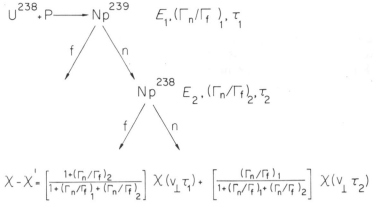

$$U^{238} + P \longrightarrow Np^{239} \quad E_1, (\Gamma_n/\Gamma_f)_1, \tau_1$$

$$Np^{238} \quad E_2, (\Gamma_n/\Gamma_f)_2, \tau_2$$

$$\chi - \chi' = \left[\frac{1+(\Gamma_n/\Gamma_f)_2}{1+(\Gamma_n/\Gamma_f)_1+(\Gamma_n/\Gamma_f)_2}\right] \chi(v_\perp \tau_1) + \left[\frac{(\Gamma_n/\Gamma_f)_1}{1+(\Gamma_n/\Gamma_f)_1+(\Gamma_n/\Gamma_f)_2}\right] \chi(v_\perp \tau_2)$$

Figure 12.6. The reaction sequence for fission fragments produced by proton bombardment of ^{238}U. For proton energies above 9 MeV fission fragments from both the initial compound nucleus $^{239}N_p^*$ (with excitation, E_1; branching ratio $(\Gamma_n/\Gamma_f)_1$ and lifetime τ_1) and from the secondary compound nucleus $^{238}N_p$ (with excitation energy, E_2; branching ratio; $(\Gamma_n/\Gamma_f)_2$; and lifetime, τ_2) are observed. The contributions of each component to the change in minimum yield is determined by the branching ratios and mean recoil distances as shown by the expression for $\chi - \chi'$

decay of the $^{238}Np^*$ compound nucleus produced by neutron emission from $^{239}Np^*$ begin to contribute significantly to the observed fission yield. At this 'threshold energy' the $^{238}Np^*$ compound nucleus is at low excitation and its decay is slow enough to give measurable lifetime. As the proton energy is increased, the excitation energy of this compound nucleus goes up and the lifetime again becomes too short to be observed. This also shows why only a lifetime limit was possible from the 12 MeV proton measurements of Brown et al.[6] or the measurements of 25 MeV helium-ion-induced fission of Melikov et al.[10]

Observation of blocking dips along $\langle 111 \rangle$ axes in the UO_2 crystals which are wider and deeper than along $\langle 110 \rangle$ axes[8] graphically illustrates the importance of multiple scattering effects in such measurements. The $\langle 110 \rangle$ axial directions contain oxygen atomic rows which act to scatter fission fragments emitted from the stronger uranium rows whereas the $\langle 111 \rangle$ directions have rows of only one type which contain both oxygen and uranium atoms.

Uncertainty in the excitation energy of the fissioning system and the complicated reaction sequence limits the usefulness of charged particle fission studies of heavy elements. This has led to attempts to measure the lifetime of fission induced by monoenergetic neutrons by Melikov et al.[10,23] and by Anderson et al.[24] Problems associated with production of an intense, reasonably well collimated neutron beam with well defined energy have thus far hampered this application but initial reports of success have been indicated by both the Russian and Danish groups.

Bombardment of lower Z targets with heavy ions offers another attractive possibility for fission studies at excitation energies close to the fission barrier. Using the apparatus shown in Figure 12.3(a) Karamyan *et al.*[16] initially reported an upper limit of 10^{-17} s for the mean decay time of polonium compound nuclei as studied by observing fission fragments from the $W(^{22}Ne, f)$ reaction induced by 174 and 98 MeV neon ions incident on a tungsten crystal. In further studies, the same workers have refined the measurement and reported monotonically increasing decay times for fission induced by 174, 146 and 116 MeV neon ions as well as the absence of any observable effect for 137 MeV ^{18}O ions.[25] The minimum yields, χ, at 90° and 160° to the incident ion beam and the difference, $\Delta\chi = \chi_{90°} - \chi_{160°}$, are shown in Table 12.1. $\Delta\chi$ is related to the lifetime of the compound nuclei as will be discussed in Section 12.4.1. Applying the analysis of Gibson and Nielsen,[8] for example, gives a mean decay time of $2·5 \times 10^{-18}$ s and $4·0 \times 10^{-18}$ s for the 174 and 116 MeV ^{22}Ne bombardments respectively.

Table 12.1 Minimum yield changes for heavy-ion-induced fission† (from reference 16)

Reaction	Ion energy	Minimum yield		$\Delta\chi$	$\overline{\Delta\chi}$
		$\chi_{90°}$	$\chi_{160°}$		
$W(^{22}Ne, f)$	174	0·216	0·180	0·036	
		0·159	0·121	0·038	0·040 ± 0·003
		0·276	0·234	0·042	
		0·324	0·337	0·047	
		0·168	0·132	0·036	
		0·410	0·371	0·039	
$W(^{22}Ne, f)$	146	0·219	0·168	0·051 ± 0·008	
$W(^{22}Ne, f)$	116	0·222	0·160	0·062	0·070 ± 0·004
		0·253	0·174	0·079	
$W(^{16}O, f)$	137	0·152	0·157	−0·005 ± 0·010	
$W(^{84}Kr, ^{84}Kr)$	74		0·292	−0·006 ± 0·012	
$W(^{84}Kr, ^{84}Kr)$	47		0·298		

† $\Delta\chi$ for the ^{84}Kr scattering measurements is taken from the two ion energies as $\Delta\chi = \chi_{74} - \chi_{47}$.

12.3.2 Inelastic Scattering

Inelastic scattering of charged particles offers the possibility of studying the lifetime of excited states of a wide variety of compound nuclei. This is limited only by the target isotopes which can be obtained in suitable thin crystal form and by the available type and energy of ion beams. In general, because of low reaction yield and current flux limitations in the measurements, it is necessary that the target crystal be thicker than about 1 μm. In many cases, thicknesses less than 10 μm are impracticable. Except for very light target nuclei, this means that many compound nuclear states will be excited over the energy thickness of the crystal. This region of very closely spaced or even overlapping levels is just where resonance fluorescence or other level width measurements cannot be applied and where statistical theories of level densities, spacing and orderings become most valid and useful in interpreting the results. The average nature of the measured lifetime for such cases is similar to that encountered in the fission lifetime case and results in similar problems in extracting meaningful lifetime values from blocking angular distributions. This will be discussed in Section 12.4.2. By making use of characteristics of the excitation function, such as sharp analogue resonances, and of the detailed energy spectrum of the scattered particles, it is possible to make the studies somewhat more specific than implied above. The situation for proton scattering from germanium is illustrated schematically in Figure 12.7.

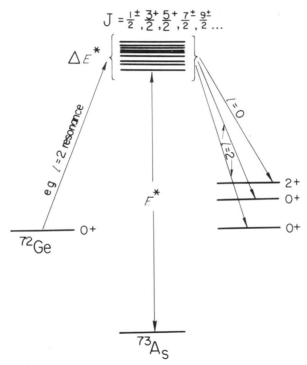

Figure 12.7. Schematic representation of excitation of ^{73}As compound nuclei by proton bombardment and decay by proton emission. The range of excitation energies ΔE^* is determined by the target thickness. The case shown involves excitation of analogue states in the compound system via an $l = 2$ resonance and decay from these states to particular states of the ^{72}Ge daughter. Nonresonant levels are also involved in the excitation and decay

Figure 12.8 shows typical energy spectra of protons scattered at 90° from a thin germanium crystal of 1·5 μm thickness at incident proton energies of 5·035 and 5·110 MeV.[26] The very strong group at high energy is due to elastic scattering from ^{70}Ge, ^{72}Ge and ^{74}Ge nuclei in the natural germanium target crystal. Each of the other peaks at lower energy corresponds to proton decay of the excited ^{71}As*, ^{73}As* and ^{75}As* compound nuclei to the specific excited states of the respective germanium daughter nuclei as indicated in the figure. These spectra were taken with a cooled 5 cm × 1 cm × 3 mm one-dimensional position-sensitive silicon semiconductor detector made by ion implantation[27] used in the apparatus shown in Figure 12.3(b). The detector, masked to 1·5 mm width along its long axis, was positioned 130 cm

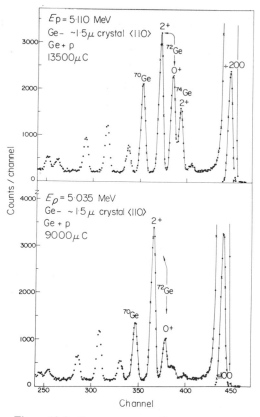

Figure 12.8. Energy spectra of protons scattered at 90° from a 1·5 μm thick germanium crystal for incident energies of 5·035 and 5·110 MeV. The proton energies were chosen to correspond with the resonant energies for excitation of analogue states in ^{71}As* and ^{73}As* compound nuclei (after Gibson *et al.*, reference 26)

from the target and carefully positioned across a ⟨110⟩ blocking minimum in such a way so as to avoid planar blocking dips. This allows the blocking distribution to be measured simultaneously for each of the energy groups as well as the background between them. Figure 12.9 shows a photograph of

Position against energy spectra
5·11 MeV protons on germanium

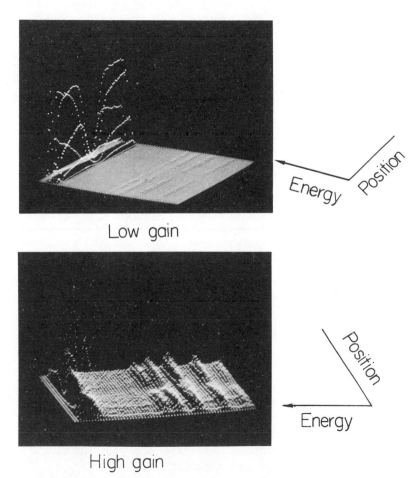

Figure 12.9. Position against energy spectral distributions photographed from cathode tube displays of the response of a one-dimensional position-sensitive semiconductor detector for 5·110 MeV proton bombardment of a thin (1·5 μm) germanium single crystal. The detector, at 90° to the incident beam direction, was positioned to intersect a ⟨110⟩ axial blocking dip. The high energy proton group (on the left) arises from elastic proton scattering in the crystal (after Gibson *et al.*, reference 17)

the cathode ray tube display from a computer-controlled two-dimensional energy against position spectrum in which only the energy range for the four highest energy groups are included. Axial $\langle 110 \rangle$ blocking dips measured in this way for proton groups corresponding to the ^{72}Ge(p, p')^{72}Ge* (2 + 2nd excited state, 0·835 MeV) are shown in Figure 12.10 for incident proton

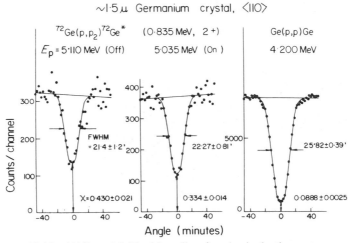

Figure 12.10. $\langle 110 \rangle$ axial blocking dips for the inelastic proton groups corresponding to the ^{72}Ge (p, p') ^{72}Ge* (2$^+$ 2nd excited state, 0·835 MeV) reaction for incident proton energies of 5·110 MeV (off the resonance for excitation of the 1/2$^+$ analogues resonance in ^{73}As*) and 5·035 MeV (at the resonance energy). Also shown for comparison is the blocking dip for elastically scattered protons at an incident energy of 4·20 MeV which corresponds to the mean emergent particle energy for the inelastic groups (after Gibson *et al.*, reference 26)

energies of 5·110 and 5·035 MeV and also a reference distribution of protons elastically scattered at the same mean energy (corresponding to an incident energy of 4·2 MeV). The blocking dips for the inelastic groups show an increased minimum yield as well as decreased width relative to the reference elastic case. At an incident proton energy of 5·035 MeV, the yield of this inelastic proton group is strongly enhanced by an $l = 2$ analogue resonance.[28] Reduction of the mean compound nucleus lifetime by this resonance is apparent. In the same experiments lifetimes for decay to other excited states in ^{72}Ge, ^{70}Ge and ^{74}Ge were also investigated.[26]

The very thin crystals required in studies of these resonance effects introduce special problems because of thermal, radiation-induced[29] or electrostatic crystal distortions during the measurements. The blocking pattern is extremely sensitive to even very small distortion effects and constant monitoring of the pattern with a detector or a phosphor viewing screen is necessary. The above measurements were taken with the crystal heated to 450°C, and the beam current was limited to 100 nA to reduce radiation and beam heating effects. Unfortunately, each material has different properties and requires careful determination of the optimum conditions. For germanium crystals the distortion effects are markedly reduced for crystals with thickness greater than 5 μm.

The first measurements of this type were those of Maruyama *et al.*[7] who also investigated inelastic proton scattering from germanium. In that work, planar blocking dips were investigated. The influence of nuclear processes on the lifetime was demonstrated by measurement of the change of the area M of the planar blocking dip as a function of incident proton energy. This is shown in Figure 12.11 for Ge(p, p)Ge, ^{70}Ge(p, p')^{70}Ge* (1st excited state,

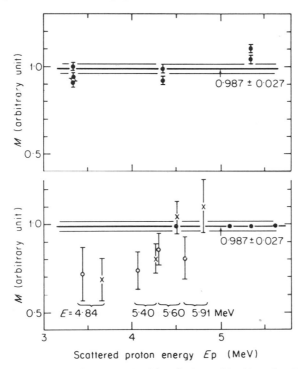

Figure 12.11. The area (M) of $\{111\}$ planar blocking dips for proton bombardment of a $10\,\mu$m germanium crystal. The observed areas have been normalized to the value for elastically scattered protons having the same energy on emergence from the crystal. The upper part shows the energy dependence of the area of the elastic group. The lower part shows the results of simultaneous measurements for both elastic and inelastic groups. Closed circles are for ^{70}Ge (p, p) ^{70}Ge elastic group; open circles, ^{70}Ge (p, p') ^{70}Ge* (1st excited state, 1·04 MeV, 2$^+$); and crosses, ^{72}Ge (p, p') ^{72}Ge* (2nd excited state, 0·835 MeV, 2$^+$) (after Maruyama *et al.*, reference 7)

1·04 MeV, 2 +) and ^{72}Ge(p, p')^{72}Ge* (2nd excited state, 0·835 MeV, 2 +)
reactions. The elastic scattering shown in the upper part of the figure was
used as a reference. The same value of M as the elastic value 0·987 corresponds to a very short compound nucleus lifetime. At incident proton energies less than 5·40 MeV, values of M for both inelastic groups are lower than
the elastic value showing lifetimes of about 3×10^{-17} s for ^{71}As* and ^{73}As*
compound nuclei. However, at higher incident energies an increase of M
is observed for the ^{72}Ge(p, p')^{72}Ge* reaction while M for the ^{70}Ge(p, p')^{70}Ge*
reaction remains about the same. This is explained by the opening of a neutron
decay channel for the compound nucleus ^{73}As* at the threshold energy of
5·21 MeV.

The pioneering Japanese work was confirmed by a combined English–
Italian team[11] who used axial blocking and were the first to utilize dynamic
two-dimensional position-sensitive and energy-sensitive detection of the
scattered protons. Figure 12.12 shows an example of two-dimensional

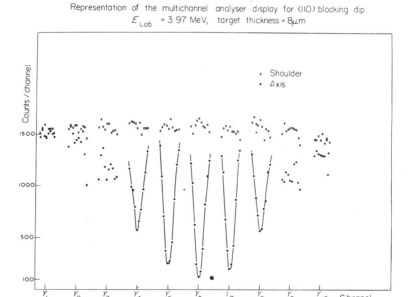

Figure 12.12. $\langle 110 \rangle$ axial blocking pattern for 5·00 MeV protons elastically scattered from a 8 μm thick germanium crystal (the energy of 3·97 MeV
indicated on the figure corresponds to the mean energy of scattered protons
emergent from the crystal). The Y's of the abscissa indicate the ten strips
in the Y direction on the front face of the checkerboard detector. The
indicated points within each Y value corresponds to the counting rate in
each of the $10X$ strips on the rear of the detector in coincidence with a
signal in the appropriate Y strip. The counts in the shoulder of the blocking
dip were obtained by translating the counter an amount corresponding to
a change in emission direction of 2° (after Clark *et al.*, reference 11)

position spectra measured by these workers using a 10×10 checkerboard detector.[30] In this figure a spectrum above Y_i denotes a position spectrum along the X-direction for a given position Y_i along the Y direction. The blocking dip in this figure is for a $\langle 110 \rangle$ axis from 3·97 MeV protons elastically scattered in an 8 μm thick germanium crystal.

It is of interest to examine in detail the energy against position spectrum from this type of measurement. Figure 12.13 shows a spectrum of 5·42 MeV

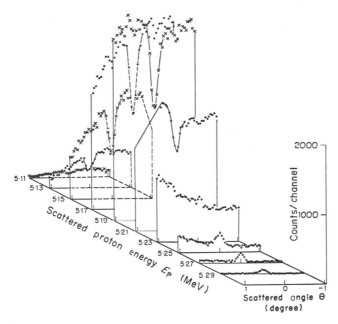

Figure 12.13. Angular against energy distributions of 5·42 MeV protons elastically scattered from a 10 μm thick germanium crystal. The abscissa angles are given relative to a $\{111\}$ planar direction (after Maruyama *et al.*, reference 7)

protons scattered at 90° from a 10 μm thick germanium crystal. Scattered protons in the energy range 5·11 to 5·29 MeV are shown as a function of position in a one-dimensional position-sensitive semiconductor detector placed to intersect a $\{111\}$ planar blocking dip. On the low energy side of this scattered energy group, the position spectra show clearly the $\{111\}$ blocking patterns. As the scattered proton energy increases, the blocking dips become less deep and on the very high energy side of the group, the dip changes to a channeling peak. In this elastic scattering example, the channeling peak corresponds to protons which have scattered from atoms far removed from the normal lattice site by thermal vibration or by lattice imperfection. The channeling peak is always observed to increase in the presence of a lifetime effect (as for the inelastic groups) because of emission of some particles at large distances ($\geqslant 0.5$ Å) from lattice sites. Although quantitative analysis of this effect is not yet possible, it is a sensitive indicator of very slowly decaying components and may be especially important in cases involving complicated decay functions (see Section 12.4.2).

Thus far, germanium remains as the only material studied in detail by inelastic scattering. The generality of the technique, however, makes it very attractive and as thin crystals of other materials become available, a large variety of such studies are expected. Already preliminary accounts of inelastic proton scattering in molybdenum,[31] copper[32] and gold[33] have been reported.

12.3.3 Resonance Reactions

The cases discussed in the preceding sections all involved excitation of many levels of the compound nuclei and in the fission case also many possible decay channels. Such results, therefore, are statistical averages and their analysis requires knowledge or assumptions about the appropriate form of the decay function. Important simplications arise for a system in which only one state in the compound nucleus is excited and the decay through only one decay channel is measured. In this case, the decay function is a simple exponential. If at the same time, information on the lifetime of the state is independently available (as from resonance analysis), then simple testing and calibration of the lifetime measurement technique becomes possible. Preliminary results from two such measurements have been reported. These are for resonant ^{27}Al$(p, \alpha)^{24}$Mg[19,20] and ^{31}P$(p, \alpha)^{28}$Si[19] reactions. Figure 12.14 shows $\langle 111 \rangle$ blocking dips for the latter case using the experimental arrangement shown in Figure 12.4. The change in the depth and width of the blocking dip when recoil is at 81° to the axis compared with that obtained for recoil almost parallel(10°) to the axis corresponds to a lifetime of approximately 8×10^{-15} s for the 9·486 MeV excited state of the ^{32}S compound nucleus. This corresponds to a level width of $\Gamma = 8 \pm 4$ eV which can be compared with the $\Gamma = 8 \pm 2$ eV calculated from the measured strengths to the ^{31}P(p, α), ^{31}P(p, γ) and ^{28}Si(α, γ) resonance reactions feeding the same 9·486 MeV level in ^{32}S.

This case is especially convenient because of the absence of α-particle production at incident proton energies away from resonance. As a result thick (GaP) crystals could be used, and the incident proton energy could be chosen to cause the α-particle producing reactions to take place at a well defined, shallow (1500 Å) depth in the crystal. Detailed, systematic investigation of such a system will be extremely important in solving many of the analysis problems discussed in the next section.

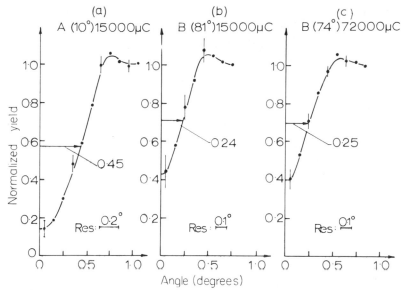

Figure 12.14. Blocking dips for α-particles emitted in the ^{31}P (p, α) ^{28}Si reaction at 0·642 MeV. The dips (a) and (b) were obtained by counting the tracks in $\frac{1}{4}$ mm wide circular rings around the $\langle 111 \rangle$ direction. In (c) the film was at twice the distance (30 cm) and tracks were counted in $\frac{1}{2}$ mm rings (after Sharma, Andersen and Nielsen, reference 19)

12.4 ANALYSIS

12.4.1 Simple Decay Function

Two parameters often used to characterize blocking (or channeling) dips are the minimum yield (emission yield directly along the axial or planar direction) and the width (taken as the half width of the dip at the half minimum point).[34] Other parameters which can also be used are the 'area' of the dip (for planar blocking) or the 'volume' of the dip (for axial blocking). Any one of these parameters is, in principle, sufficient to determine the compound nucleus lifetime for the case in which the decay is assumed to be a simple exponential function of time $A/A_0 = e^{-\lambda t}$.

(1) Minimum Yield

(a) *Continuum model.* A major effect in the emission angular distribution of charged particles from a crystal compared with emission from a random material is the large decrease in emission yield along crystallographic directions. This decrease (the ratio of the aligned to the random yield is designated as χ and called the 'minimum yield') is used to indicate the strength of the blocking effect and is therefore expected to be one of the most sensitive measures of the lifetime. For a continuum approximation to the row potential for a crystal axis and for small displacements from mean lattice positions, Lindhard[34] has derived a relation between the minimum yield and the mean square displacement $\langle r^2 \rangle$

$$\chi = \langle r^2 \rangle / r_0^2.$$

The effective area of each row in a plane perpendicular to the row is given by $\pi r_0^2 = 1/Nd$, where N is the atomic density in the crystal and d is the atomic spacing along the row. In normal emission from lattice atoms, $\langle r^2 \rangle$ can be replaced by the mean square thermal vibrational amplitude ρ^2. The continuum approximation in the derivation results in underestimation of the minimum yield by this relationship. This effect has been investigated for planar[35] and axial[36,37] channeling. On the basis of Monte Carlo calculations of particle trajectories, Barrett[36] has proposed the form

$$\chi = C \langle r^2 \rangle / r_0^2 = CNd\pi\rho^2 \tag{12.1}$$

for emission from axes and has found for 400 keV to 5 MeV protons in a variety of crystals over a wide temperature range that $C \approx 2$–3. In the case of a finite lifetime, the displacement x, caused by recoil of the compound nucleus, must be taken into account. For simple exponential decay with a mean recoil distance $(v_\perp \tau)$ (where $v_\perp$ is the recoil velocity perpendicular to the row and τ is the mean lifetime). This gives a minimum yield:

$$\chi = CNd\pi(\rho^2 + 2v_\perp^2\tau^2). \tag{12.2}$$

The factor of two in the recoil displacement term comes from integration of the exponential decay function from zero to the centre of the channel, r_0.

In general, for cases in which the lifetime effect is absent, observed minimum yields are larger than indicated by equation (12.1) because of scattering in the crystal and at the crystal surface. This is influenced by the depth of emission, the initial perfection of the crystal, radiation damage during the measurement, crystal distortion and by the presence of impurity films on the crystal surface. Since these effects contribute independently to the minimum yield, they are generally combined with the thermal effects into an additive constant. The change in minimum yield owing to the lifetime effect is given as

$$\Delta\chi = \chi - \chi' = 2CNd\pi v_\perp^2\tau^2 \tag{12.3}$$

which assumes that, except for lifetime effects, all other contributions to the minimum yield for the reference measurement (χ') are the same as the primary measurement (χ). Independence of $\Delta\chi$ for values of χ' varying over a factor of three (from 0·12 to 0·37) in ^{187}W(Ne, f) studies (see Table 12.1)[25] and over a factor of about two (0·23 to 0·45) in ^{238}U(p, f) studies[8] supports this assumption.

Scattering inside the crystal involving changes of transverse energy have yet another effect on the minimum yield analysis. Equation (12.3) assumes that steering of the particle trajectory by the continuum potential takes place at arbitrarily large values of r (out to r_0). The appropriate screened Coulomb potentials are known, however, to decrease rapidly in both magnitude and gradient as the distance is increased. Depending on the depth of emission and the type and quality of the crystal, changes in angle owing to such scattering should become equal to the angular changes induced by the potential at some distance of emission from the row, r_c, called the critical cutoff distance. Particles emitted at distances larger than r_c are not blocked by the row potential and are therefore able to move in the direction of the row. The contribution of such particles to the minimum yield is given by their fractional probability, which for simple exponential decay is $\exp(-r_c/v_\perp \tau)$. Inclusion of this into the minimum yield analysis gives[8]

$$\Delta \chi = \chi - \chi' = \overbrace{\frac{2Cv_\perp^2 \tau^2}{r_0^2} \left\{ 1 - \left(1 + \frac{r_c}{v_\perp \tau} \right) \exp\left(-\frac{r_c}{v_\perp \tau} \right) - \frac{1}{2} \left(\frac{r_c}{v_\perp \tau} \right)^2 \exp\left(-\frac{r_c}{v_\perp \tau} \right) \right\}}^{\text{I}}$$

$$\overbrace{+ \exp\left(-\frac{r_c}{v_\perp \tau} \right)}^{\text{II}} \tag{12.4}$$

where I includes those particles emitted inside r_c and II those emitted at distances larger than r_c. The relative contributions of these two terms to the minimum yield is shown in Figure 12.15 which shows $\Delta \chi$ as a function of the mean displacement for fission fragments emitted along the $\langle 111 \rangle$ axes in uranium dioxide crystals for three values of the cutoff parameter r_c. It is evident from this figure that even for large values of the cutoff (corresponding to very little multiple scattering), the second term in equation (12.4) is dominant except for very small mean displacements.

These calculations assume statistical equilibrium for particle motion in the transverse plane. This assumption becomes valid only after penetration of a thickness which depends on the crystal, the emitted particle Z and the emitted particle's longitudinal and transverse energy. Transmission studies through thin silicon and germanium crystals[38] indicate that the necessary thickness for 4 MeV protons of relatively low transverse energy may be almost one micrometre in germanium and almost two micrometres in silicon.

However, a much more serious assumption lies in the cylindrically symmetric single string continuum potential used to determine the particle motion through the lattice. For phenomena which take place within about $0.5 \, \text{Å}$ of the atomic row, this approximation is valid. At larger distances, however, the potential contours deviate markedly from cylindrical symmetry

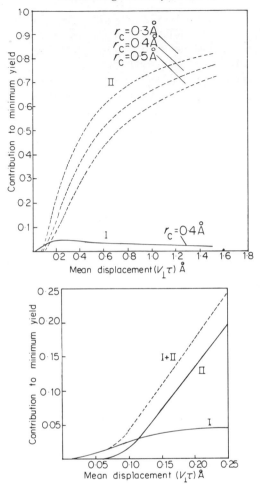

Figure 12.15. Equation (12.4) plotted for emission along a [111] axis in uranium dioxide for $C = 2·5$. The top part of the figure shows the second term of the equation for three different values of the cutoff parameter r_c and the first term for $r_c = 0·4$ Å). The lower part is an expanded plot for small values of the mean displacement for $r_c = 0·4$ Å (after Gibson and Nielsen, reference 8)

as shown in Figure 12.16 for a $\langle 110 \rangle$ axis in germanium. It is apparent from the importance of the second term of equation (12.4) that the long tail on the exponential decay function plays an important role in determining the minimum yield. Consequently, the assumption of cylindrical symmetry must be questioned even for short mean recoil distances.

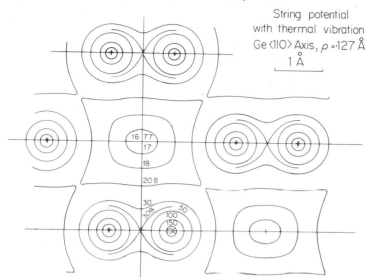

Figure 12.16. Potential contours in a plane perpendicular to a ⟨110⟩ axis is germanium based on a Lindhard standard free atom potential. The potential values for the various contours are given in volts

(b) *Computer simulation.* An alternative to analytical calculations based on the single-string continuum potential approximation is to consider charged particle motion through single crystals as independent binary scatterings of the particle with individual lattice atoms. Such calculations have been carried out in considerable detail in investigations of particle channeling and blocking phenomena[39–41] and have been shown to give good correlation with continuum calculations in the single string limit.[42] Application of this approach to analysis of blocking determinations of nuclear lifetimes was suggested by Massa[14] and Sona[15] and has been further developed by Clark *et al.*[11]

Since individual atomic scattering events are included and since the atoms are allowed to vibrate thermally, multiple scattering owing to lattice atoms is included in these calculations in a natural way. Unfortunately, the huge amount of computer time required often restricts such simulations to limited emission depths, numbers of particles and emission conditions. In addition, numerous recalculations are required to investigate the effect of different decay times, decay functions, recoil directions etc. The computation time may be substantially reduced in many cases by turning the calculation around, calculating the particle trajectories for a channeling geometry[41] and using the 'rule of reversibility'[34] to apply it to the blocking case. In this way, the same particle trajectories can be used over and over for different

lattice vibrations, compound nuclear decay functions and recoil directions
as well as for investigations of particle flux distributions, depth effects and
other parameters.

A large scale Monte Carlo computer simulation of this type has recently
been carried out for the case of proton scattering in a thin germanium crystal
illustrated in Figures 12.9 and 12.10.[43] Computed values of the minimum
yield, χ, against the mean recoil distance for three different recoil directions
are shown by the solid lines in Figure 12.17. The dashed lines are continuum

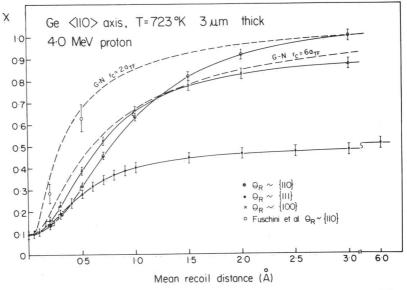

Figure 12.17. Calculated values of the minimum yield, χ, as a function of the
mean recoil distance, $v_\perp\tau$. The solid curves were calculated by the Monte
Carlo program described in the text for three different recoil directions
tilted about $1\cdot5°$ from the planar direction indicated in each case. The dotted
curves calculated by equation (12.4) for two different values of the critical cutoff
distance, r_c and the open circles from the Monte Carlo calculations of Fuschini
et al. (reference 44) were shifted to coincide with the solid curves at zero mean
recoil distance (after Hashimoto, Barrett and Gibson, reference 43)

model calculations for two different values of the critical cutoff radius (equation (12.4)), and two points are shown from the Monte Carlo calculation of
Fuschini *et al.*[44] All cases correspond to a $\langle 110 \rangle$ blocking pattern measured
at $90°$ to the incident beam direction. The solid curves are not normalized
and the excellent agreement at zero recoil with the minimum yield measured
for the elastic dips of Figure 12.10 shows that the multiple scattering effects
through the $3\ \mu$m emission depth in the crystal are well reproduced in the

calculation. The analytical curves and the results of Fuschini *et al.* are normalized at zero recoil because they are calculations of $\Delta\chi$ and do not contain thickness effects.

The most striking and important feature of Figure 12.17 is the strong dependence of the minimum yield on the recoil direction. This is a direct result of the potential contour distribution shown in Figure 12.16 and is closely related to the flux peaking effects discussed in Chapters 2 and 3 of this volume. In fact the same particle trajectory calculations used to obtain the results of Figure 12.17 can be used to obtain the flux distribution of channeled particles. This is shown in Figure 12.18 for particles that have

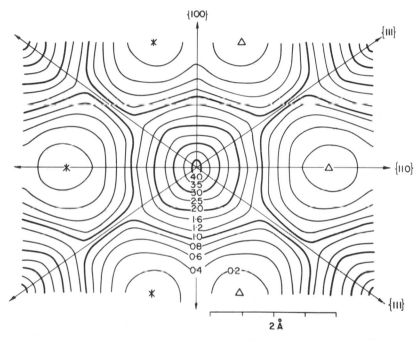

Figure 12.18. Flux density contours for 4·0 MeV protons incident parallel to a $\langle 110 \rangle$ axis in germanium evaluated for the depth range of 0 to 3·0 μm into the crystal by a large scale Monte Carlo calculation. The triangles and stars correspond to mean lattice positions (in two different horizontal planes) for atoms in the $\langle 110 \rangle$ axial rows of germanium

penetrated 3 μm into the crystal.[43] The regions of high particle flux correspond to points from which particles emitted in the row direction can move easily through the crystal. Recoils originating at atomic lattice sites which move nearly parallel to $\{111\}$ planes avoid regions of high flux density which accounts for the low value of χ for that recoil direction compared with recoil along either $\{110\}$ or $\{100\}$ planes.

The results of Monte Carlo calculations of Fuschini *et al.*[44] are in marked disagreement with the {110} curve of Figure 12.17 which corresponds to the same recoil direction and proton energy. This may be a result of the shallow emission depth used in the Fuschini *et al.* calculations (2000 Å). It has been shown by Barrett[36] that the minimum yield for elastically scattered particles ($v_\perp \tau = 0$) undergo fluctuations as the emission depth is increased and this effect appears to be enhanced both in amplitude and depth for larger recoil distances.[43] The discrepancy may also arise in part from a cutoff at 0·5 Å in the interaction potential used in the Fuschini *et al.* calculation. The effect of a cutoff is apparent from the two analytical curves shown in Figure 12.17.

(2) *Dip Width*

The influence of the emission position on the width of blocking dips has been nicely demonstrated by measurements and calculations of the effect of thermal vibrations by Andersen.[45] Andersen and Feldman[42] have shown that changes in the width of observed blocking dips owing to thermal lattice vibration is almost entirely due to the position of the emitting nucleus (as opposed to variation in the subsequent steering effect owing to vibration of other atoms in the row). This means that the quantitative results of Andersen's calculations can, to a good approximation, be used directly to determine the mean displacement owing to recoil from measured changes in the dip width. Although the gaussian displacement distribution caused by thermal vibration decreases more rapidly with distance than the exponential distribution expected from the lifetime effect, the width is little influenced by the tail of the distribution but is determined by the mean value. Andersen's calculated displacement dependence is shown in Figure 12.19.

In applying this type of analysis, the observed width should not be compared quantitatively to the calculated ψ_1 values. Instead it is preferable to determine the width change relative to a control measurement in which lifetime effects are absent. Measured widths in the absence of lifetime effects are generally 20–30 % smaller than calculated values[46] even for emission (or channeling) depths of less than 1000 Å. This is apparently because of multiple scattering.[47] For larger depths the discrepancy is even larger. It is not clear how scattering effects should be incorporated in analysis of width changes, and it is felt that the ratio of the measured width to that obtained for the calibration measurement is perhaps the most justifiable technique to use.

The width of the blocking minimum is not as sensitive to the long tail in the exponential decay distribution as is the minimum yield. Consequently the continuum string approximation should be valid to larger values of the mean recoil distance when considering the dip width as compared to the limiting values when the minimum yield is considered. The usual method of specifying the experimental width by the half angle at half minimum[34]

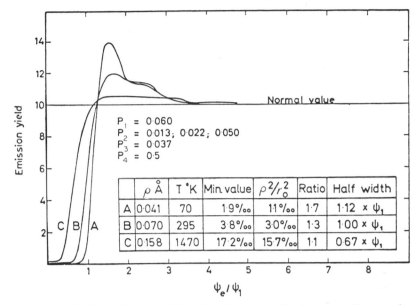

Figure 12.19. Influence of the vibrational amplitude, ρ, on the angular distribution of 500 keV protons emitted from a $\langle 100 \rangle$ axis in tungsten. The P_1, P_2, P_3, P_4 parameters correspond to physically reasonable conditions. The unit 0/00 stands for 'parts per thousand'. These calculated curves can also be used to evaluate width changes owing to exponential displacement within the framework of the single string approximation (after Andersen, reference 38)

needs to be reconsidered however since it still depends on the measured minimum yield. Perhaps a determination based on the width at some fixed fractional yield (such as 70% of the random yield for example) would be more useful. This has not been investigated but qualitative examination of the contours of Figures 12.16 and 12.18 and some calculated results for 4 MeV protons emitted along $\langle 110 \rangle$ axes in germanium[43] suggests that such an analysis should not be too bad for mean recoil distances up to about 0·7 Å.

(3) *Dip Area and Volume*

Related to both the minimum yield and the width of a blocking dip is the area (for planes) or the volume (for axes). When an appreciable change in width is observed, this will tend to dominate changes in the area and especially the volume. Measurement of this parameter has an advantage in statistical accuracy since the entire measured distribution can be used. It should be noted however that comparable statistical accuracy can be achieved in minimum yield and width determinations by fitting the entire observed distribution with a general function composed of a combination of a straight line and a gaussian dip. Such a function would be of the form:

$$\{a + b(\chi - \chi_0)\}\left[1 - \left(\exp\left\{-d(Q)\left(\frac{\chi - \chi_0}{\sigma}\right)^Q\right\}\right)\right].$$

The width or minimum yield can then be extracted from the parameters (a, b, d, Q) of the fitted function.[48] Another possible advantage of the area or volume determination is that it allows relatively poor angular resolution to be used in the measurement.[11] One disadvantage of this type of measurement is that, because of the inverted bell shape of observed blocking distributions, it is generally more sensitive than minimum yield or width determinations to choice of the random level.

Calculation of the angular distribution pattern of emitted particles can be made in terms of the continuum potential approximation as illustrated by the calculations of Andersen.[45] From such calculations it is straightforward to obtain the area or volume of the blocking dips. Such calculations for planar blocking have been carried out by Komaki and Fujimoto,[13] who obtain the following expression for the ratio of the planar dip area in the presence of a lifetime effect to the reference dip:

$$M(v_\perp \tau)/M(v_\perp \tau = 0) = \{\alpha(v_\perp \tau) - \beta\}/\{\alpha(0) - \beta\}$$

where

$$\alpha(v_\perp \tau) = \int_{-\infty}^{\infty} g(x; v_\perp \tau, \rho)U(x)\,dx$$

and

$$\beta = \frac{1}{d}\int_0^d U(x)\,dx$$

in which

$$g(x, v_\perp \tau, \rho) = \frac{1}{\sqrt{2\pi}\rho v_\perp \tau}\int_0^{\infty} \exp\{-(\chi - \chi')^2/2\rho^2\}\exp(-\chi'/v_\perp \tau)\,dx'$$

where $U(x)$, ρ and d are the average planar potential, the transverse component of the rms amplitude of the thermal vibration and the lattice spacing respectively.

The calculated ratios for blocking of 5 MeV protons by $\{111\}$ planes in germanium as a function of $v_\perp \tau$ are shown in Figure 12.20.

A similarly detailed analytical calculation has not been carried out for the volume change in an axial blocking dip. Clark et al.[11] have fitted Monte Carlo computer calculations of the dip for 5 MeV protons emitted along a [110] axis in germanium to a function of the form $R(v_\perp \tau)/R(v_\perp \tau = 0) = 1 - \exp(r_c/v_\perp \tau)$, where R corresponds to the volume of the dip and r_c is a

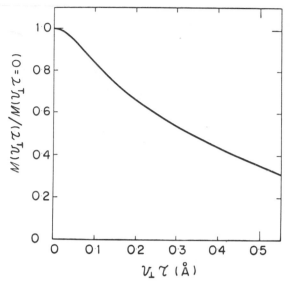

Figure 12.20. Variation of the area of planar blocking dips (normalized to the dip obtained for zero displacement) as a function of mean displacement for protons emitted along {111} planar directions in germanium (after Komaki and Fujimoto, reference 13)

fitting parameter originally envisaged as an effective cutoff distance beyond which steering owing to the string potential no longer takes place. The value $r_c = 0.879\, a_{TF}$ obtained from the fitting is anomalously low in view of the shallow emission depths used in the computer simulation and the range of effectiveness of the string potential into the axial channel as determined from flux peaking[49] or particle transmission[38] studies so it should be regarded more as an arbitrary fitting parameter than a physically significant cutoff distance.

It is assumed in the dip area or volume analysis as in the width change analysis that scattering effects involving changes of transverse energy are either negligible or that they affect both the primary and the reference measurements by the same multiplication constant. This was investigated by Fuschini *et al.* who folded assumed gaussian dispersion functions into blocking dips obtained from computer simulation calculations.[44] The effect of different amounts of multiple scattering on the $\langle 110 \rangle$ axial dip for 4.4 MeV protons in germanium is shown in Figure 12.21. They conclude that, although the dip areas for the primary measurement ($R(v_\perp \tau)$) and for the control ($R(v_\perp \tau = 0)$) are each dependent on multiple scattering, the ratio $R(v_\perp \tau)/R(v_\perp \tau = 0)$ is virtually independent of such effects.

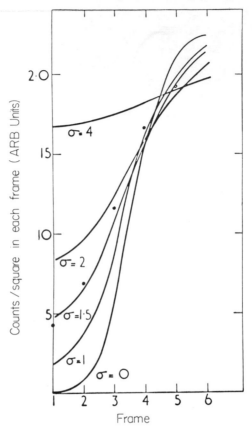

Figure 12.21. Calculations of the effect of multiple scattering on $\langle 110 \rangle$ axial blocking dips in germanium crystals, $\sigma = 0$ corresponds to a computer simulated (Monte Carlo) blocking dip with no added scattering. The other solid curves have gaussian distributions of various width folded into the computer simulated curve. $\sigma = 1$ corresponds to 0·087 degrees. The points were measured for 4·4 MeV protons scattered from a 25 μm thick germanium crystal (after Fuschini *et al.*, reference 44)

(4) *Complete Blocking Distribution*

(a) *Single-string continuum model.* As shown in Figure 12.19, complete emission angular distributions can be constructed in a straightforward fashion under the assumption of a continuum string (atomic row) potential. It is instructive to consider two limiting cases for such a calculation. The first is an assumption of statistical equilibrium for particle motion in the transverse plane. This means that the direction of the momentum vector in the transverse plane has equal probability in any direction independent of its original direction and leads to an axially symmetric emission distribution centred on the axial direction. Under this assumption the so-called 'single string' calculation[45] is carried out. The influence of the other strings is included only in a surface transmission term. In this limit the recoil direction is not taken to have a unique azimuthal direction but is assumed to be azimuthally symmetric. Since an axially symmetric distribution is obtained, a single cut through the axis is sufficient to describe the result.

As noted previously the statistical assumption becomes valid only after penetration of a thickness which depends on the crystal, the emitted particle Z, and the emitted particle's longitudinal and transverse energy. A sensitive test of the assumption of statistical equilibrium is the symmetry of the observed blocking pattern. Asymmetry or a shift of the centroid of the dip relative to a control measurement indicates that statistical equilibrium has not been achieved.

The origin of asymmetry and centroid shift in the blocking dips is easily seen by considering the second limiting case. In this case we retain the well defined recoil direction of the compound nucleus and assume *no* statistical averaging of the direction of the momentum vector in the transverse plane. Such a calculation is of little direct relevance to most experimental situations since passage through a thickness of only a few hundred ångströms of crystal will appreciably modify the resultant distribution. The results of such a calculation for 4 MeV protons emitted along a $\langle 110 \rangle$ axis in germanium is shown in Figure 12.22. The mean recoil distance was taken as 0·2 Å. Thermal vibration of the emitting atoms was included but not the effect of surface transmission which would have the major result of giving a nonzero yield parallel to the axis but will not significantly change the widths or shapes of the emission dips. A strong shadow effect owing to the string of atoms from which the recoiling nucleus originated is seen in the tailing contours at 180° to the recoil direction. The half width of the blocking dip is in fact almost four times as large at 180° to the recoil direction as is at 0° to the recoil direction. If it were possible to make use of this asymmetry, it would perhaps be the most sensitive lifetime indicator. Unfortunately, the initial distribution represented by Figure 12.22 is rapidly modified by scattering of emitted particles from other rows (or strings) of atoms as they penetrate the crystal. In addition to averaging of the intensity contours around the axial direction, blocking effects owing to the ordered array of atomic strings in the transverse plane leads to azimuthal structure in the emitted particle distribution. Both of these effects depend sensitively on the thickness of crystal through which the emitted particles must penetrate before emerging and on the initial emission position and direction in the lattice.

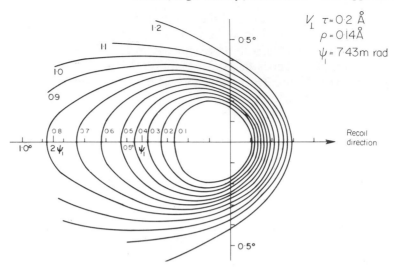

Figure 12.22. Single string continuum calculation of emergent particle intensity contours for 4·0 MeV protons emitted along a $\langle 110 \rangle$ axis of germanium. This is a pure single string calculation including no azimuthal averaging or surface transmission effects.

(b) *Computer simulation.* As discussed previously an alternative to analytical calculations based on continuum potential approximations is to consider charged particle motion through single crystals as independent binary scatterings of the particle with individual lattice atoms.

Such calculations can be used not only to determine single parameters of the blocking distribution such as the minimum yield or dip width or area but at the expense of computing time can also give the entire shape of the blocking distribution. In spite of severe limitations imposed by the computing time required, Sona has used this technique to explore the shift in the centre of mass of the blocking dip as a function of the mean recoil distance, the target thickness and target temperature for 12 MeV protons in silver crystals.[15] Massa[11,14] has computed blocking dips for 4·3 MeV protons emitted along $\langle 110 \rangle$ axes in Ge for $(v_\perp \tau) = 0$ and $(v_\perp \tau) = 0.2$ Å. To reduce the computation effort required, it was necessary in the latter case to fold the distribution around the axial direction. In addition, the emission depth was limited to 2000 Å and a cutoff in the interaction potential of 0·5 Å was applied. Unfortunately, these limitations and approximations severely restrict application of the results and serve to point out the major problem in this approach to analysis of experiments. It is not uncommon for a Monte Carlo calculation to take longer, be more expensive and have higher statistical uncertainty than the associated experiment.

As pointed out previously, turning the calculation around and making use of reversibility can in some cases lead to a significant improvement in efficiency. The most complete Monte Carlo computer simulation calculations reported to date are for 4 MeV protons emitted from germanium crystals along a $\langle 110 \rangle$ axis for thickness up to $3\ \mu m$.[43] Emission patterns corresponding to an azimuthal slice through the axis along a line tilted 16.5 degrees from a $\{100\}$ plane for different recoil directions and mean recoil distances are shown in Figure 12.23. This azimuthal direction corresponds

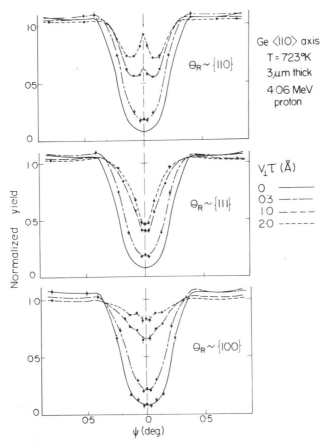

Figure 12.23. Computer simulated blocking angular distributions for emission of 4.06 MeV protons from a $3\ \mu m$ thick germanium crystal for three different compound nuclear recoil directions, θ_R, and four different mean recoil distances, $v_\perp \tau$. These calculations are all for particles emergent in a plane containing a $\langle 110 \rangle$ axis and inclined 16.5 degrees to a $\{100\}$ plane to avoid planar channeling effects (after Hashimoto, Barrett and Gibson, reference 43)

to the orientation of a one-dimensional position sensitive detector used in an extensive series of proton scattering measurements in germanium[26,50] and was chosen to avoid interference from low index planes. The strong dependence of the minimum yield on recoil direction noted previously is apparent as well as dependence of the width, area and other features of the blocking dips.

The curves of Figure 12.23 indicate the necessity to include the multistring potential of Figure 12.16 in analysis of any case in which the mean recoil distance exceeds about 0·3 Å. Of course changing the recoil angle away from 90 degrees relative to the string is a possible technique to limit the perpendicular component of the velocity and therefore the mean recoil distance ($v_\perp \tau$). This may be important in some cases to obtain quantitatively reliable results from a single-string continuum analysis. However, this approach forces the size of the measured effects to be small and therefore more difficult to obtain accurately and has the more serious disadvantage of not allowing the striking differences shown in Figure 12.23 to be used to gain information about the lifetime and, in particular, about the shape of the decay function. The blocking dips have also been found to be sensitive to emission depth (at least up to depths of 3 μm for 4 MeV protons in germanium) and to the use of cutoffs in the interaction potential used.[43] Although this type of analysis apparently will allow sensitive and quantiative interpretation of experimental results, it is impracticable to carry out for every case of interest. Solution of this problem remains as an important goal.

(c) *Multistring continuum calculations.* One approach to inclusion of the crystal geometry in a less extensive and expensive calculation has been suggested by Hashimoto.[48,51,52] In these calculations he assumes a continuum potential for each row and calculates the particle trajectory through the two-dimensional multistring potential. Although computer analysis is necessary to solve the equation of motion and individual particle trajectories are computed as in the Monte Carlo case, the calculation is considerably less expensive than a full binary Monte Carlo calculation. In fact this approach is a compromise between the single string and the binary scattering calculation and contains some of the advantages and disadvantages of both. As discussed for the Monte Carlo calculations, the trajectories can be computed for either the emission or channeling geometry. Using a channeling geometry, calculated blocking dips for 4 MeV protons emitted along the $\langle 110 \rangle$ axis in germanium for recoil direction nearly parallel (1·5° tilt) to a $\{111\}$ plane are shown by the solid curves of Figure 12.24[52] and compared to the Monte Carlo computations for different values of the mean recoil distance, $v_\perp \tau$. At $v_\perp \tau = 0$ the experimental result obtained for the elastically scattered protons of Figure 12.10 is also shown.

All of the recoil direction and depth effects noted for the Monte Carlo calculations are observed in the multistring continuum calculations except for those arising from multiple scattering due to lattice atom thermal vibrations. This is the reason for the differences between the blocking distributions calculated by the two methods for small values of $v_\perp\tau$ as shown in Figure 12.24. As the mean recoil distance increases, the effects of multiple scattering decrease (since nuclear multiple scattering is most important) and two calculations give more nearly the same result. It may be possible to include multiple scattering effects in the multistring continuum

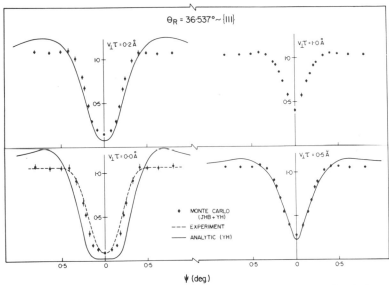

Figure 12.24. Blocking distributions for 4·20 MeV protons emitted from a 3 μm thick germanium crystal along a $\langle 110 \rangle$ axis calculated by a large scale Monte Carlo computer simulation (solid points) and by a multistring continuum model (solid lines) for recoil 1·5 degrees from a $\{111\}$ planar direction for various mean recoil distances, $v_\perp\tau$. An experimental measurement for $v_\perp\tau - 0$ taken from the elastic proton scattering curve of Figure 12.10 is also shown (dashed curve) (after Hashimoto and Barrett, reference 52)

calculation by using a formalism of the type used for calculations of dechanneling.[53] This is currently under investigation.[48]

It would also be of considerable interest to include energy loss effects in such calculations (or in Monte Carlo calculations). At the high particle energies of interest for this application, the energy loss is almost entirely due to electron excitation and is therefore separable from the steering effect of the rows or planes of atoms in the crystal. Retention of information on the particle trajectories through the lattice should allow estimation of the energy loss[54] and would hopefully permit quantitative use of the 'channeling peaks' observed on the high energy side of inelastic proton groups as noted in

Section 12.3.2. This 'channeling effect' should be strongly enhanced for recoil directions of the type shown in the upper part of Figure 12.23.

Making use of the increased efficiency of a multistring continuum calculation, it may be possible to obtain the entire blocking distribution for a particular case. This is about two orders of magnitude more time consuming than the single slice computations of the type shown in Figure 12.23 and 12.24 and is at the present time impractical by Monte Carlo techniques. A preliminary multistring continuum calculation for 4 MeV protons emitted from a mean depth of 800 Å in germanium from nuclei with a mean recoil distance of 0·4 Å is shown in Figure 12.25.[51] It is apparent that strong asymmetry is still present for this depth and that planar blocking effects have

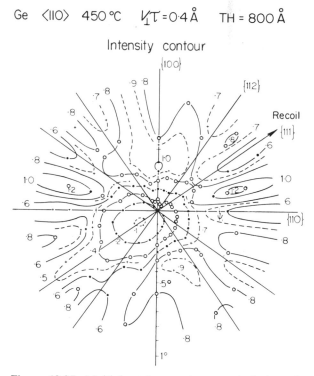

Figure 12.25. Multiple string continuum calculation of emergent particle intensity contours for 4·0 MeV protons emitted along a ⟨110⟩ axis of germanium for a crystal thickness of 800 Å (Y. Hashimoto, private communication)

already begun to appear. As the emission depth increases, the asymmetry begins to disappear but preliminary results indicate significant asymmetry still remaining after a thickness of 2000 Å.[48]

12.4.2 Complex Decay Function

A serious limitation of all of the analyses discussed up to this point is the assumption of a simple, single component, exponential decay function. Such an assumption is justified in the case where a single isolated nuclear state is excited,[19,20] but this is limited to special cases in light nuclei. Gibson and Nielsen justified this assumption in their fission lifetime studies by pointing out that the width changes observed were consistent with the minimum yield changes.[8] These two features show different responses to the shape of a complex decay function. Clark *et al.* recognized the danger of a single parameter analysis in their inelastic proton lifetime studies in germanium but gave theoretical arguments to support the assumption. The differences ($\sim$ factor of two) between lifetimes obtained from blocking dip volume changes and minimum yield changes for this case[44] suggest, however, the possibility of some problem with this assumption. Detailed measurements of the angular distributions of the inelastic proton groups at different bombardment energies[26] also show systematic inconsistencies between minimum yield changes and width changes which suggests a more complicated decay behaviour. For this reason it is always desirable to determine more than one parameter in the blocking distribution. For example, the minimum yield can be very much affected by the presence of a long-lived component of relatively low yield which will not significantly change the width of the dip. The width depends in a more direct way on the *mean* displacement (the dip volume or area is a less simple parameter to evaluate since it depends on both the width and minimum yield). In principle, the detailed shape of the blocking distribution including any possible asymmetry or distortion on the edges[9] contains considerable information about the shape of the decay function. The recoil direction dependence shown in Figure 12.23 may also be a powerful tool for this purpose. At the present time the analysis of such distributions is not sufficiently advanced to extract such information. This represents an important and difficult challenge to the theory of particle blocking in single crystals. Even relatively simple model calculations in which two or three components with different decay constants and which were present in different amounts would be useful guides to qualitative indicators. At the same time, calculations of the decay functions expected from appropriate nuclear models can also act as guides to help interpret the results and to improve the analysis. The first such calculations are those of Malaguti *et al.*[55] who analysed the case of ^{70}Ge bombarded with 5.4 MeV protons on the basis of a statistical model using a Porter–Thomas level distribution. The resulting decay function is shown as the solid line in Figure 12.26. Also shown is an exponential distribution normalized at two points near the beginning of the calculated distribution (dotted line). It can be seen that the calculated decay function has a long, slowly decaying tail when compared to the exponential.

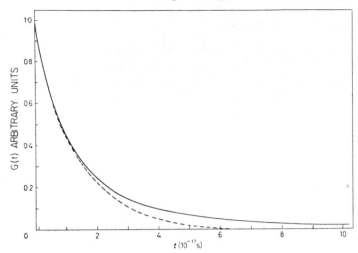

Figure 12.26. Calculated time dependence of the [71]As compound nucleus decay probability function $G(t)$ at 9·7 MeV excitation energy (solid line) based on a statistical nuclear model compared with a simple exponential decay (dashed line) normalized at short times (after Malaguti, Uguzzoni and Verondini, reference 55)

This is in the right direction for explaining the differences noted in Section 12.3.2 between decay times determined from area or width analyses and those determined from minimum yield changes.

12.5 CONCLUSIONS AND PROGNOSIS

It is clear that blocking studies of nuclear lifetimes is a demonstrated tool, capable of application to important and detailed nuclear questions concerning level widths, level densities, nuclear structure effects, fission barrier parameters, effects of specific perturbations such as analogue resonances, etc. Even the current accuracy (which is in the range of 20% to a factor of three, depending on the experimental problems and the confidence with which the decay function can be predicted) is high enough for meaningful determinations. It is also clear, however, that improvement of the accuracy, extension of the range, and extraction of information about the decay function represent some of the most important and demanding challenges to detailed understanding of particle channeling and blocking effects in single crystals. In particular, better understanding is required of changes in the particle trajectory distribution as a function of depth and effects of crystal impurities and imperfections as well as surface disorder and contamination. Detailed analysis of lifetime effects is closely related to considerations of flux peaking in impurity ion location studies and of emission angular distributions of particles channeled through very thin crystals and is also related to the general questions of dechanneling. Continued active developments in these areas give promise that one can expect early improvement in the lifetime analysis situation. In addition, more complete and realistic decay functions predicted from nuclear model calculations will serve as important guides and checks on the measurement and analysis. The recent calculation for inelastic proton scattering in germanium[55] is an important step in this direction.

From the experimental standpoint, detailed measurements of simple or well understood systems are required to provide real tests of analytical and experimental techniques. Cases in which a single, isolated state is excited such as ^{31}P(p, α) 5·28 [19] or ^{27}Al(p, α)^{24}Mg [20] are especially important, particularly if the level width can be accurately determined by other techniques. It would also be very useful to make a systematic investigation of a system with a known or expected nonsimple decay function. Variation of the recoil direction relative to an axis would be informative in investigating the effect of different decay components on the shape of the emission distribution.

It is probable that refinement of present techniques and experiments will extend to lifetime measurements to less than 10^{-18} s. This may involve cooling of the crystal to reduce thermal vibrations and use of heavy ions or higher energies to increase recoil velocity as has already been demonstrated.[25] In this same vein G M Temmer has pointed out the possibilities of using very high energy particles.[56] Relativistic time dilation could extend the time limit one or two orders of magnitude and the determination of lifetimes of single particles such as π^0, η^0 and Σ^0 represents an exciting possible extension of this technique. At lower energies, measurement of compound elastic scattering strengths, fission barrier heights and shapes and nuclear level broadening effects are only examples of some of the challenging and important possibilities which will continue to expand as this tool becomes more widely recognized and used and more completely developed.

References

1. Domeij, B. and Bjorkqvist, K., *Phys. Lett.*, **14**, 129 (1965).
2. Gemmel, D. S. and Holland, R. E., *Phys. Rev. Lett.*, **14**, 945 (1965).
3. Tulinov, A. F., Kulikauskas, V. S. and Malow, M. M., *Phys. Lett.*, **18**, 304 (1965).
4. See C. S. Barrett, Chapter 11 in this volume.
5. Tulinov, A. F., *Doklady Akademii Nauk SSSR*, **162**, 546 (1965). *Sov. Phys.–Doklady*, **10**, 463 (1965).
6. Brown, F., Marsden, D. A. and Werner, R. D., *Phys. Rev. Lett.*, **20**, 1449 (1968).
7. Maruyama, M., Tsukada, K., Ozawa, K., Fujimoto, F., Komaki, K., Mannami, M. and Sakurai, T., Conf. on Nuclear Physics in Energy Region of Tandem Accelerators, Tokyo, Japan (December 1968) JAERI-1184, p. 141; *Phys. Lett.*, B, **29**, 414 (1969); *Nucl. Phys.*, A, **145**, 581 (1970).

8. Gibson, W. M. and Nielsen, K. O., *Bull. Am. Phys. Soc.*, **14**, 629 (1969); Int. Symp. on the Physics and Chemistry of Fission, Vienna (July 1969) p. 861, IAEA-SM-122/129; *Phys. Rev. Lett.*, **24**, 114 (1970).
9. Melikov, Yu. V., Otstavinov, Yu. D., Puzanov, A. A. and Tulinov, A. F., *Zh. E. Ksp. Teor. Fiz.*, **55**, 1690 (1968); *Sov. Phys.–JETP*, **28**, 888 (1969).
10. Melikov, Yu. V., Otstavnov, Yu. D. and Tulinov, A. F., *Zh. Eksp. Teor. Fiz.*, **56**, 1803 (1969); *Sov. Phys.–JETP*, **29**, 968 (1969); *Yad. Fiz.*, **12**, 50 (1970); *Sov. J. of Nucl. Phys.*, **12**, 27 (1971).
11. Clark, G. J., Poate, J. M., Fuschini, E., Maroni, C., Massa, I. G., Uguzzoni, A. and Verondini, E., *Nucl. Phys.*, A, **173**, 73 (1971).
12. Treacy, P. B., *Australian J. of Sci.*, **22**, 334 (1960).
13. Komaki, K. and Fujimoto, F., *Phys. Lett.*, A, **29**, 544 (1969); *Phys. stat. sol. (a)*, **2**, 875 (1970).
14. Massa, I., *Lett. al Nuovo Cimento*, **III**, 186 (1970).
15. Sona, P., *Il Nuovo Cimento*, **LXVI**, 663 (1970).
16. Karamyan, S. A., Melikov, Yu. V., Normuratov, F., Otgonsuren, O. and Solov'eva, G. M., JINR preprint R7-5300 (1970); *Yad. Fiz.*, **13**, 944 (1971); *Sov. J. Nucl. Phys.*, **13**, 543 (1971).
17. Gibson, W. M., Maruyama, M., Mingay, D. W., Sellschop, J. P. F., Temmer, G. M. and Van Bree, R., *Bull. Am. Phys. Soc.*, **16**, 557 (1971).
18. Fleischer, R. L., Price, P. B. and Walker, R. M., *Ann. Rev. Nucl. Sci.*, **15**, 1 (1965).
19. Sharma, R. P., Andersen, J. U. and Nielsen, K. O., to be published.
20. Komaki, K., Fujimoto, F., Hakayama, H., Ishii, M. and Hisatake, K., to be published.
21. Augustyniak, W. M., Brown, W. L. and Lie, H. P., *IEEE Trans. on Nucl. Sci.* **NS-19** 196 (1972).
22. Leagsgaard, E., private communication.
23. Melikov, Yu. V., Otstavnov, Yu. D., Tulinov, A. F. and Chetchenin, N. G., *Nucl. Phys.*, to be published.
24. Andersen, J. U., Nielsen, K. O., Stak-Nielsen, J. and Hellborg, R., to be published.
25. Karamyan, S. A., Oganesyan, Yu. Ts. and Normuratov, F., JINR reprint P7-5512; *Yad. Fiz.*, **14**, 499 (1971); *Sov. J. of Nucl. Phys.*, **14**, 279 (1972).
26. Gibson, W. M., Hashimoto, Y., Keddy, R. J., Maruyama, M. and Temmer, G. M., *Phys. Rev. Lett.*, **29**, 74 (1972).
27. Laegsgaard, E., Martin, F. W. and Gibson, W. M., *Nucl. Instrum. and Meth.*, **60**, 24 (1968); *IEEE Trans. on Nucl. Sci.*, **NS-15**, 239 (1968).
28. Temmer, G. M., Maruyama, M., Mingay, D. W., Petruscu, M. and Van Bree, R., *Phys. Rev. Lett.*, **26**, 1341 (1971).
29. Meek, R. L., Gibson, W. M. and Sellschop, J. P. F., *Rad. Effects*, **11**, 139 (1971).
30. Hofker, W. K., Oosthoek, D. P., Hoeberechts, A. M. E., van Dantzig, R., Mulder, K., Oberski, J. E. J. and Koerts, L. A. Ch., *IEEE Trans. on Nucl. Sci.*, **NS-13**, 208 (1966).
31. Clark, G. J., Fuschini, E., Maroni, C., Massa, I. G., Malaguti, F., Uguzzoni, A., Verondini, E. and Wilmore, D., Int. Conf. on Stat. Properties of Nuclei, Albany, New York, August 1971, to be published.
32. Lassen, N. O. and Andersen, P. B., unpublished.
33. Maruyama, M., Hashimoto, Y., Alvar, K. and Temmer, G. M., unpublished.
34. Lindhard, J., *Dan. Vid. Selsk.*, *Mat. Fys. Medd.*, **34**, No. 14 (1965).
35. Altman, M. R., Feldman, L. C. and Gibson, W. M., *Phys. Rev. Lett.*, **24**, 464 (1970).
36. Barrett, J. H., *Phys. Rev.*, **166**, 219 (1968).
37. Komaki, K., Fujimoto, F., Ishii, M. and Nakayama, H., *Phys. Lett.*, A, **37**, 271 (1971).

38. Armstrong, D. D., Gibson, W. M., Goland, A., Golovchenko, J. A., Levesque, R. A., Meek, R. L. and Wegner, H. E., *Rad. Effects*, **12**, 143 (1972).
39. Morgan, D. V. and van Vliet, D., *Can. J. Phys.* **46**, 503 (1968).
40. Morgan, D. V., Chapter 3, this volume.
41. Barrett, J. H., *Phys. Rev.* B, **3**, 1527 (1971).
42. Andersen, J. U. and Feldman, L. C., *Phys. Rev.* B, **1**, 2063 (1970).
43. Hashimoto, Y., Barrett, J. H. and Gibson, W. M., *Phys. Rev. Lett.*, **30**, 995 (1973).
44. Fuschini, E., Malaguti, F., Maroni, C., Massa, I., Uguzzoni, A., and Verondini, E., *Il. Nuovo Cimento* **10A**, 177 (1972).
45. Andersen, J. U., *Dan. Vid. Selsk., Mat. Fys. Medd.*, **36**, No. 7 (1967).
46. Picraux, S. T., Davies, J. A., Eriksson, L., Johansson, N. G. E. and Mayer, J. W., *Phys. Rev.*, **180**, 873 (1969).
47. Campisano, S. U., Foti, G., Grasso, F., Quercia, I. F. and Rimini, E., *Rad. Effects*, **13**, 23 (1972).
48. Hashimoto, Y., private communication.
49. Andersen, J. U., Andreasen, O., and Davies, J. A., *Rad. Effects*, **7**, 1 (1970).
50. Alvar, K., Gibson, W. M., Hashimoto, Y., Kanter, E., Keddy, R. J., Levesque, R. A., Leuca, I., Maruyama, M., Mingay, D. W., Petruscu, M., Sellschop, J. P. F., Temmer, G. M. and Van Bree, R., to be published, see also *Bull. Am. Phys. Soc.* **16**, 557 (1971), **17**, 560 (1972); **18**, 119 (1973).
51. Hashimoto, Y., *Bull. Am. Phys. Soc.*, **17**, 560 (1972), to be published.
52. Hashimoto, Y. and Barrett, J. H., *Bull. Am. Phys. Soc.*, **18**, 119 (1973).
53. Campisano, S. U., Grasso, F. and Rimini, E., *Rad. Effects*, **9**, 153 (1971).
54. Appleton, B. R., Erginsoy, C. and Gibson, W. M., *Phys. Rev.*, **161**, 330 (1967).
55. Malaguti, F., Uguzzoni, A., and Verondini, E., *Lett. al. Nuovo Cimento*, **2**, 629 (1971).
56. Temmer, G. M., Rutgers University, private communication.

CHAPTER XIII

Foreign Atom Location

J. A. DAVIES

A phenomenon such as channeling, which produces large and predictable fluctuations in so many observable quantities (nuclear reaction yields, Rutherford scattering, X-ray emission, etc.), usually finds application in several fields. Indeed in recent years, such applications have begun to attract even more interest than the basic channeling studies themselves.

The preceding chapter summarizes one important application to nuclear physics—the measurement of extremely short nuclear lifetimes—in which channeling (or blocking) effects are used to determine the exact lattice location of a recoiling nucleus at the instant that it emits a charged particle. In the present chapter, we discuss how this same 'atom-locating' aspect of channeling may be applied to foreign atom location. Particular emphasis is given to its use in studying ion-implanted semiconductors, since the channeling-effect technique provides simultaneous information on at least two important implantation questions: the amount of radiation damage present in an implanted crystal and the type of site on which the implanted foreign atom is located.

For a convenient entry into the current literature of these applications, the reader is referred to the proceedings[1-3] of several ion-implantation conferences, and also to chapters 3 and 4 of a recent monograph[4] on ion implantation in semiconductors.

These solid state applications may be subdivided into three main categories: foreign atom location, radiation damage, and surface studies. All three rely upon the same basic principle—the ability of a channeled beam to recognize atomic displacements as small as approximately 0·1 Å with respect to an aligned set of lattice rows or planes. In favourable cases, this feature enables the location of specific atoms within the unit cell to be accurately specified, thereby providing unique *geometrical* information concerning their lattice sites. Such geometrical information often serves as a valuable complement to the energy-level or bonding information obtainable by other means (electron spin resonance, hyperfine field, Hall-effect studies, etc.).

Let us consider briefly the basic principles involved in all atom-location studies. We will then discuss in more detail the specific problems of foreign-atom detection, illustrating them with appropriate examples from the current literature.

391

13.1 BASIC PRINCIPLES

For the present discussion, the most important aspect of the channeling phenomenon is the fact that channeled particles are unable to approach closer than approximately 0·1 Å (the mean vibrational amplitude ρ_{rms} of the lattice atoms) to the aligned rows or planes of atoms. The crystal is thus divided into allowed and forbidden regions as shown in Figure 13.1. For such channeled particles, all physical processes requiring an impact parameter

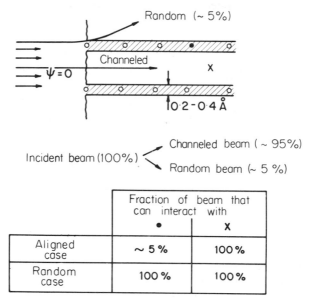

Figure 13.1. The splitting of a perfectly aligned beam into channeled and random components in passing through the crystal surface

smaller than about 0·1 Å (wide-angle elastic scattering, nuclear reactions, inner-shell X-ray production) are completely prohibited and extremely strong attenuations (up to 100-fold) in the yields of such processes can therefore be achieved, as has been demonstrated in Chapter 1.

These large attenuations in yield occur not only for interactions with the lattice atoms, but also for those foreign atoms located within the forbidden region such as the substitutional (●) site in Figure 13.1. On the other hand, those foreign atoms located within the *allowed* region for the channeled beam (such as the site marked X in Figure 13.1) will be able to interact with both the channeled and the random components of the beam and so will not exhibit any attenuation in yield. Indeed, as we shall see later (Figures 13.7, 13.9 and

13.10), the yield for such interstitial atoms must actually *increase* somewhat as the beam becomes channeled, since (in order to conserve the total number of particles in the beam) one cannot have a decrease in flux in the forbidden region near the atomic rows without producing an increase in flux elsewhere.

This marked difference in behaviour between shadowed and nonshadowed atoms forms the basis of the channeling-effect technique of atom location. One simply compares the angular dependence of the interaction yield with the foreign atoms to that with the lattice.

13.2 TRIANGULATION AND LATTICE SYMMETRY

The existence of a strong yield attenuation with the foreign atoms (such as that observed with gold in Figure 13.4) is not in itself sufficient proof that the foreign atoms are on substitutional sites in the crystal lattice, but only that they must lie somewhere within the shadow of the aligned set of atomic rows. However, successive measurements along two or more different directions (triangulation) can be used to locate the exact position of the foreign atom. Furthermore, in lattices having a high degree of symmetry (such as the cubic lattice of Figure 13.4), it is often possible to invoke symmetry arguments as well.

Both these procedures are illustrated schematically, for a simplified two-dimensional lattice, in Figure 13.2. If, for example, the foreign atom is substitutional (●), it lies within the shadow of all close-packed rows in the lattice and therefore large attenuations will be observed along both the ⟨01⟩ and the ⟨11⟩ directions. On the other hand, if the atom lies at the intermediate position marked by X in Figure 13.2, 50 % of the atoms are located along the particular [01] direction, while the other half falls on the completely equivalent set of [10] rows. In such a case, the maximum attenuation that can occur along any selected direction of incidence is only 50 %. In a three-dimensional cubic lattice there would of course be *three* equivalent sets of axes, [100], [010], and [001], and hence the maximum attenuation at the edge-interstitial (X) position would be only 33 %. Neither of the X atoms in Figure 13.2 lies in the shadow of the ⟨11⟩ direction and, therefore, along this axial direction, no attenuation* would be observed. A third possible site, the position marked □, lies in the middle of the [01] and the completely equivalent [10] channels, but is in the shadow of the [11] and [1̄1] rows. In this case, therefore, we would expect no attenuation* in yield along ⟨10⟩ and a strong attenuation along ⟨11⟩. Thus, by observing orientation effects along two or more directions, it is often possible to locate the exact site of specific foreign atoms.

---

* For such mid-channel atoms, there is in fact an increase in yield when the beam becomes well channeled, as will be discussed in Section 13.5.

Location of foreign atoms by channeling

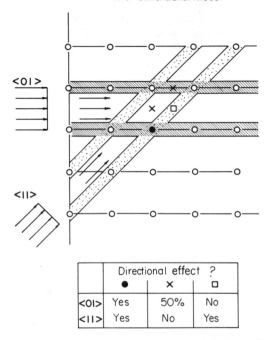

Two-dimensional model

Figure 13.2. A two-dimensional model illustrating how the channeling effect may be used to locate foreign atoms in a crystal. As shown by the table, three possible sites for a foreign atom (●, □ and ×) can be readily distinguished by comparing the channeling behaviour along the ⟨01⟩ and along the ⟨11⟩ directions

13.3 FOREIGN ATOM DETECTION

For the method to achieve widespread application, we must be able to detect a small concentration of foreign atoms in the presence of a large excess of lattice atoms. Fortunately, there are many close-impact processes from which to choose and, provided that the atomic concentration of foreign atoms exceeds approximately 10^{-3}–10^{-4}, a satisfactory one can usually be found.

To date, the most commonly used process in atom-location studies has been Rutherford backscattering. This is a semi-universal technique in that it can be readily applied to all cases in which the foreign atom is significantly heavier than the substrate atom (Figure 13.3). It has the advantage of measuring simultaneously the amount of lattice disorder present at various depths in

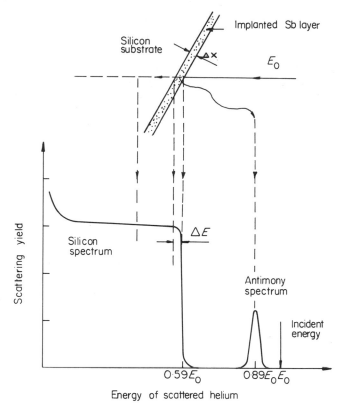

Figure 13.3. A typical energy spectrum for the scattering yield of helium ions ($E_0 \sim 1$ MeV) from an amorphous silicon target containing a heavy impurity atom (antimony) implanted to a depth of approximately 0.1 μm beneath the crystal surface

the crystal. It also provides information on the depth distribution of the foreign atoms. Whenever the foreign atom is lighter than the substrate, however, some other process is required, since the impurity peak in Figure 13.3 would then be superimposed on the continuum region produced by scattering from the heavier substrate atoms.

For Li, B, C, N, O, F, S, and a few other low-Z foreign atoms, specific nuclear reactions are often suitable[5]—such as (p, α), (p, γ), or (d, p). In these cases, Rutherford scattering can still be used to monitor the interaction of the beam with the lattice atoms.

Alternatively, one can sometimes detect the yields of characteristic inner-shell X-rays[6] resulting from proton- or helium-ion bombardment, particularly if a high-resolution solid state spectrometer is available to distinguish the foreign-atom X-rays from those of the substrate. For elements with $Z \lesssim 25$, this has the added advantage that the cross section for X-ray production is appreciably greater than that for Rutherford backscattering and so it may be applicable to even lower atomic concentrations. This aspect has yet to be fully exploited. The use of heavier ion beams (O, Na, Ar, etc.) to generate selectively the desired foreign-atom X-ray has been investigated by Cairns.[7] In general, large bombardment doses (and radiation damage) are involved and the sensitivity is rather poor. Hence it is of limited value in most atom-location studies.

Finally, the angular distribution (blocking) of emitted alpha particles[8,9] or positrons[10] from radioactive implants has also been successfully used in certain atom-location studies, but the number of suitable α- or β^+-emitting isotopes is not very great. A much larger number of β^--emitting isotopes are available, but unfortunately the channeling behaviour of *negative* particles does not yet provide a simple quantitative atom-locating technique such as the one to be discussed here.

Most of the following examples involve foreign atoms that are heavier than the substrate and so Rutherford scattering was selected as the appropriate close-impact process. It should be remembered, however, that similar channeling studies may be performed with any of the other close-impact processes, such as nuclear reactions and inner-shell X-ray production.

13.4 SUBSTITUTIONAL CASE

The case of an exactly substitutional atom (Figure 13.4) is quite straight-forward: along all axial and planar directions, the yield attenuation for the foreign atoms should be identical (both in magnitude and in angular width) to that for the host lattice.* In this case the full angular scan in Figure 13.4 is not really necessary, since two yield measurements—perfectly aligned and random—are sufficient to confirm that the foreign-atom and host-atom yield attenuations along a given axis are identical.

Indeed, whenever a *large* attenuation in yield is observed, even the tri-angulation procedure can sometimes be eliminated. Lattice-symmetry considerations require most of the scattering centres to be located at the intersection of all equivalent axes in the crystal, e.g. at the intersection of the six $\langle 110 \rangle$ axes in the copper lattice of Figure 13.4, since obviously the helium beam in these experiments cannot be aligned with more than one axis at a time.

In consequence, the fraction of substitutional foreign atoms in the crystal can be specified quite accurately. For instance, the observed ($\sim 94\%$) attenuation in the scattering yield from the foreign (gold) atoms in copper

---

* This would not be quite true if the foreign atoms had a significantly larger vibrational amplitude than that of the host lattice.

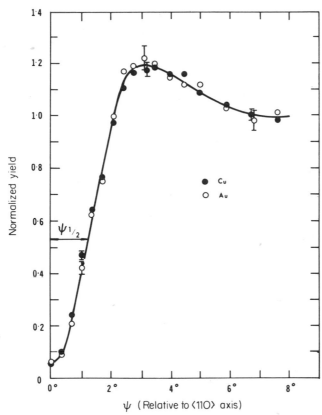

Figure 13.4. Angular dependence of the normalized backscattered yields of 1·2 MeV ^{4}He ions from Au and Cu atoms
in a single crystal containing approximately 2 at. % Au. The
energy windows correspond to a depth region 400–900 Å
below the surface (Alexander and Poate 1972[11])

(Figure 13.4) indicates that at least 94% are located along this particular
$\langle 110 \rangle$ row. Presumably a similar fraction is located along each of the other
five sets of $\langle 110 \rangle$ rows. Since in the fcc lattice of copper the intersection of two
or more $\langle 110 \rangle$ rows occurs only at the substitutional sites, we therefore
deduce that no more than 1/5 of the missing ($\sim 6\%$) fraction of the foreign
atoms are located in a nonsubstitutional position along any particular set.
Hence, at least 93% of the foreign atoms must be on the substitutional sites.

Strictly speaking, since the gold and copper yield curves in Figure 13.4 are
indistinguishable from one another, we can conclude that both types of atoms
are equally well shadowed along each set of $\langle 110 \rangle$ rows. Hence, virtually
100% of the gold atoms must be located within approximately 0·1 Å (the
mean vibrational amplitude) of the substitutional sites. In practice, this

perfect shadowing is only seldom observed. Usually, the yield attenuation for the foreign atoms is considerably weaker than that for the host atoms, indicating that some of them are not substitutional (cf the Bi in Si example in Section 13.6.4).

13.5 FLUX DISTRIBUTION WITHIN THE CHANNEL

Whenever the foreign atoms are not all substitutional, a quantitative interpretation of the measurements requires some knowledge of the flux profile of a channeled beam as a function of angle of incidence, depth into the crystal, etc. This problem is a complex one and has been discussed already in detail in Chapters 2 and 3. Here we shall merely summarize those aspects that are significant in atom-location studies.

13.5.1 Parallel Incidence ($\psi = 0$) Case

When positive ions enter a crystal exactly parallel to a channel axis, their initial transverse energy $E_\perp$ is determined solely by the value of the potential $U(r_{in})$ at the point of entry. Assuming that conservation of transverse energy is maintained, each ion is then confined to that region of the channel where the potential energy is less than or equal to $U(r_{in})$. Thus, in Figure 2.3, helium ions entering the $\langle 100 \rangle$ channel of copper on (for example) the 0·5 eV equipotential contour can only move within the area of the channel bounded by this contour. We know from the work of Lindhard[12] that ions of a given $E_\perp$, after penetrating sufficiently deep (typically ~ 1000 Å) to achieve statistical equilibrium, have an equal probability of being found anywhere within their accessible area A_i. This leads to a simple analytical expression[13] for the normalized flux F_i along the ith equipotential contour:

$$F_i = \int_{A_i}^{A_0} \frac{dA}{A} = \ln \frac{A_0}{A_i} \qquad (13.1)$$

A_0 is the total area of one channel. Note that this expression does not involve any assumption about the form or magnitude of the potential $U(r)$ across the channel. In the limit of $A_i \to 0$, the mid channel flux should, according to equation (13.1), increase logarithmically towards infinity. At such large distances from atomic rows, however, $U(r)$ becomes sufficiently small and flat that conservation of transverse energy is no longer a valid assumption. Instead, various factors such as beam collimation, electron scattering, and surface disorder introduce a finite spread in $E_\perp$, and this causes the flux near the centre of a channel to level out at some limiting value given by

$$F(\bar{r}_0) = 1 + \ln \frac{A_0}{A_1} \qquad (13.2)$$

where A_1 is defined as the area of this flat central region.

In order to evaluate the magnitude of these multiple scattering effects on the flux distribution, Van Vliet[14] has employed the technique of computer simulation. In the absence of multiple scattering, the flux profile is very similar to the analytical result of equation (13.1)—except that the finite area unit used in the computer simulation (in Figure 3.15) restricts $F(\mathbf{r}_0)$ to a finite value. In addition, Van Vliet's work shows: (i) that $F(\mathbf{r}_0)$ is sensitive to multiple scattering effects; (ii) that for relatively realistic conditions $F(\mathbf{r}_0)$ can still be at least three times the random value; and (iii) that the flux distribution near the strings is quite insensitive to multiple scattering effects.

The above analytical estimates (equations (13.1) and (13.2)) are valid only after statistical equilibrium has been achieved, i.e. at depths greater than or approximately equal to 1000 Å. Within approximately the first 1000 Å, computer simulations as a function of depth (Figure 3.16) indicate that $F(\mathbf{r}_0)$ exhibits a significant oscillatory dependence and that the amplitude of these oscillations is strongly damped by various multiple-scattering effects. Experimental evidence confirming the existence and approximate magnitude of such depth oscillations has recently been obtained by Eisen and Uggerhøj,[15] but many more computer simulations and experimental tests are required before they can be accurately predicted.

Most atom-location experiments have been performed using ion-implanted crystals, where the depth of the foreign atoms is often less than 1000 Å. In such cases, a quantitative treatment of the flux-distribution question still requires detailed computer simulations for each individual system.

13.5.2 Angular Dependence of The Flux Profile

Although the flux for perfect alignment ($\psi_{in} = 0$) is an important parameter in atom-location studies, the *variation* of flux with angle of incidence can be equally significant. Thus, in determining the site of the scattering atom, we often rely on the angular half width ($\psi_{\frac{1}{2}}$ or ψ_p) of the dip or peak in yield as well as on the magnitude (χ_{min} or χ_{max}) of the yield at perfect alignment. We should expect the flux in the mid-channel region (where the potential is small and almost constant) to fall rapidly as ψ_{in} is increased from zero. Accurate analytical expressions for the angular halfwidth of the mid-channel flux peak (ψ_p) are not available, but estimates by Andersen *et al.*[13] and by Van Vliet[14] for 1 MeV ^{4}He in $\langle 110 \rangle$ silicon indicate it to be about a factor of five smaller than the critical angle $\psi_{\frac{1}{2}}$ for channeling. This marked difference between ψ_p and $\psi_{\frac{1}{2}}$ is valuable in complex atom-location cases, since it enables us to distinguish clearly the individual yield contributions of mid-channel atoms and of atoms lying on the atomic rows (cf Figure 13.7(b)).

Let us now consider the case of an atom lying fairly close to an atomic row (e.g. somewhere on the 50 eV contour of Figure 2.3). Only those ions having a transverse energy ($E_\perp = U(r_{in}) + E\psi_{in}^2$) greater than 50 eV can encounter such an atom. Since for most of the beam the initial potential energy term $U(r_{in})$ is much less than 50 eV, we can with reasonable accuracy ignore the distribution of $E_\perp$ in estimating the 'critical angle' for scattering off such an atom. This enables us to identify $E_\perp$ ($=50$ eV) with the initial transverse kinetic energy term $E\psi_{in}^2$. In the general case, Lindhard's continuum potential

$$U_1(r) = \frac{Z_1 Z_2 e^2}{d} \ln\left\{\left(\frac{Ca}{r}\right)^2 + 1\right\} \tag{13.3}$$

may be inserted in place of this 50 eV example, thus providing a simple relationship between the displacement r of an atom from a row and the critical half angle $\psi_{\frac{1}{2}}$ for interacting with it.

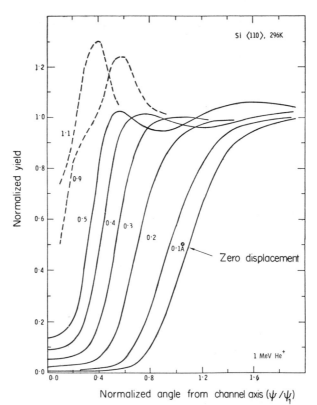

Figure 13.5. Calculated angular distributions as a function of equilibrium displacement distance r (in Å) for 1 MeV helium in $\langle 110 \rangle$ silicon at 296 K. (From Picraux et al.[16])

Where greater accuracy is required, Van Vliet[14] has given a more detailed and careful analysis of the problem. Also, for the $\langle 110 \rangle$ channel in silicon, Picraux *et al.*[16] have recently computed an extensive set of angular distributions of yield for various *r* values. As can be seen in Figure 13.5, the angular width of the dip decreases strongly and χ_{min} increases with increasing displacement.

It is interesting to note that at fairly large displacements (~ 1 Å in Figure 13.5) the distribution starts to exhibit a small peak at the edge of the dip in yield. This is presumably the beginning of the mid-channel flux-peaking effect which, at still larger displacements, moves to zero angle and completely replaces the dip. Morgan and Van Vliet[17] have indeed suggested that the existence of these intermediate yield peaks at nonzero angles of incidence may in certain cases provide useful atom-location information. Unfortunately, these off-axis peaks occur in a flux region that is rather sensitive to multiple scattering and depth effects. This is presumably the reason why no clear-cut experimental case has yet been found to exhibit them (cf examples 13.6.1 and 13.6.2).

13.6 SOME NONSUBSTITUTIONAL EXAMPLES

From the discussion in Section 13.5, the channeling-effect technique may seem a rather complicated and inaccurate tool for locating *non*substitutional atoms, since the flux distribution in the mid-region of the channel is so difficult to predict. However, interstitial foreign atoms are often located on well defined sites, e.g. on the tetrahedral interstitial site of the diamond-type lattice, or on the octahedral (face-centred) site of the body-centred cubic lattice. In such cases, they will be perfectly shadowed along some of the low-index directions (cf Figure 13.2) and their location can thus be determined without a detailed knowledge of the flux distribution across the channel. Let us look at a few examples illustrating how this is accomplished.

13.6.1 Boron in Tungsten: an Octahedral Interstitial Impurity

A particularly simple and clear-cut example of the use of channeling to locate an interstitial foreign atom is provided by the recent work of Andersen *et al.*[5] on boron-implanted tungsten crystals. The two commonly expected interstitial sites in a bcc lattice, the octahedral site and the tetrahedral site, are illustrated in Figure 13.6. Comparison of the axial projections in the lower part of this figure with the angular scans in Figure 13.7 shows that only the octahedral site is consistent with the data. Although both sites would give a peak at approximately 0° incidence* for each of the three axes, only the

---

* In principle, since the foreign atoms are not on the mid-channel position along either $\langle 111 \rangle$ or $\langle 110 \rangle$, the peak in yield should be displaced slightly from 0°; however, as noted in Section 13.5, such an effect would be easily masked by multiple scattering.

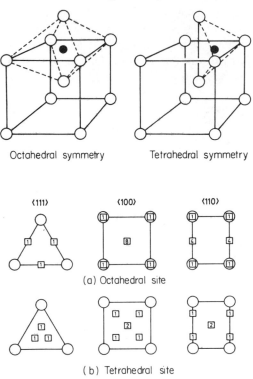

Figure 13.6. Projections showing the octahedral and tetrahedral interstitial positions in tungsten, as viewed along each of the three principal axial channels. The numbers ①, ②, etc., indicate the relative probability of the interstitial site having its projection at this particular position. Thus, along the ⟨100⟩ for example, 2/3 of the octahedral sites fall at the mid-channel position and 1/12 are shadowed by each of the four atomic rows

octahedral site should exhibit a $\frac{1}{3}$ 'perfectly shadowed' component along the ⟨100⟩ and ⟨110⟩ axes. Both angular scans display this $\frac{1}{3}$ dip, and its angular width is in reasonable agreement with the width of the tungsten host dip.* Moreover, the high and narrow peak along ⟨100⟩ agrees well with $\frac{2}{3}$ of the boron being at the optimum flux-peaking site in the centre of this channel.

Further information on an interstitial site can often be obtained from planar scans. For example, to verify that the octahedral site was correct, Andersen *et al.* also performed the {100} planar scan shown in Figure 13.8.

---

* The boron dips appear to be slightly broader than those of the host, but this may be due to the fact that the nearest-neighbour B–W distance along the row is only half the W–W distance.

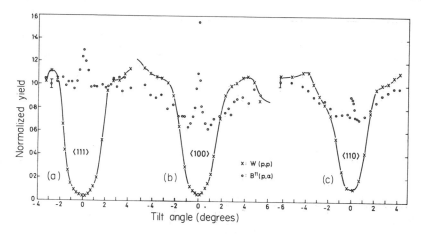

Figure 13.7. Angular scans across the principal axial channels for 700 keV protons on a boron-implanted tungsten crystal: (a) ⟨111⟩, (b) ⟨100⟩, (c) ⟨110⟩. (Anderson *et al.*[5])

Both the angular width and size of the boron dip are in good agreement with the host. The octahedral site is the *only* interstitial site in tungsten which can exhibit this perfect shadowing behaviour along the {100} plane.

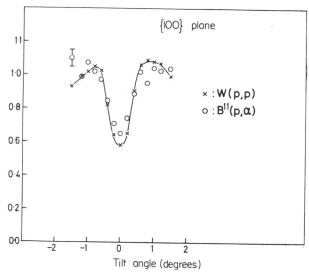

Figure 13.8. Angular scan through the {100} plane for 700 keV protons incident on a boron implanted tungsten crystal (Andersen *et al.*[5])

13.6.2 Zr and Tl in Silicon: Tetrahedral Interstitial Impurities

A similar investigation of Zr-implanted silicon has enabled Domeij *et al.*[18] to locate the Zr atoms as being on the tetrahedral interstitial site of the diamond-type lattice. Such a site is perfectly shadowed along both $\langle 111 \rangle$ and $\langle 100 \rangle$; hence, along these directions the scattering yield from Zr exhibits a broad, deep dip similar to that from the silicon. Along the $\langle 110 \rangle$ channel, however, the tetrahedral site lies fairly close to the mid-channel position, thus producing an extremely narrow peak in the Zr yield (Figure 13.9). The magnitude of this flux peak (approximately twice the random level) is consistent with the value of 1·75 calculated by Van Vliet[14] for the tetrahedral site.

In a closely related series of measurements on Tl-implanted silicon, Domeij *et al.*[18] found that Tl atoms are partly substitutional and partly on the tetrahedral interstitial sites. Unfortunately, their data were not sufficiently precise to determine accurately the fractional distribution between the two sites. More recently, they have reported the use of double-alignment techniques (Section 13.7) to enhance the distinction between substitutional, random, and interstitial contributions to the $\langle 110 \rangle$ yield.

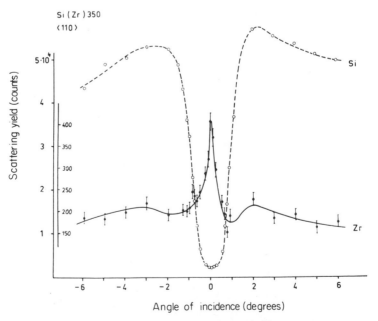

Figure 13.9. Scattering yield for 1·8 MeV ^{12}C ions in Zr-implanted silicon as a function of angle of incidence relative to the $\langle 110 \rangle$ direction. (Domeij *et al.*[18])

13.6.3 Yb in Silicon: a More Complex Site

One of the first experimental evidences for flux peaking in the mid-channel region was obtained by Andersen *et al.*[13] in attempting to determine the lattice location of Yb in silicon. Their detailed angular scans across each of the low-index directions are shown in Figure 13.10. The magnitude and width of the narrow peak along ⟨110⟩ is similar to that for the Zr implant (Figure 13.9). However, the behaviour along the other two axes clearly indicates that the Yb atoms are not on the *tetrahedral* site. In both cases the Yb dip is much weaker than that for the Si lattice and furthermore in the ⟨111⟩ case it is about a factor of two narrower than the Si dip. Substitution of this (⟨111⟩) reduced angular width into equation (13.3) indicates that the Yb atoms must be located about 0·5 Å off the ⟨111⟩ axes. On the other hand, along ⟨100⟩, the angular width and depth of the Yb dip suggests that $\frac{1}{3}$ of the sites are located almost exactly in the shadow of each set of ⟨100⟩ rows. These considerations are sufficient to locate the site of the Yb atoms as being along the ⟨100⟩ row, approximately midway between two adjacent tetrahedral interstitial sites. Again, it is interesting to note that the normalized flux at this site (for incidence parallel to the ⟨110⟩) is estimated by Van Vliet[14] to be 1·85, in good agreement with the observed yield of 1·6 in Figure 13.10.

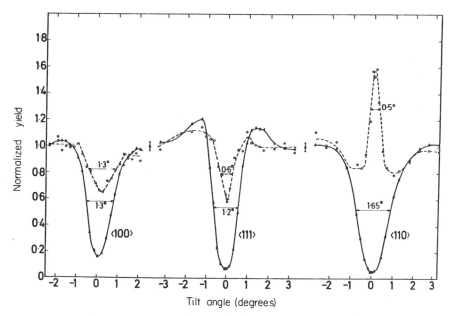

Figure 13.10. Scattering yield from the Yb atoms (O) and from the Si atoms (△) in the implanted region, as a function of the angle between the incident beam direction and various low-index axes. (Andersen *et al.*[13])

13.6.4 Bi in Silicon: an 'Almost Substitutional' Case

In some of the early atom-location studies[19] on Bi-implanted silicon, large (5–10-fold) attenuations in scattering yield off the Bi atoms were observed along all the low-index directions and so it was concluded that at least 80–90% of the Bi atoms must be substitutionally located in the lattice. Presumably, the other 10–20% are randomly distributed in the lattice, e.g. at dislocations, precipitation sites, etc. Subsequently, more careful measurements[16,20] showed that the angular width of the Bi dip (Figure 13.11)

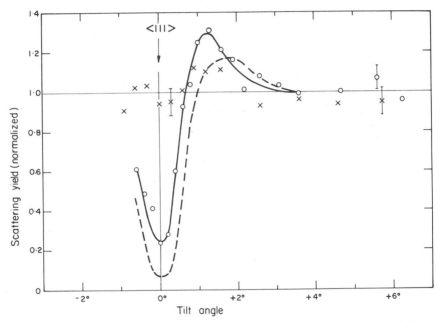

Figure 13.11. Orientation dependence of the scattering yield of 1·0 MeV helium ions from Bi (O), Au (X), and Si (- - -) atoms in ion-implanted silicon crystals (Eriksson *et al.*[20])

is significantly narrower than that of the silicon lattice, indicating that the Bi atoms are not perfectly shadowed. By inserting this reduced angular half width ($\psi_{\frac{1}{2}}$) into equation (13.3) one can estimate the mean displacement of the Bi atoms. From the data in Figure 13.11, the displacement turns out to be approximately 0·25 Å. For such a heavy atom as Bi it does not seem reasonable to attribute this displacement to an enhanced vibrational amplitude. Hence, it presumably indicates a permanent (static) displacement of the Bi atoms.

Picraux *et al.*[16] have recently shown that additional information can be obtained by fitting the entire shape of the dip, rather than merely using the half angle. As seen in Figure 13.12, somewhat better agreement with their channeling data is obtained by assuming that half the Bi atoms are displaced 0·45 Å from the Si lattice sites and the other half are exactly substitutional. They find that the calculated distribution is sensitive to the magnitude but not to the direction of the foreign-atom displacement from the substitutional site. This is an interesting refinement of the channeling-effect technique.

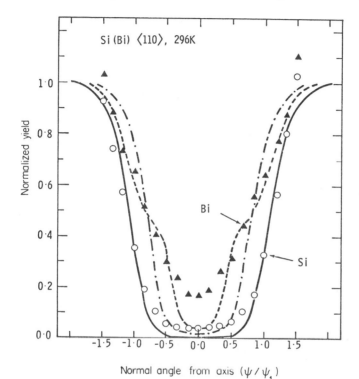

Figure 13.12. Backscattering yield of 1·0 MeV helium ions from Bi (▲) and Si (○) atoms in a Bi-implanted silicon crystal. The solid line is the calculated distribution (see Figure 13.5) for the silicon lattice, the chain line for Bi atoms displaced 0·2 Å along the ⟨110⟩ directions, and the broken line for Bi atoms 50% on lattice sites and 50% displaced 0·45 Å along the ⟨110⟩ directions. (Picraux *et al.*[16])

It is clear from Figure 13.12 that the observed perfectly aligned yields (χ_{min}) from both the Bi and the Si atoms lie considerably above the calculated distributions. For Bi, this discrepancy in the normalized yield is approximately 0·15, indicating that perhaps 15% of the Bi atoms are randomly distributed in the lattice. On the other hand, Barrett[21] has recently shown by Monte Carlo calculations that χ_{min} at the lattice site should be increased by about a factor of three above the simple $(\chi_{min} \simeq \pi N d\rho_{rms}^2)$ estimate of Lindhard;[12] if a similar factor applies also for the slightly displaced Bi atoms, then this χ_{min} discrepancy would almost completely disappear (cf Figure 11 of reference 16).

An experimental test of whether or not a small fraction of the bismuth is randomly distributed is discussed in Section 13.7.

13.6.5 Randomly Distributed Foreign Atoms

In many systems, such as the Au in Si case included in Figure 13.11, no significant* fluctuation in scattering yield from the foreign atom is observed along any axial or planar direction. These atoms must therefore be on some sort of randomly distributed site within the lattice. Channeling measurements are unable to distinguish the exact location of such 'randomly distributed' atoms. Consequently, we cannot specify whether they are pinned on dislocations, situated in locally disordered regions, contained in precipitates or merely distributed over a large number of different sites.

13.7 DOUBLE ALIGNMENT

Another interesting refinement utilized by Picraux *et al.* in studying Bi-implanted silicon[16] and also by Domeij[22] in studying interstitial dopants (Section 13.6.2) is the technique of double alignment.[23] This requires both the incident and outgoing (detected) beams to be aligned simultaneously with major crystal axes, or planes. First, the incident beam is aligned with one axis in the crystal, and then a well collimated detector is aligned with another† axis in order to observe only those scattered particles that emerge parallel to a channeling direction. Particles originating from a lattice site are prohibited (i.e. 'blocked'—see Chapter 11) from emerging parallel to any lattice row or plane, whereas particles originating from a mid-channel position will have an enhanced emission probability parallel to the channel. According to the rule of reversibility, the blocking phenomenon is closely related to that of channeling. It therefore produces a further 30–100-fold attenuation in the observed yield from lattice sites and an approximately threefold *increase* from mid-channel sites.

For atoms on lattice rows, calculations[24] of the double-alignment minimum yield χ' indicate that

$$\chi' = \nu(\alpha)\chi_{min}^2 \tag{13.4}$$

---

* Yield attenuations of less than about 20% are usually of doubtful significance in atom-location studies; the reasons for this are discussed in Section 13.8.

† Alternatively, by using an annular detector in the so-called 'uniaxial' technique of Appleton and Feldman,[24] both alignments can be achieved simultaneously along the same axial direction.

where χ_{min} is the single-alignment yield for channeling or blocking, and $v(\alpha)$ is 1–2, depending on the angle α between the incoming and outgoing channel axes. It is interesting to note that, in their Bi in Si investigation (Figure 13.13), Picraux *et al.* observed (χ'/χ_{min}^2) ratios of 1·5–1·8 for the scattering off Bi atoms, in good agreement with those predicted by equation (13.4), thus confirming that almost all (i.e. $>95\%$) of the bismuth atoms must be located close to the substitutional sites. This clearly indicates that the discrepancy between the measured and calculated single-alignment yields in Figure 13.12 is *not* due to 15% of the Bi atoms being randomly distributed through the lattice.

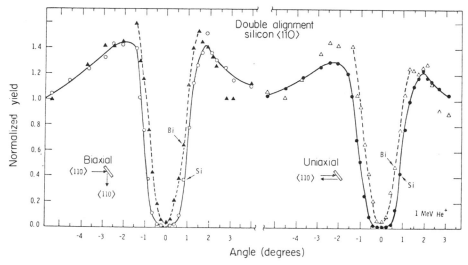

Figure 13.13. Uniaxial and 90° biaxial double-alignment angular distributions for ⟨110⟩ axes, showing the Bi and Si dips for 1 MeV helium scattering from the same bismuth-implanted silicon crystal as in Figure 13.12 (Picraux *et al.*[16])

For atoms on mid-channel sites, the channeling and blocking effects should again involve a geometrical combination (i.e. $\chi' \simeq \chi_{max}^2$), so that the observed two-fold increase in single-alignment yield in (for example) Figure 13.9 should be enhanced in double-alignment to an approximately four-fold effect. This could be a valuable tool for distinguishing the relative contribution of mid-channel atoms from that of randomly distributed ones. Domeij[22] has recently reported observing such an increase in his channeling studies of Tl-implanted silicon.

There is one serious limitation to the use of double alignment. Because of the need of a well collimated detector, the yield or count rate per micro-coulomb of beam is much less than in similar single-alignment studies and much larger beam currents are therefore required. In many cases, the increased radiation damage completely overrides any advantages that may be involved.

13.8 LIMITATIONS

In most of the above cases, it was possible to locate rather accurately the site of the foreign atoms. However, our examples all involved high quality single crystals with a detectable amount of foreign atoms embedded at a fairly shallow depth (typically ~ 1000 Å) beneath the surface. The crystal types were always cubic (either fcc, bcc, or diamond), thus possessing a high degree of lattice symmetry. Furthermore, in almost every case, the interaction yield with the foreign atoms exhibited fairly large ($\gtrsim 30\%$) yield attenuations along at least some of the low-index directions or planes.

Whenever one or more of these conditions is not fulfilled, serious complications may arise.

13.8.1 Crystal Perfection

Obviously, the technique cannot be applied to polycrystalline or amorphous substrates. Even in single-crystal material, however, considerable caution must be exercised whenever the aligned interaction yield (χ_{min}) with lattice atoms remains considerably greater than its theoretically predicted level.* In such cases, anomalous scattering processes (surface contamination, radiation damage, lattice defects, etc.) must be significantly perturbing the channeled flux distribution. Since the relative importance of such perturbations at interstitial and substitutional sites may be quite different, it would be dangerous to apply any simple subtraction procedure to correct for an anomalously large χ_{min} value. In ion-implanted crystals, there is often a large increase in χ_{min} owing to radiation damage introduced by the implantation (cf Chapter 14). In such cases, it is desirable to anneal the implanted crystals properly *before* attempting the atom-location studies. This, of course, introduces the complication that the implanted atoms may change their lattice location during the anneal.

It should also be stressed that a proton or helium-ion beam at MeV energies is *not* a particularly gentle probe for investigating a crystal lattice and consequently significant radiation damage may sometimes occur during the atom-location measurements. In metallic or semiconductor crystals such effects are usually quite small; on the other hand, in organic or ionic crystals they can be extremely serious. Even in those cases where the total amount of lattice damage is small, some of the radiation-induced defects may interact with specific foreign atoms in such a way as to change their lattice location. Thus, boron,[25] thallium[26] and arsenic[27] atoms in silicon have been observed to move significantly from substitutional to nonsubstitutional sites under the influence of proton or helium-ion bombardment.

---

* For a low-index axial direction, χ_{min} should typically be no greater than 1–5 % of the random value.

13.8.2 Detection Sensitivity

As noted in Section 13.3, it is usually possible to detect foreign atoms at concentration levels down to about 10^{-4} atom fraction; for many solid state investigations, this sensitivity is quite adequate. However, in the semi-conductor field, the concentration range of greatest interest starts at the 10^{-4} level and extends downwards through several orders of magnitude. Consequently, an increase in detection sensitivity would be highly desirable; at present, the most promising prospect is probably the use of inner-shell X-ray production resulting from proton or helium-ion bombardment (cf Section 13.3).

13.8.3 Depth Distribution

If the distribution of foreign atoms extends deep into the crystal, one can (in principle) use channeling measurements to study the depth dependence of their lattice location. However, this introduces two serious complications. Firstly, multiple scattering produces a significant dechanneling of the beam (cf Chapter 7) and so the observed angular dependence of the interaction yields will be considerably weaker and less predictable than near the surface. Secondly, the beam undergoes significant energy loss in traversing the region containing the foreign atoms. Since the cross sections for Rutherford scattering, X-ray production, etc., are strongly energy dependent, and since the stopping powers for random and for channeled trajectories are quite different, it then becomes necessary to apply a correction for this energy loss. Unfortunately, it is difficult to decide what stopping power is appropriate for the aligned case and thus the correction cannot always be accurately evaluated. Hence, the analysis of deep ($\gtrsim 0.5 \ \mu$m) implants[28] or of uniformly doped crystals[27] is more difficult than the shallow-implant examples considered in this chapter.

13.8.4 Low-symmetry Crystals

The value of triangulation and lattice symmetry arguments (Section 13.2) in determining the site of foreign atoms in highly symmetrical lattices has already been clearly illustrated. Thus, for example, in the Au in Cu case (Figure 13.4), a scan through one major axis was sufficient to establish that the Au atoms were virtually 100 % on substitutional sites. However, in a low-symmetry crystal such as monoclinic or triclinic, a similar angular scan would merely indicate that the foreign atoms were located somewhere within the shadow of the aligned set of rows. Triangulation procedures then become much more difficult and it is no longer possible to determine unambiguously the position of the foreign atoms, except perhaps when they are located on a single, well defined site.

13.8.5 Small Yield Attenuations

Whenever one observes only a small attenuation in the interaction yield with the foreign atoms, several complications arise. Firstly, lattice symmetry arguments no longer offer much help. Even in a cubic crystal, a 16 % attenuation along $\langle 110 \rangle$ (with the same angular width as that for the lattice atoms) could arise either from 16 % of the atoms being on lattice sites, or from 100 % of them being at some intermediate site along the $\langle 110 \rangle$ row, or various combinations of these two alternatives. Secondly, triangulation between two or more axes does not contribute much information either, since the foreign atoms contributing to a small dip along one axis may not be on the same site as those involved in another small dip along a different axis. Thirdly, there are no truly *random* directions in a single crystal and thus the 'random yield' required for normalization cannot usually be specified to better than about 5 %. This uncertainty in the random level is of little consequence whenever the foreign-atom and lattice yields both exhibit large attenuations. For example, in the case of a 90 % dip in foreign-atom yield, a $\mp 5 \%$ uncertainty in the choice of random level corresponds to less than a 1 % error in the fraction of shadowed atoms. On the other hand, if the foreign-atom yield attenuation is small, a significant error may be introduced. Thus, in the above example of a 16 % dip in foreign-atom yield, a $\mp 5 \%$ uncertainty in the choice of random level would set upper and lower limits of 21 % and 11% for the shadowed fraction. This large spread is almost useless for quantitative atom-location purposes.

Atom-location studies therefore require the foreign-atom yield to exhibit a fairly large attenuation along at least some of the low-index axes or planes. In highly symmetrical (e.g. cubic) lattices, we have seen examples where approximately 35–50 % dips in yield were quite sufficient (cf Figures 13.7, 13.8 and 13.10). In less symmetrical lattices, however, considerably larger attenuations would probably be necessary.

13.9 SUMMARY

Our basic understanding of the channeling behaviour of MeV low-Z ions is already sufficiently well established to provide a simple technique for pinpointing the geometrical location(s) of specific foreign atoms within a crystal lattice. Provided the interaction yield of the channeled beam with the foreign atoms exhibits a strong angular dependence, then measurements of the angular width and magnitude of this dip (or peak) in yield along several different crystallographic directions, together with lattice symmetry and triangulation arguments, often permit the site of the foreign atom to be specified to within approximately 0·1 Å.

For interstitially located atoms, a more accurate description of the channeled flux distribution and its dependence on depth, $E_\perp$, etc., would certainly be desirable. Several groups are tackling this problem experimentally and with computer simulations and we hope soon to see considerable progress.

References

1. Proc. 1st Int. Conf. on Ion Implantation (Thousand Oaks, California, May 4–9, 1970), *Rad. Effects* **6, 7** (1970–1).
2. Proc. European Conf. on Ion Implantation (Reading, Sept. 7–9, 1970). Stevenage, England, Peter Peregrinus (1970).
3. Proc. 2nd Int. Conf. on Ion Implantation (Garmisch, May 24–28, 1971). Berlin, Springer-Verlag (1971).
4. Mayer, J. W., Eriksson, L. and Davies, J. A., *Ion Implantation in Semiconductors*, New York, Academic Press (1970).
5. Andersen, J. U., Laegsgård, E. and Feldman, L. C., Proc. Gausdal Conf. on *Atomic Collisions in Solids, Rad. Effects*, **12**, 219 (1972).
6. van der Weg, W. F., den Boer, J. A., Saris, F. and Onderdelinden, D., *Proc. European Conf. on Ion Implantation, Reading, Sept. 1970*, Stevenage, England, Peter Peregrinus, p. 198 (1970).
7. Cairns, J. A., *Nucl. Instrum. and Meth.*, **92**, 507 (1971).
8. Domeij, B., *Nucl. Instrum. and Meth.*, **38**, 207 (1965).
9. Matzke, Hj. and Davies, J. A., *J. appl. Phys.*, **38**, 805 (1967).
10. Uggerhøj, E. and Andersen, J. U., *Can. J. Phys.*, **46**, 543 (1968).
11. Alexander, R. B. and Poate, J. M., Proc. Gausdal Conf. on Atomic Collisions in Solids, *Rad. Effects*, **12**, 211 (1972).
12. Lindhard, J., *Dan. Vid. Selsk. Mat. Fys. Medd.*, **34**, No. 14 (1965).
13. Andersen, J. U., Andreasen, O., Davies, J. A. and Uggerhøj, E., *Rad. Effects*, **7**, 25 (1971).
14. Van Vliet, D., *Rad. Effects*, **10**, 137 (1971).
15. Eisen, F. and Uggerhøj, E., Proc. Gausdal Conf. on Atomic Collisions in Solids, *Rad. Effects*, **12**, 233 (1972).
16. Picraux, S. T., Brown, W. L. and Gibson, W. M., *Phys. Rev.* B, **6**, 1382 (1972).
17. Morgan, D. V. and Van Vliet, D., Proc. Gausdal Conf. on Atomic Collisions in Solids, *Rad. Effects*, **12**, 203 (1972).
18. Domeij, B., Fladda, G. and Johansson, N. G. E., *Rad. Effects*, **6**, 155 (1970).
19. Mayer, J. W., Davies, J. A. and Eriksson, L., *Appl. Phys. Lett.*, **11**, 365 (1967).
20. Eriksson, L., Davies, J. A., Johansson, N. G. E. and Mayer, J. W., *J. appl. Phys.*, **40**, 842 (1969).
21. Barrett, J. H., *Phys. Rev.*, B, **3**, 1527 (1971).
22. Domeij, B., Proc. Gausdal Conf. on Atomic Collisions in Solids.
23. Bøgh, E., Proc. Int. Conference on Solid State Physics Research with Accelerators, BNL-50083, p. 76 (1967).
24. Appleton, B. R. and Feldman, L. C., *Proc. Sussex Conference on Atomic Collision Phenomena in Solids*. Amsterdam, North Holland, p. 417 (1970).
25. Davies, J. A., Denhartog, J., Eriksson, L. and Mayer, J. W., *Can. J. Phys.*, **45**, 4053 (1967).
26. Fladda, G., Mazzoldi, P., Rimini, E., Sigurd, D. and Eriksson, L., *Rad. Effects*, **1**, 243 (1969).
27. Haskell, J., Rimini, E. and Mayer, J. W., *J. appl. Phys.*, **43**, 3425 (1972).
28. Gibson, W. M., Martin, F. W., Stensgaard, R., Palmgren Jensen, F., Meyer, N. I., Galster, G., Johansen, A. and Olsen, J. S., *Can. J. Phys.*, **46**, 675 (1968).

CHAPTER XIV

Application to Radiation Damage

F. H. EISEN

14.1 INTRODUCTION

Channeled ions are sensitive to the position of the atoms of the host crystal, just as they are sensitive to the location of foreign atoms. Channeling may, therefore, be applied to the study of radiation damage in crystalline solids. The principles outlined in Chapter 13 for its application to foreign atom location in crystals also apply to the study of radiation damage in crystals. However, there are additional problems which are peculiar to the study of radiation damage. One of these is the inability to distinguish experimentally between reactions of the analysing beam with displaced atoms and with those atoms of the crystal which have not been displaced. The channeled fraction of the beam interacts only with displaced atoms whereas the ions in the random fraction of the beam interact with all the atoms of the crystal. A separation between these two contributions must be included in the calculation of the number of displaced atoms. This is one of the most important problems connected with applying channeling techniques to the study of radiation damage.

A further difference between foreign atom location and radiation damage studies is that the channeling technique is not as sensitive in the radiation damage case. This is associated with the fact that a finite scattering yield is observed from an undamaged crystal. The magnitude of this yield then determines the smallest amount of damage which can be observed with the channeling technique. Using single alignment, this will usually amount to the displacement of at least 1 % of the atoms of the crystal.

A close impact process is required for the detection of displaced atoms by channeled ions. Backscattering has been used almost exclusively for this purpose in radiation damage studies utilizing the channeling technique. The use of backscattering makes it possible to obtain information concerning the depth distribution of the radiation damage, as well as the amount of the damage. This depth information is also helpful in extracting the amount of the damage.

The depth scale for the damage distribution can, in principle, be established from a knowledge of the channeled and random stopping powers of the analysing beam in the material being investigated. Often these data are not available. If stopping power data are available, there is still some uncertainty as to what stopping power should be used for the channeled ions. These factors limit the precision with which the depth scale for a damage distribution can be calculated.

The backscattering spectra give no direct information concerning the nature of the defects producing the scattering. It is possible to reach some tentative conclusions about the types of defects which are most likely to produce backscattering events. However, one of the current problems with the application of the channeling-backscattering technique is that there is not a clear understanding of what is detected by its use.

These aspects of the application of channeling to the study of radiation damage will be discussed in detail below, together with some of the practical problems encountered in applying this technique.

14.2 DECHANNELING

A typical energy spectrum for channeled protons backscattered from an ion implanted silicon sample is shown in Figure 14.1, together with spectra for protons channeled along the same axis in an undamaged silicon crystal and for random incidence of the proton beam on a silicon crystal. The disordered region of the implanted sample is buried below the surface as indicated schematically in the inset of the figure. The backscattering yields observed in the labelled parts of the spectrum for the implanted sample are due to scattering from the regions of the inset with the corresponding label. The peak in the scattering yield observed in region B is due to the scattering of the channeled protons from the displaced silicon atoms in the crystal. The scattering yield from region C of the crystal, which is below the damaged region, is higher than the scattering yield observed from the same depth (i.e. for about the same channel numbers) in the undamaged crystal. This is due to the spreading of the channeled beam which occurs in the disordered region of the crystal and which results in the dechanneling (see F. Grasso, Chapter 7) of some of the channeled ions. These dechanneled ions can be backscattered by all of the atoms of the crystal, and so they contribute to the backscattered yield in region B of the spectrum where the damage peak occurs, as well as to the yield in region C. Assuming that the analysing beam can be divided between a channeled component (also termed the aligned beam) which can only be scattered from atoms which are not on lattice sites (scattering centres) and a random component which is scattered by all the atoms of the crystal, we have the following relation,[1]

$$y'(t) = y_n(t)\left[\left\{1 - \chi'(t)\right\}\frac{N'(t)}{N} + \chi'(t)\right].\qquad(14.1)$$

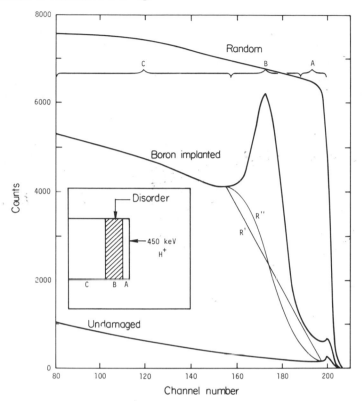

Figure 14.1. Aligned ($\langle 110 \rangle$) and random backscattering spectra for a 450 keV proton beam incident on silicon. The spectrum showing a damage peak was obtained from a sample implanted with 8×10^{15} boron ion/cm^2. The line labelled R′ is the linear approximation to $\chi'(t)$ discussed in the text and the curve labelled R″ was calculated using the Keil plural scattering theory. The labelled regions of the spectrum for the boron-implanted sample correspond to the regions with the same label in the schematic disorder distribution shown in the inset (after Eisen and Welch, 1971)

$y'(t)$ is the backscattered yield from depth t, y_n is the normal yield (i.e. the yield with the beam incident in a random direction), χ' is the random fraction of the analysing beam, N' is the density of scattering centres and N is the atomic density of the crystal. The first term on the right is due to the scattering of channeled ions from scattering centres and the second is due to the scattering of the random portion of the beam. If equation (14.1) is solved for $N'(t)$ we get

$$N'(t) = N \frac{\{[y'(t)/y_n(t)] - \chi'(t)\}}{\{1 - \chi'(t)\}}. \tag{14.2}$$

Equations (14.1) and (14.2) neglect the possible difference in scattering cross section for the aligned and random beams at a given depth, which can occur because the aligned and random beams may have different energies at the same depth. (See Section 14.4.)

It can be seen from equation (14.2) that in order to evaluate the concentration of scattering centres at a given depth, the random fraction of the beam at that depth, $\chi'(t)$, must be known. The simplest method which has been used to approximate the random fraction of the beam in the damaged region has been to draw a straight line from a point on the aligned curve for the undamaged sample, near the beginning of the damaged region, to a point on the spectrum for the damaged sample, just behind the damaged region. This is illustrated in Figure 14.1. A comparison of the results of applying this approximation with those obtained using more sophisticated calculations of $\chi'(t)$ will be made below. Often in applying this linear approximation, the area between this line and the backscattered energy spectrum is measured and taken as proportional to the total concentration of scattering centres. This effectively ignores the term in the denominator of equation (14.2) and can result in errors as high as 20 to 30 % for large amounts of damage.

Since dechanneling is produced by the same scattering centres that cause the backscattering of the aligned beam, it should be possible to relate the random fraction of the analysing beam to the concentration of scattering centres, assuming that a scattering centre makes the same proportionate contribution to the backscattered yield as it does to the dechanneling. (This will not be true for atoms which are only a small distance from lattice sites, as is discussed in Section 14.3.) Several workers[2-5] have done this using a single scattering, plural or multiple scattering treatment of the spreading of the aligned beam by the scattering centres. In all cases, it has been assumed that if a channeled ion is scattered through an angle greater than ψ_{crit}, the critical angle for channeling, it is dechanneled and becomes a part of the random beam. Scattering from the random beam into the channeled beam is neglected.

If the dechanneling is due to single deflections through an angle greater than ψ_{crit}, then $\chi'(t)$ is given by[1]

$$\chi'(t) = 1 - \{1 - \chi_1(t)\}\, e^{-\gamma(t)} \tag{14.3}$$

and

$$\gamma(t) = \frac{\pi Z_1^2 Z_2^2 e^4}{E^2 \psi_{crit}^2} \int_0^t N'(t')\, dt' \tag{14.4}$$

where $\chi_1(t)$ is the normalized aligned yield from an undamaged crystal, Z_1 and Z_2 are the atomic numbers of the projectile and target atoms respectively, e is the electronic charge and E the energy of the projectile ions. These equations may be obtained by solving the dechanneling problem if it is assumed that the dechanneling which normally occurs owing to multiple scattering processes in undamaged crystals and the disorder dechanneling are independent of each other. This is a reasonable assumption if the disorder dechanneling is indeed entirely due to single scattering.

Single scattering accounts for dechanneling in a reasonable manner for low amounts of disorder, but as the amount of disorder increases, plural or multiple scattering of the channeled ions should become important. The dechanneling problem, assuming that plural or multiple scattering dominates the dechanneling, has not been rigorously solved. The following estimate[2] for $\chi'(t)$ has been used:

$$\chi'(t) = \chi_1(t) + [1 - \chi_1(t)]F(t), \qquad (14.5)$$

where $F(t)$ is the fraction of the aligned beam which has been scattered beyond ψ_{crit} at depth t. This estimate essentially treats the normal and disorder dechanneling as independent. In applying (14.5), $F(t)$ has been calculated by finding the fraction of ions which would be scattered beyond ψ_{crit} by a randomly distributed number of atoms/cm^2 given by $\int_0^t N'(t')\,dt'$. This ignores the fact that ions which have been scattered through an angle greater than ψ_{crit} are dechanneled, and so are no longer actually part of the scattering distribution. This approximation and that of the independence of normal and disorder dechanneling should become progressively worse as the amount of disorder increases. However, there seems to be no better estimate of $\chi'(t)$ available at present.

The most common method of calculating damage distributions from the above relations has been to do an iterative calculation in which it is assumed that $\chi'(0) = \chi_1(0)$. $N'(0)$ is then calculated and used to obtain the value of χ' for the next portion of the damage distribution. This process is repeated for each point in the measured scattering distribution. The validity of the results can be judged by a kind of self-consistent condition, i.e. that the calculated value of the scattering yield behind the damage peak should agree with the observed value. Or, stated another way, the calculated distribution of scattering centres must return to zero behind the damage peak.

The single scattering treatment of backscattering data has been used by several workers.[2,4,5] This has usually given satisfactory results only when applied to samples with a small amount of disorder. In a heavily disordered sample, the amount of dechanneling calculated from a single scattering treatment is usually lower than that observed. However, Hart[5] has obtained good (i.e. self-consistent) results from channeling analyses of ion-implanted silicon samples, carried out with 120 keV protons and 280 keV helium ions, using a single scattering treatment.

Better success has been obtained with multiple scattering and plural scattering treatments. This includes the use[2,3] of an approximate multiple scattering formula given by Lindhard,[6] though this does not seem to be a good enough approximation in all cases. In this treatment, $F(t) = \exp\{-\psi_{crit}^2/\Omega^2\}$, where

$$\Omega^2 = 2\pi\left(\frac{Z_1 Z_2 e^2}{E}\right)^2 \ln(1{\cdot}29\varepsilon) \int_0^t N'(t')\,dt'.$$

ε (the reduced energy used by Lindhard[6]) is given by

$$\varepsilon = \frac{aE}{Z_1 Z_2 e^2} \cdot \frac{M_2}{M_1 + M_2}.$$

M_1 and M_2 are the masses of the projectile and target atoms respectively and $a = a_0 0{\cdot}8853(Z_1^{\frac{2}{3}} + Z_2^{\frac{2}{3}})^{-\frac{1}{2}}$, where a_0 is the Bohr radius.

Actually, for the number of scattering centres observed in most cases where channeling has been applied to radiation-damaged samples, a plural scattering treatment would seem to be more approrpriate. That is, the number of scattering events is usually not high enough to justify the use of a multiple scattering treatment. The plural scattering work of Keil *et al.*[7] has been applied by Westmoreland[4] and co-workers to damage produced by boron ions in silicon. This procedure seems to result in self-consistent results (as defined above) over a wider range of damage and analysing beam types and energies than any of the other scattering treatments which have been used. For this case, $F(t)$ is evaluated from tables of the fraction of the beam scattered beyond a given angle ($\hat{G}^*(m, \theta)$) in the paper of Keil *et al.* Recently, Meyer[8] has published a plural scattering theory based on a scattering cross section which seems more appropriate than that used by Keil *et al.* This has not yet been applied to the calculation of disorder distributions, but the calculations of Rimini *et al.*,[9] for scattering in aluminium films on silicon suggest that the results would not differ greatly from those obtained using the Keil theory, at least in the case of silicon.

A comparison of the results of applying single scattering, plural scattering, and the Lindhard multiple scattering formula to a particular backscattered energy spectrum is shown in Figure 14.2. It can be seen that the single scattering treatment results in a scattering centre distribution which does not return to zero behind the peak, whereas the multiple scattering treatment results in a distribution with negative scattering centre concentrations, which are of no physical significance. Use of the Keil plural scattering theory resulted in a calculated scattering centre distribution which behaved in a satisfactory manner behind the damage peak.

A comparison of the disorder distribution calculated from a set of back-scattering data for boron-implanted silicon using the straight line approximation to $\chi'(t)$ and the calculation based on the Keil plural scattering theory is shown in Figure 14.3. A comparison of the total amount of disorder calculated by these two approaches for silicon samples implanted with different doses of boron is shown in Figure 14.4. In both figures, the two

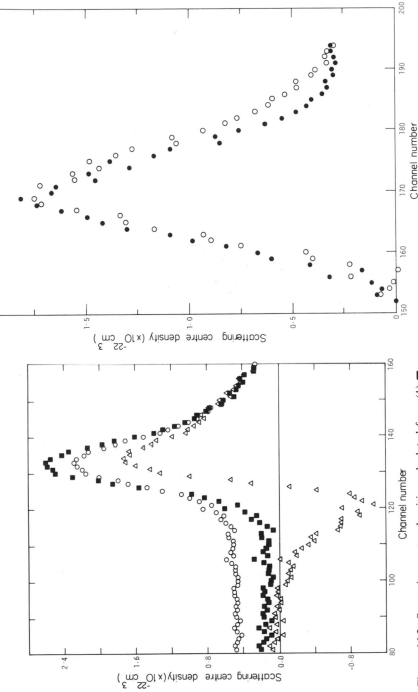

Figure 14.3. Comparison of disorder distributions calculated using the straight line approximation (●) and the Keil plural scattering theory (○) to obtain $\chi'(t)$[10,13]

Figure 14.2. Scattering centre densities calculated from (1) ■, plural scattering procedure outlined in the text, (ii) ○, single scattering treatment using equations (14.4) and (14.5) and (iii) △, multiple scattering formula used in references 2 and 3[4]

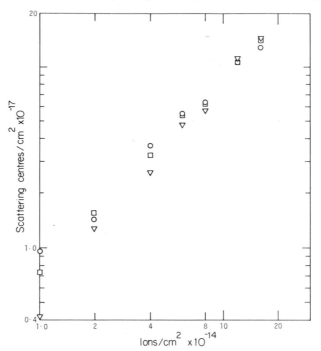

Figure 14.4. Comparison of the results of calculating the total disorder in a sample implanted with different doses of boron by (i) ▽, using the straight line approximation, (ii), ☐, using the Keil plural scattering theory to estimate the dechanneling, and (iii) ○, using the level behind the damage peak as discussed in the text. The backscattering measurements were carried out with 450 keV protons[10]

approaches to estimating $\chi'(t)$ give very similar results. In Figure 14.1 we have shown the value of $\chi'(t)$ which results from the application of the Keil plural scattering theory. Near the surface of the damaged sample, the value of $\chi'(t)$ calculated from the plural scattering approach is lower than the value obtained from the straight line approximation; whereas the plural scattering approach gives values of $\chi'(t)$ which are higher than the straight line approximation deeper in the damaged sample. These results suggest that the straight line approximation may give good results in cases where a distinct damage peak is evident in the backscattered energy spectrum.

Such a peak is not always observed in radiation-damaged samples. One example is shown in Figure 14.5(a).[11] Here it is not at all obvious where the damage distribution in the carbon-implanted gallium arsenide sample ends, so that the straight line approximation cannot be applied. When the plural scattering theory is used to calculate the damage distribution, a

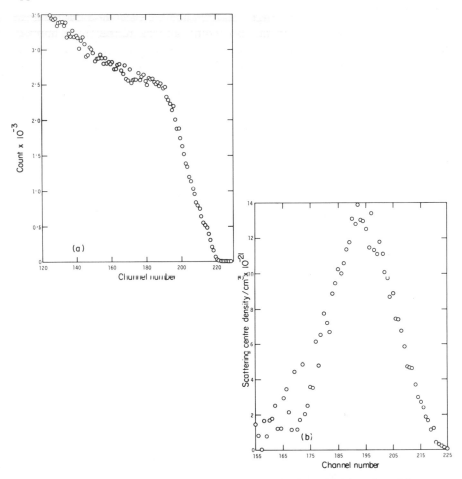

Figure 14.5. (a) Aligned spectrum for a GaAs sample implanted with carbon. Measurements with 450 keV protons. (b) Scattering centre distribution calculated from the data in (a) using Keil plural scattering theory[11]

definite peak results as shown in Figure 14.5(b). The application of the channeling effect technique to the study of disorder in epitaxial semiconductor layers is discussed by Mayer in Chapter 16. Here again there is no peak in the disorder distribution which would enable one to apply the straight line approximation. The experience gained in applying the plural scattering theory to the calculation of disorder distributions from data which show a distinct peak in the backscattered yield is of considerable value in approaching the problem of calculating disorder distributions in those cases in which a peak is not observed.

Bøgh[12] has approached the analysis of channeling effect data for damaged crystals in a slightly different manner. In his approach, the entire damage distribution is calculated on the assumption that there has been no dechanneling. This distribution is then used to calculate the dechanneling that would occur assuming a single scattering dechanneling mechanism. This process is iterated until the calculated scattering yield behind the damage peak agrees with that observed experimentally. One test of the validity of this procedure is shown in Figure 14.6, where measurements made with 0·5 MeV helium ions and 0·5 MeV protons on the same damage distribution are compared.

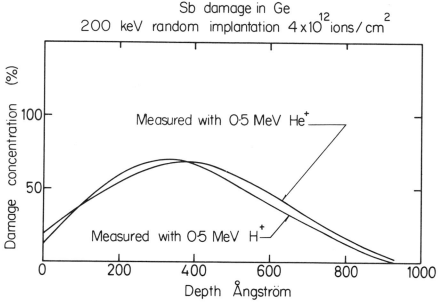

Figure 14.6. Comparison of calculated damage distributions obtained from data for a germanium sample implanted with antimony, taken with 500 keV protons and 500 keV He^+ (from reference 12)

The dechanneling correction for the helium ion measurements would be greater than that for the proton measurements. The fact that the two distributions are in close agreement is an indication that the procedure used to calculate the dechanneling is reasonably accurate.

It is sometimes possible to obtain a measure of the total damage introduced in an irradiated sample from the level of the scattering yield behind the damage peak. One way of doing this is to calculate from the Keil scattering theory the number of scattering centres which would be required to produce this amount of dechanneling. Eisen and Welch[13,14] have used this procedure in treating data for boron and carbon-implanted silicon. The results obtained for the total amount of damage in the sample using this procedure are also shown in Figure 14.4. The agreement with the other methods is good.

Some caution must be used in taking the yield behind the peak as a measure of total disorder. Figure 14.7 shows backscattering results for a sample implanted with boron at a low temperature and a sample implanted with boron at room temperature. It can be seen that the sample implanted at low temperature shows a lower amount of dechanneling even though the total disorder is clearly higher than for the sample which was implanted with boron at room temperature. This experimental difference probably indicates that factors related to defect structure are important in determining the dechanneling (see Section 14.3). Such factors have not been taken into account in any of the dechanneling treatments. This is a serious limitation upon the validity of these treatments at the present time.

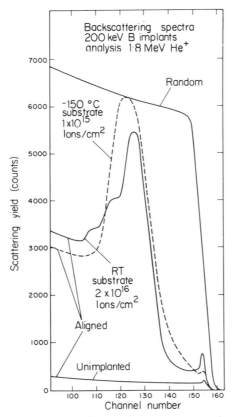

Figure 14.7. Spectra from 1·8 MeV He⁺ room temperature analyses of samples implanted with boron at − 150 °C (dashed line) and room temperature (solid line)[4]

14.3 SENSITIVITY TO VARIOUS KINDS OF DEFECTS

We now turn to a discussion of the kinds of defects which should be detected by the channeling effect technique. We know from Chapter 13 that the channeled beam will interact only with atoms which are more than about 0·1 Å from the lattice sites of the crystal. It might therefore be expected that the channeling effect technique would detect the interstitial atoms produced by radiation damage and also those atoms which are slightly displaced from their normal lattice site owing to strain from neighbouring defects such as single or multiple interstitials or vacancies. The number of scattering centres calculated from equation (14.2) is actually given by

$$N'(t) = \int N(x, y, t)P(x, y, t)\,\mathrm{d}x\,\mathrm{d}y \qquad (14.6)$$

where $N(x, y, t)$ is the distribution of scattering centres across the channel at depth t and $P(x, y, t)$ is the distribution of the channeled beam over the channel at depth $t(\int P(x, y, t)\,\mathrm{d}x\,\mathrm{d}y = 1)$. It is clear from Figure 13.6 that $P(x, y, t)$ will be much lower near the strings surrounding a channel than it is in the remainder of the channel. This results in a reduced contribution to $N'(t)$ owing to atoms which are only a few tenths of an ångström off their normal lattice position. It is worth noting that if the damage consists of interstitial atoms uniformly distributed across the channel with density $N(t)$, then from equation (14.6) $N'(t) = N(t)$. Therefore the scattering yield from such an interstitial distribution is independent of the spatial distribution of the channeled ions.

As an example, we compare the scattering expected from an interstitial silicon atom which is near the centre of the channel with that expected from the strain around a divacancy in silicon. Watkins[15] has estimated the displacement of the six silicon atoms surrounding a divacancy to be about 0·25 Å. Near a string the flux of channeled ions is approximately $Nd\pi\rho^2$ where ρ is the distance from the string and d is the atomic spacing along the string.[16] Using this expression, we obtain a value of 0·038 for the flux at 0·25 Å from the $\langle 110 \rangle$ strings in silicon. On average, only half the six silicon atoms will be displaced 0·25 Å perpendicular to the $\langle 110 \rangle$ strings. The others are displaced only about half this distance from the string. Assuming that the flux is about unity at the position of the interstitial we see that the contribution of the divacancy to the backscattering yield would be only about 15% of the contribution of the interstitial.

The flux distribution of the channeled ions will, of course, be affected by scattering in the damaged region. This will significantly reduce the flux-peaking near the centre of the channel but the flux near the strings will still be lower than in the rest of the channel. We can use Van Vliet's results on the flux distribution near a string[16] (his equation 5.9) to estimate the change which takes place in a damaged sample. For example, assume that $\chi' = 0.20$ behind the damage peak in a $\langle 110 \rangle$ silicon sample. From Keil's results we can estimate the beam spreading which would produce this amount of dechanneling. When this is used with Van Vliet's results, we find that the flux at a point 0.25 Å from the $\langle 110 \rangle$ strings would be increased to about 0.11 because of the spreading of the channeled beam in the disordered region.

These conclusions indicate that interstitials and interstitial clusters are the defects which are primarily responsible for the backscattering observed in channeling effect measurements. This is especially true when the initial flux distribution has not been greatly perturbed in the damaged layer, as will be the case for a thin disordered layer or for measurements in the low damage portion of a damage against dose curve.

The disorder produced in ion-implanted semiconductors, as measured by the channeling effect technique, is often appreciably greater than one would predict from collision theory. This problem has been discussed by Sigmund[17] and by Marsden and co-workers.[18] They have suggested that the reason may be due to the sensitivity of the technique in detecting small displacements from lattice sites. The considerations presented above suggest that this is unlikely and that some other explanation for the discrepancy must be found.

While atoms which are located only a small distance from their equilibrium lattice position have a lower probability of producing backscattering of channeled ions, they can produce more dechanneling per backscattering event than do atoms located more nearly in the centre of the channel. This is because all the channeled ions which have access to such slightly displaced atoms have a high transverse energy; whereas ions with a high transverse energy are only a small fraction of the total number of ions which can interact with an interstitial in the centre of the channel. The additional energy which must be transferred in order to cause dechanneling in a single scattering event is appreciably lower for the channeled ions which can come close to the atomic rows than for those with trajectories limited to the central part of the channel. The cross section for such an energy transfer to a channeled ion, in a collision with an interstitial atom located on the equipotential contour corresponding to the transverse energy of the ion, can be simply estimated as a function of the distance of the interstitial from an atomic row, using the Lindhard string potential.[6] The results show that this cross section is twice as high for interstitials 0.2 Å from $\langle 110 \rangle$ strings in silicon as for interstititals in the centre of the channel and four times as high if the distance from the string is 0.15 Å. The consequence of this can be illustrated by considering the changes which would take place in the backscattering spectrum for a sample with interstitial atoms distributed fairly uniformly across the channel, if additional displaced atoms located very close to the strings were introduced into the sample. The resultant backscattering spectrum should show an increase in the dechanneling relative to the scattering from the displaced atoms. Such an effect may contribute to the difference in dechanneling between silicon samples implanted with boron at liquid nitrogen temperature and at room temperature which was shown in Figure 14.7.

In principle, it should be possible to use the triangulation techniques described in Chapter 13 to obtain some information about the location of interstitial atoms when studying a radiation-damaged sample by the channeling technique. In practice, however, there seem to be no cases in which it has been possible to obtain information in this way. Where a comparison of the backscattering yields measured with the beam channeled along different axes has been reported,[19] no significant difference in the number of scattering centres has been observed. It seems likely that this is because the interstitial atoms exist mainly in clusters at the high levels of disorder which are detectable by the channeling effect technique. This would preclude the existence of unique interstitial positions which could be detected by triangulation.

A comparison of the number of scattering centres observed using single alignment and double alignment (see Section 13.7) may give information concerning the location of the interstitial atoms. The expression for the number of scattering centres observed in double alignment is similar to that for single alignment (equation (14.6)) except that the integrand contains the square of the distribution of channeled ions. This will be greater than one near the centre of the channel and less than one near the strings. Therefore, if the interstitials were primarily at the centre of the channel, double alignment should give a higher yield than single alignment and the reverse should be true if they lie near the strings. Hirvonen *et al.*[20] have used a comparison of single alignment and double alignment yields in carbon-implanted silicon to conclude that for low carbon doses the mean square displacement distance of the interstitials from $\langle 110 \rangle$ strings ranged from 0·4 to 0·56 Å. A possible difficulty with this result is the fact that these authors also observed some beam annealing (see below) in their work. Since double alignment measurements usually require a longer exposure to the analysing beam than single alignment measurements, the possibility of beam annealing effects which could lower the observed scattering yield is greater in the double alignment case.

14.4 DEPTH SCALE

The energy $E_2(0)$ of particles of initial energy E_1, which are backscattered from the surface atoms of a crystal, is given by $E_2(0) = k^2 E_1$, where k is the kinematic factor for the scattering (see Grasso, Section 7.2.1). Particles which are backscattered from a depth t in the crystal have an energy which differs from $E_2(0)$ by[1]

$$\Delta E = k^2 E_1 - E_2(t) = t\left\{k^2 S(E_1) + S(E_2)\frac{\cos\theta_1}{\cos\theta_2}\right\} \qquad (14.7)$$

where $S(E_1)$ and $S(E_2)$ are the stopping powers of the projectile ions in the crystal at energies E_1 and E_2, respectively, and θ_1 and θ_2 are the angles between the normal to the sample surface and the incident and backscattered beam directions, respectively (see Figure 7.4).

There are two problems connected with using equation (14.7) to convert the energy of the backscattering data to a depth scale. The first is that it is often not possible to find measured values of the stopping powers to substitute in the equation. This then necessitates that the stopping powers be estimated from existing data and will naturally be of limited accuracy. For the aligned spectra $S(E_1)$ is the stopping power of the channeled beam. Even if the appropriate channeling energy loss spectra have been measured it is not clear exactly what number should be used for this stopping power. It is presumably somewhat smaller than the random stopping power but it is clearly not as small as the stopping power of the best channeled ions. It is also possible that the channeled stopping power will vary as the amount of disorder in the crystal changes. In many cases, the approximation has been made that $S(E_1)$ is one-half the random stopping power. It is clear, however, from recent measurements of helium ion stopping in silicon[21] that this is often incorrect. These measurements showed that the ratio of channeled stopping to random stopping is a function of energy which has a maximum at about the same energy as that at which the maximum in the random stopping power is observed.

In some cases, the magnitude of the uncertainty in the depth scale caused by uncertainty in the value of the channeled stopping power, is minimized because the value of k^2 is small compared with one. In single alignment measurements $S(E_2)$ is the random stopping power. It can be seen from equation (14.7) that the error in the depth scale caused by uncertainty in the value of the channeled stopping power can be decreased by observing the backscattered ions at a large value of θ_2, which increases the distance that the outgoing beam travels in the sample relative to the distance travelled by the channeled beam. For double alignment measurements $S(E_2)$ is also a channeled stopping power with the same uncertainties as to what value should be used as discussed above for $S(E_1)$. It is therefore particularly difficult to establish a depth scale for double alignment measurements with good accuracy.

Equation (14.7) is actually an approximation. The exact relation[1] involves the integration of the stopping powers over the distance travelled by the particles in the crystal. This results in a variation in the conversion factor between energy and depth in the crystal as a function of the depth in the crystal. However, the error caused by using a constant depth scale is probably small compared with the error resulting from uncertainties in the values of stopping powers to be used in equation (14.7). This probably limits the accuracy with which the depth scale can be established to about $\pm 10\%$ in most cases.

A more direct method has been employed by Westmoreland *et al.*[4] to establish the depth scale in boron-implanted silicon. They used the damage peak itself as a marker in the crystal and removed layers of known thickness from the crystal, observing the shift in the energy position of the damage peak when this was done. This enables one to calculate a value of $\Delta E/t$ without the necessity of considering separately the values of the stopping powers involved. The accuracy of the layer removal method is limited by the accuracy with which the thickness of the layers which are removed can be measured. The value determined by Westmoreland *et al.* for the depth scale in $\langle 110 \rangle$ silicon with the analysis performed with 1·8 MeV helium ions and $\theta_1 = 0$, $\theta_2 = 16°$, was 45 ± 4 eV Å^{-1}.

If the measured channeled and random stopping powers[21] for helium in silicon are substituted into equation (14.7), the value of $\Delta E/t$ obtained is 44 eV Å^{-1}. The channeled stopping power used in this calculation of $\Delta E/t$ was the value obtained from the position of peak of the energy spectra for transmission of helium ions through thin silicon crystals, i.e. it was calculated from the most probable energy loss of the channeled ions. The agreement of the value of $\Delta E/t$ calculated from equation (14.7) with the value determined by layer removal, suggests that it may be appropriate in other cases to use as the value of the channeled stopping power that stopping power which corresponds to the most probable energy loss of the channeled ions.

In evaluating the values of $y'(t)$, $y_n(t)$, and $\chi_1(t)$ to be used in calculating $N'(t)$, it has been common to use values of these quantities at a given energy in the backscattered spectra as though they were the yields from a given depth in the crystal. This is not entirely correct, since the quantity $S(E_1)$ in equation (14.7) may be different for aligned and random orientations and may depend on the amount of disorder in the sample in the aligned case. Ziegler[22] has calculated disorder distributions by converting the energy scale to a depth scale. In doing this he has integrated the appropriate stopping powers over 200 Å layers of the sample. He has examined the effects of various assumptions about the channeled stopping power, including the assumption of a linear dependence on the scattering centre density. He has calculated disorder distributions using a straight line approximation for $\chi'(t)$ and included the effects of the variation of the scattering cross section with energy. The effect of these corrections depends on the energy of the analysing beam. An example of the results of Ziegler's calculation compared with the results obtained using Lindhard's multiple scattering formula to evaluate $F(t)$ (see Section 14.2) without making any of the corrections outlined above, is shown in Figure 14.8. The area under the two curves and the peak depths differ by less than 10% but there is a significant difference in the shapes of the two distributions.

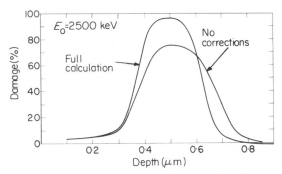

Figure 14.8. A comparison of the results of the calculation of a damage distribution from a particular backscattering spectrum, with and without the corrections suggested by Ziegler (reference 22)

14.5 SOME PRACTICAL CONSIDERATIONS

Both proton and helium ion beams have been used for analysing radiation damaged samples by the channeling effect technique. At a given energy, the helium ion beam will provide greater depth resolution. However, the dechanneling produced by a given amount of disorder may be greater for the helium beam than for the proton beam. The choice between the two kinds of ions will then be a compromise between requirements for depth resolution and for low dechanneling. A helium ion beam may be preferred for a thin, heavily damaged layer because of the need for better resolution, and the dechanneling problem would be reduced in such a layer. For a thicker layer a proton beam might provide the best results. It is usually desirable to work at as high an analysing beam energy as possible since this will also reduce the amount of dechanneling.

The possibility that the disorder distribution may be affected by the analysing beam must often be considered. Effects of the analysing beam upon the disorder distribution have been observed by several workers. Beam annealing has been found to be caused by both proton and helium ion beams in silicon samples implanted with boron ions at low temperatures.[4] Beam annealing has also been observed in silicon samples implanted with carbon or antimony at room temperature.[12,20] Bøgh and co-workers have also observed an increase in the original damage, particularly in germanium.[12] Such effects are not observed in all cases, however. For example, silicon samples implanted with boron at room temperature do not appear to be affected by the analysing beam.[4] Such samples may be less sensitive to beam annealing effects because of the large amount of annealing which takes place during the boron implantation.[23]

The sensitivity of the channeling effect technique to radiation damage may be increased by employing double alignment. The increase in sensitivity will be proportional to the decrease in the minimum yield observed in an undamaged sample and should be at least a factor of ten. If double alignment measurements are employed one must be particularly careful to check the possibility of beam annealing affecting the results, since the double alignment measurements usually result in a greater exposure of the sample to the analysing beam.

Both the random yield and the aligned yield from a damaged sample must be measured in order to determine the scattering centre density from equation (14.1). It is difficult to obtain good reproducibility in random spectra because of the difficulty in finding an orientation of the crystal which eliminates channeling effects. Ziegler and Crowder[24] have proposed an operational definition of 'random' which they find is reproducible to $\pm 1 \%$. This consists of setting θ (see Figure 7.1 for a schematic of the experimental setup) so that the beam makes an angle with the channeling axis which is at least ten times the critical angle. The crystal is then rotated about the ϕ axis while the random spectrum is recorded, so that the random is an average over all azimuthal angles.

Ziegler and Crowder give values for two corrections to be applied to a random spectrum measured in this manner. The first was evaluated for helium ions in silicon by comparing such a random spectrum with the spectrum from amorphous silicon (produced by bombardment with silicon ions). It is a function of which major axis of the crystal is aligned with the ϕ axis of the goniometer, and the energy of the helium ion beam. The correction factors they obtained were less than 3%. They may depend on such experimental factors as the accuracy with which the crystal axis is aligned with the ϕ axis of the goniometer; and the possibility of double alignment occurring at some values of ϕ, which would depend on the placement of the detector for the backscattered ions relative to the goniometer.

The magnitude of the second correction can be reduced by proper placement of the detector for the backscattered particles relative to the goniometer. It results from the fact that the spectra are recorded with a constant energy width per channel, so that the value of the random yield depends upon the value of $\Delta E/t$ as determined from equation (14.7). In changing the sample orientation to go from an aligned to a random direction, the values of θ_1 and θ_2 may be changed. This can result in a change in $\Delta E/t$ which will make a significant change in the observed random yield. Since θ_1 is usually about $0°$ a small change in its value will not be significant. The important factor is the possibility of a change in θ_2. This change can be minimized if the detector for the backscattered particles is placed in the plane of the beam and the θ

tilt axis (see Figure 7.1). On the other hand, if the detector were placed in the plane lying 90° to this plane the variation in θ_2 would be maximized and significant effects on the random yield may result.

14.6 SUMMARY

The channeling effect technique can be used to give information on both the amount of disorder and the disorder distribution in a radiation damaged sample. The principal problems in applying the technique are associated with : (i) the dechanneling produced in the disordered material, (ii) the effects of the spatial distribution of the channeled ions, and (iii) the conversion of the energy scale of the backscattered spectra to a depth scale. As a result of these factors the accuracy of the technique in determining the depth and width of the disorder distribution in ion-implanted samples is probably not better than about 10 % at the present time. In order to improve upon this accuracy it will probably be necessary to take into account the structural details of defects in calculating dechanneling and to attain a better knowledge of stopping powers, particularly as regards the appropriate value to use for channeled stopping and the effects of disorder upon channeled stopping.

GLOSSARY OF SYMBOLS

a	Thomas–Fermi screening radius
a_0	Bohr radius
d	Atomic spacing along a symmetry axis
e	Electronic charge
E	Energy of ions in analysing beam
E_1	Incident energy of analysing beam
$E_2(t)$	Energy of particles scattered from depth t in the crystal
ΔE	Energy difference between particles scattered from surface of crystal and from depth t
$F(t)$	Fraction of aligned beam scattered beyond ψ_c
k^2	Kinematic scattering factor
M_1, M_2	Masses of projectile and target atoms
N	Atomic density of damaged crystal
$N'(t)$	Scattering centre density at depth t
$N(x, y, t)$	Distribution of scattering centres across a channel at depth t
$P(x, y, t)$	Distribution of channeled ions across a channel at depth t
$S(E_1), S(E_2)$	Stopping powers of analysing beam at energies E_1 and E_2
$y'(t)$	Scattering yield from depth t in damaged crystal
$y_n(t)$	Random scattering yield from depth t
Z_1, Z_2	Atomic numbers of projectile and target atoms
ε	Lindhard reduced energy

θ	Tilt angle about vertical axis of goniometer
θ_1	Angle between normal to crystal and beam direction
θ_2	Angle between normal to crystal and beam scattered into detector
ϕ	Rotation angle about horizontal axis of goniometer
ρ	Distance from an atomic row
$\chi'(t)$	Random fraction of analysing beam at depth t
$\chi_1(t)$	Normalized aligned scattering yield from depth t in an undamaged crystal
ψ_{crit}	Critical angle for channeling

References

1. Bøgh, E., *Can. J. Phys.*, **46**, 653 (1967).
2. Feldman, L. C. and Rogers, J. W., *J. appl. Phys.*, **41**, 3776 (1970).
3. Eisen, F. H., Welch, B., Westmoreland, J. E. and Mayer, J. W., *Atomic Collision Phenomena in Solids*, Ed. D. W. Palmer, M. W. Thompson and P. D. Townsend, Amsterdam, North-Holland, p. 111 (1970).
4. Westmoreland, J. E., Mayer, J. W., Eisen, F. H. and Welch, B., *Rad. Effects*, **6**, 161 (1970).
5. Hart, R. R., *Rad. Effects*, **6**, 51 (1970).
6. Lindhard, J., *Dan. Vid. Selsk. Matt. Fys. Medd.*, **34**, No. 14 (1965).
7. Keil, E., Zeitler, E. and Zinn, W., *Z. für Naturforsch.*, **15a**, 1031 (1960).
8. Meyer, L., *phys. stat. sol. (b)*, **44**, 253 (1971).
9. Rimini, E., Lugujjo, E. and Mayer, J. W., *Phys. Rev. B*, **6**, 718 (1972).
10. Eisen, F. H., unpublished work.
11. Harris, J. S., unpublished work.
12. Bøgh, E., Høgild, P. and Stensgaard, I., *Rad. Effects*, **7**, 115 (1971).
13. Eisen, F. H. and Welch, B., *Rad. Effects*, **7**, 143 (1971).
14. Eisen, F. H. and Welch, B., *Proc. European Conf. on Ion Implantation*, Stevenage, England, Peter Peregrinus, p. 227 (1970).
15. Watkins, G. D., *Radiation Effects in Semiconductors*, Ed., F. L. Vook, New York, Plenum Press, p. 67 (1968).
16. Van Vliet, D., *Rad. Effects*, **10**, 137 (1971).
17. Sigmund, P., *Appl. Phys. Lett.*, **14**, 114 (1969).
18. Marsden, D. A., Bellavance, G. R., Davies, J. A. and Sigmund, P., *phys. stat. sol.*, **35**, 269 (1969).
19. Eriksson, L., Davies, J. A., Johansson, N. G. E. and Mayer, J. W., *J. appl. Phys.*, **40**, 842 (1969).
20. Hirvonen, J. K., Brown, W. L. and Glotin, P. M., *Ion Implantation in Semiconductors*, Ed. I. Ruge and J. Graul, Berlin, Springer-Verlag, p. 8 (1971).
21. Eisen, F. H., Clark, G. J., Bøttiger, J. and Poate, J. M., *Rad. Effects*, **13**, 93 (1972).
22. Ziegler, J. F., *J. appl. Phys.*, **43**, 2973 (1972).
23. Westmoreland, J. E., Mayer, J. W., Eisen, F. H. and Welch, B., *Appl. Phys. Lett.*, **15**, 308 (1969).
24. Ziegler, J. F. and Crowder, B. L., *Appl. Phys. Lett.*, **20**, 178 (1972).

CHAPTER XV
Application to Surface Studies

E. Bøgh

15.1 INTRODUCTION

One of the most fruitful methods for studying channeling of, for example, energetic protons or He ions in crystals is to measure the directional dependence of the probability of hitting the crystal atoms with the well collimated ion beam as proposed by Lindhard.[1] In many cases, this is most conveniently done by measuring the wide-angle elastic-scattering yield, using the experimental technique introduced by Bøgh and Uggerhøj.[2] Already the first investigations, where this technique was used, indicated that channeling could be applied to studies of several properties of crystalline solids, one of them being the study of crystal surfaces.

It is well known that the yield of processes such as nuclear reactions and characteristic X-ray excitation by, for example, protons, also exhibits a similar orientation dependence and may, in principle, be used in the same way as elastic scattering. For a general application the method based on elastic scattering has, however, proved to have several advantages over methods based on other processes. The following discussion therefore refers to the application of wide-angle elastic scattering, but adaption to an application of other processes should be self-explanatory. Furthermore, only measurements performed with light, energetic ions (for example protons or helium ions) are discussed; thus the wide-angle elastic scattering is described by the familiar Rutherford-scattering cross section. The exclusion of heavier ions and lower energies does not reflect a deprecation of the use of such ions; the reason is simply that the author's own experience mainly falls within measurements with light ions.

It should be mentioned that during the last 25 years Rutherford-scattering yield measurements have been used routinely for elemental analysis of the surface region of solids. Prominent examples are: determination of the composition of targets used in nuclear reaction studies, and measurements of shallow diffusion profiles in material science. In fact, the fundamental formulae upon which the earlier application of Rutherford scattering and the more recent application of the combination of Rutherford scattering and channeling are based, were developed as an appendix to a study of nuclear

435

reactions by Brown *et al.*[3] An excellent analysis of 'prechanneling' application of Rutherford scattering for surface analysis has been given by Rubin.[4]

It is easy to visualize several fields which are normally considered surface physics, where channeling may be used as a tool, and several examples have been reported in the literature.[5-8] Owing to a limited depth resolution, the results of these investigations have all been averages over approximately 300 Å. In the following, the main emphasis will be placed upon an analysis of those aspects of the combination of channeling and Rutherford scattering which are of importance when information about the first few atomic layers of a crystal is to be obtained, i.e. in applications where the technique may be complementary to LEED and Auger spectroscopy.

15.2 FOUNDATION OF THE METHOD

In the standard channeling experiment with Rutherford scattering, the crystal is bombarded with a well collimated ion beam; the Rutherford scattering yield is measured by detecting the ions scattering through a certain angle. By performing an energy analysis of the scattered ions, the yield can be measured as a function of the depth in the crystal. The directional dependence of the yield (i.e. its dependence on the direction in the crystal along which the ion beam is entering the crystal) is measured, and it is observed that whenever the beam is parallel to a major crystal axis, then the yield is only approximately 1 % of the yield observed when the beam is not parallel to any crystal axis or plane. In the following, the former yield will be referred to as the *channeled yield* and the latter as the *normal yield*. A brief interpretation is that channeling causes small regions around the atomic rows along which the ion beam is injected to be inaccessible to a large fraction (up to 99 %) of the ions; this fraction constitutes the channeled ions. The radius of these forbidden zones is of the order of the Thomas–Fermi screening distance, i.e. 0·1–0·2 Å for protons and He ions. It turns out that in the experiment where the exterior ion beam is perfectly aligned with a crystal axis, all ions which hit the surface at distances from the surface atoms larger than the radius of the forbidden zones (and only those ions) become channeled, and they stay channeled over a considerable distance inside the crystal. Bearing in mind that Rutherford scattering is an event which requires an ion to come as close as approximately 10^{-4} Å to the centre of an atom, it is immediately seen why the channeled yield is much lower than the normal yield: in short, the channeled ions cannot 'hit' the atoms as long as these are in normal lattice positions.

For some purposes, this primitive interpretation of a Rutherford-scattering channeling experiment needs some refinements. Thermal vibrations very often displace atoms a distance from their equilibrium position which is comparable to the radius of the forbidden zones; in such cases, the effect of the thermal vibrations must, of course, be considered.

In the present context it is of importance to note that, very close to the surface, the radius of the forbidden zones is smaller than deeper in the crystal. Therefore, let us consider the case of a perfect crystal being bombarded with an ion beam perfectly aligned with a crystal axis and look for the probability of hitting (i.e. the yield from) the first atom in the row, the second atom, etc. Obviously, the yield from the first atom does not depend on beam direction; all other yields will therefore be listed relative to this yield, which corresponds to the normal one mentioned above.

The trajectories of ions relatively close to a single row of atoms are illustrated in Figure 15.1. The ions are pushed away from the row by correlated

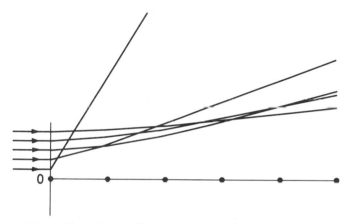

Figure 15.1. Sketch of ion trajectories close to a row of atoms

collisions with the individual atoms and, in this way, a zone where the ion flux is zero is created around the row. The flux distribution around a static row without atomic vibrations can be determined relatively easily by using power potentials[9] to describe the individual collisions. The shape of the trajectories in Figure 15.1 is that of an r^{-2} potential, but the absolute values of the deflection angles are greatly enlarged in the figure. In reality, they are typically less than 1°. In Figure 15.2 the depth dependence of the boundary of the forbidden zone is shown. The curve corresponds to a 0·5 MeV He ion beam incident along a $\langle 110 \rangle$ axis in a gold crystal. The room temperature rms vibrational amplitude perpendicular to the row is indicated in the figure.

In a real crystal, the atomic vibrations cause the forbidden-zone concept to be less meaningful close to the crystal surface because, in this region, no point is completely inaccessible to ions; even points on the row may be hit, although the probability is very small. The calculation of the ion flux distribution is rather more involved than in the static case, but can be performed numerically. The yield from an atom can then be calculated by integrating the

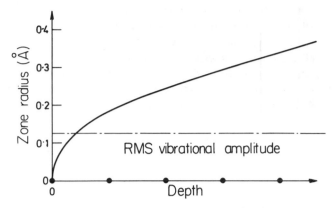

Figure 15.2. Radius of the forbidden zone for 0·5 MeV He
ions around a ⟨110⟩ axis in gold

product of the ion flux and the distribution function of the vibrating atom
over the distance from the row. In Figure 15.3 the result of a calculation for
0·5 MeV He ions in gold at room temperature is shown; an r^{-2} potential and
a gaussian-type atom-displacement distribution were applied. It is seen that
the yield decreases very rapidly with increasing crystal depth and, for most
purposes, it will suffice to take into account only the second atom.

It should be mentioned that the calculated yield from the second, the third,
etc., atom is not very accurate. This is largely due to the uncertainty of the
rms vibrational amplitude. This parameter may be determined in several
ways, for example by measurements of heat capacity, elastic constants,
velocity of sound, resistivity, Debye–Waller factor, etc., and usually the results
of these various determinations do not agree too well. The deviations often
stem from the fact that different phenomena are influenced by different parts
of the vibrational spectrum. The effect considered here primarily depends on
waves with a wavelength comparable to the nearest-neighbour distance.
Furthermore, it should be borne in mind that the vibrations of surface atoms
differ from the vibrations of the bulk atoms. In Figure 15.3 a 50 % increase of
the rms vibrational amplitude increases the yield from the second atom by a
factor of 3. The choice of an ion–atom interaction potential is, of course,
also crucial for the results. A detailed discussion of classical scattering by
screened potentials is presented in reference 9.

The above considerations refer to an isolated row of atoms. In a crystal,
atoms on a row can also be hit by ions already deflected by another row
through an angle larger than the critical channeling angle ψ_c. Close to the
surface, the most important contribution comes from the nearest-neighbour
row. Figure 15.4 illustrates the effect. Ions are deflected by row (I). Having
travelled the distance $d/\sin \psi$, where d is the space between nearest-neighbour

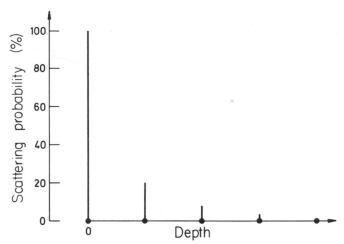

Figure 15.3. Scattering probability for 0·5 MeV He ions along a ⟨110⟩ axis in gold at room temperature. 100 % equals the normal Rutherford scattering yield

rows and ψ the total angle through which an ion is deflected by row (I), the ion approaches row (II); only if ψ is larger than ψ_c can the ion penetrate into the atoms of the row. If ψ is smaller than ψ_c, it is channeled. The yield stemming from ions already scattered by another row is most conveniently obtained by using Monte Carlo calculations as was done by, for example,

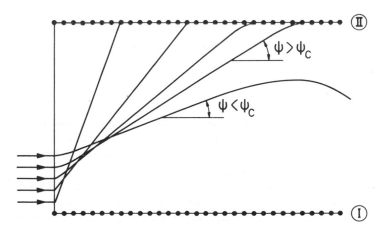

Figure 15.4. Illustration of the yield from an atomic row due to ions already deflected an angle ψ by another atomic row. ψ_c is the critical angle for channeling

Barrett.[10] Figure 15.5 shows the result of such a calculation, which also includes the effect of more distant rows for 0·5 MeV He ions in gold at room temperature. It is seen that only for atoms at depths larger than approximately d/ψ_c is there a significant probability of being hit. Actually, the first bump on the curve, representing the point where the shadow edge reaches the nearest-neighbour rows, occurs at a depth slightly larger than d/ψ_c. This is in

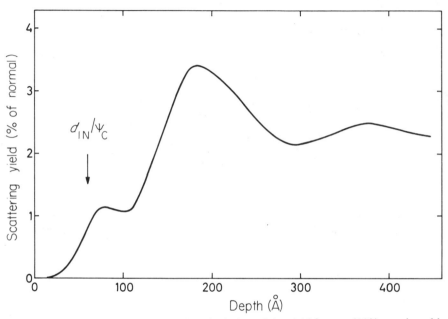

Figure 15.5. Result of a Monte Carlo calculation of the yield from a $\langle 110 \rangle$ row in gold due to ions already deflected by another $\langle 110 \rangle$ row. The primary ions are 0·5 MeV He ions. d_{1N} is the distance between nearest neighbour rows. ψ_c is the critical angle for channeling

agreement with the fact that the highest ion density in a vibrating lattice is to be found somewhat outside the critical channeling angle, cf. the shoulder in a channeling dip. The other bumps represent the points where the shadow edge reaches the second-nearest and the third-nearest neighbour rows, respectively. At larger depths, one can apply the well known channeling picture which predicts an almost constant yield.

The total yield is obtained by adding the results of Figures 15.3 and 15.5. The sum is shown in Figure 15.6, which clearly demonstrates the unique role of surface atoms in scattering experiments: *Their yield is from 10 to 100 times higher than that from other crystal atoms.*

It is from measurements of this yield that information about the crystal surface is obtained. For example, the yield from the second atom strongly

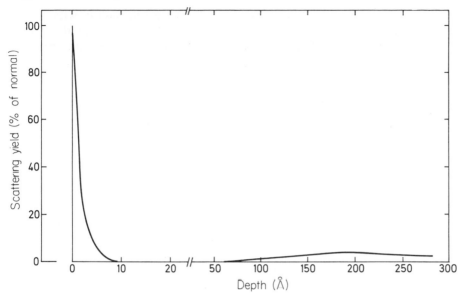

Figure 15.6. Sum of the yields from Figures 15.3 and 15.5. Note the change in the depth scale

depends on the distance to the first atom and on its vibrational amplitude; thus the spacing between the first and the second layer, and the thermal vibrations of the surface atoms may be studied. In channeling, the terms 'surface atoms' and 'first atomic layer' are often used as designations for the first atoms in the rows which are aligned with the ion beam. As illustrated in Figure 15.7, these atoms are not necessarily all situated in the uttermost

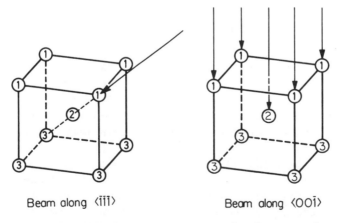

Beam along ⟨ī1̄1⟩   Beam along ⟨00ī⟩

Figure 15.7. Illustration of the concept of surface atoms in channeling

atomic plane; to measure the vibrations of the uttermost atoms on the (001) plane of tungsten, one must use the $\langle\bar{1}\bar{1}1\rangle$ axis as the channeling direction; when the $\langle00\bar{1}\rangle$ axis is used, the atoms in the second plane are also exposed to the ion beam, as illustrated in the figure.

15.3 EXPERIMENTAL EQUIPMENT

Criteria for the selection of the optimum experimental conditions may be found from an analysis of one of the major advantages of the use of the Rutherford scattering technique, namely the possibility of measuring the depth dependence of the scattering yield. This measurement can be carried out with a certain depth resolution Δt, which is indicative of the extent to which the apparatus may be considered a good tool for surface studies. That there is a possibility that the depth dependence may be measured is a consequence of the fact that the energy scale of the scattered-ion spectrum can be converted into a depth scale. The conversion formulae were worked out several years ago by Brown and co-workers[3] in a very detailed analysis of the energy relations in the scattering target; at that time, the channeling effect in single crystals was not yet known to exist. In reference 2, however, it was shown that the results of the analysis of Brown and co-workers can be applied to channeling, although some uncertainty as to the depth conversion of the channeled spectrum arises owing to an indefiniteness in the effective stopping power of the ion.

With reference to Figure 15.8, the conversion formulae are:

$$t \approx \{k^2 E_1 - E_2(t)\}\left(k^2 S_1(E) + \frac{\cos\theta_1}{\cos\theta_2}S_2(E)\right)^{-1} \tag{15.1}$$

$$\Delta t \approx \Delta E\left(k^2 S(E_1) + \frac{\cos\theta_1}{\cos\theta_2}S(E_2)\right)^{-1} \tag{15.2}$$

$$k = \frac{M_1\cos\theta}{M_1 + M_2}\left\{\left(\frac{M_1\cos\theta}{M_1 + M_2}\right)^2 + \frac{M_2 - M_1}{M_2 + M_1}\right\}^{\frac{1}{2}} \tag{15.3}$$

where M_1 is the ion mass and M_2 the mass of the scattering atom; $S_1(E)$ and $S_2(E)$ are the average stopping powers of the target material along the ion path in and out, respectively. ΔE is the sum of all factors, leading to an energy spread of ions scattered at the same depth.

Under channeling conditions, equations (15.1) and (15.2) should be applied with some caution. During the first collision with a string or a plane, for example, the energy loss of ions observed in a scattering experiment is often anomalously high[11,12] close to the surface, and equations (15.1) and (15.3) are only approximate (but for the present purpose still useful) estimates.

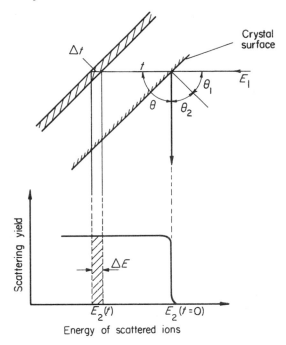

Crystal
surface

Scattering yield

ΔE

$E_2(t)$ $E_2(t=0)$

Energy of scattered ions

Figure 15.8. Principle of Rutherford
scattering experiment

As mentioned in Section 15.2, most of the information about the crystal surface is extracted from a measurement of the surface-yield peak illustrated in Figure 15.6, and it is therefore required of the apparatus that ΔE be small enough for the corresponding Δt not to be much larger than the width of this peak (i.e. a few atomic distances). Close to the surface, the instrumental energy spread is the main contribution to ΔE; at a larger depth, straggling effects are also significant. The most important instrumental contribution is usually the finite energy resolution of the ion-energy analyser which is used to measure the energy spectrum of the scattered ions.

In the following is described an apparatus used for surface investigations at the University of Aarhus. It represents an attempt to create optimum experimental conditions. In many cases, simpler equipment would suffice. Figure 15.9 is a sketch of the apparatus. The primary ion beam is usually obtained from the Aarhus 500 kV heavy-ion accelerator which is able to supply isotope-separated ion beams of most elements with energies of 30–500 keV for singly charged ions; in some cases, sufficient quantities of triply charged ions are produced in the ion source, i.e. so that the upper energy will then be 1 and 1·5 MeV, respectively.

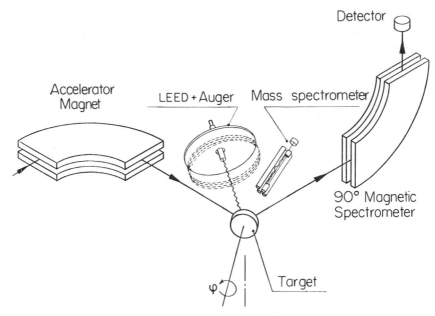

Figure 15.9. Sketch of experimental equipment

The beam collimation calculated from the beam-defining apertures is $0 \cdot 1°$. This value indicates the reproducibility after a shut-down and start-up of the accelerator. During an uninterrupted measurement, the beam divergence is considerably smaller. Angular scans through narrow channeling dips have indicated a beam divergence of less than $0 \cdot 05°$. The beam cross section is typically $1 \cdot 4 \times 2 \cdot 0 \text{ mm}^2$.

The scattering chamber is a modified Varian LEED system, which provides facilities for crystal cleaning by ion bombardment and heating, LEED investigations, Auger spectroscopy, and gas analysis (for example in connection with desorption measurements) with a Balzer quadrupole analyser. The chamber is evacuated by means of a 220 l s^{-1} ion pump, a titanium sublimation pump, and/or a liquid-helium cryopump. The chamber is connected to the accelerator by a differentially pumped chamber. During measurements, the pressure in the scattering chamber is typically 5×10^{-11} Torr. Owing to a flow of gas through the beam apertures, the effective pressure at the crystal surface is, however, somewhat higher. Desorption measurements have indicated that this effect is equivalent to an effective pressure of approximately 5×10^{-10} Torr at the crystal surface.

In most measurements, the scattering angle is 90°. The ion analyser is a 90° sector magnet with a 60 cm radius and inclined boundaries $(26 \cdot 6°)$ for two-directional focusing. The maximum-acceptance solid angle is approximately 5×10^{-4} steradians.

The ultimate energy spread of this equipment is $\Delta E/E \sim 2 \times 10^{-3}$. For 0·5 MeV He ions, this corresponds to $\Delta t \sim 20$ Å in silicon and $\Delta t \sim 10$ Å in gold.

A major disadvantage of this set-up is the small acceptance solid angle of the ion analyser and, in some cases, the integrated beam current necessary for a measurement ($\sim 250\ \mu C/cm^2$) has caused changes in the crystal surface.

The over-all performance of the apparatus may be judged from the result of a measurement of the channeled yield in gold, as shown in Figure 15.10.

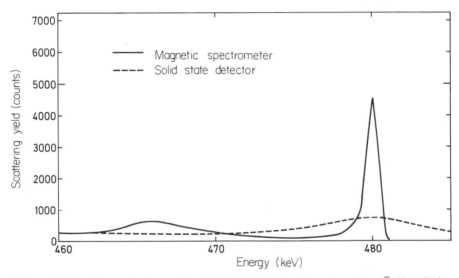

Figure 15.10. Measured channeled yield. 0·5 MeV He ions along the $\langle \overline{1}10 \rangle$ axis in gold at room temperature. Surface plane : (100). Normal yield = 8400 counts

The uncertainty of the yield curve is from 1·1 to 1·2 times the uncertainty which can be ascribed to mere counting statistics. The unique role of the surface atoms predicted from the considerations in Section 15.2 is very apparent, cf Figure 15.6. The fact that the height of the surface peak is only 55% of the random yield reflects the finite resolution of the apparatus, as does the FWHM of the peak. To illustrate the importance of a good depth resolution, the result of a measurement performed with a solid state detector, the energy resolution of which is ten times poorer than that of the magnetic spectrometer, is also shown in the figure.

According to Figure 15.6, the area of the surface peak is expected to correspond to the yield from approximately 1·3 atomic layers (cf the definition of atomic layers discussed at the end of Section 15.2), i.e. from approximately $2·2 \times 10^{15}$ atoms/cm^2. The area of the measured peak turns out to correspond to the yield from $4·2 \times 10^{15}$ atoms/cm^2. There are several causes for this discrepancy. Firstly, the gold surface is a (100) plane which has the anomalous (1 × 5) surface structure in disaccord with the bulk structure. The atoms in this over-layer structure do not shield the underlying atoms against the ion beam. Thus the yield from at least 1.7×10^{15} atoms is added to the surface peak. Secondly, the calculated yield of Figure 15.6 was based upon the thermal vibrational amplitude for bulk atoms. In itself, this value is rather uncertain and, furthermore, the vibration of an atom close to the surface is known to be larger than that of an atom in the bulk. A 50 % increase of the vibrational amplitude would make the calculated surface peak agree with the measured yield. Finally, the surface preparation may have introduced defects into the crystal; it should perhaps be mentioned here that the method is much more sensitive to point defects than is, for example, LEED.

15.4 EXAMPLES OF APPLICATION

A true evaluation of this method for the study of surfaces can, of course, be obtained only after it has been applied to specific surface problems. At the University of Aarhus, three problems are currently under investigation, namely (i) cleaning procedures used for the preparation of surfaces, (ii) surface vibrations and surface expansion, and (iii) reconstruction of surfaces. It is too early to draw any definite conclusions regarding these problems but, in order to illustrate how the method is used, two examples obtained in the course of the studies shall be mentioned in the following. It should be mentioned that the measurements to be presented were performed in a vacuum system with a pressure approximately two orders of magnitude higher than that of the system described above. Therefore the surfaces have undoubtedly been somewhat contaminated.

Figures 15.11 and 15.12 show channeled-yield curves measured at two different stages of the cleaning procedure for a silicon crystal. With the purpose of investigating the contamination from minute metallic impurities during etching, the crystal had been etched in a gold-chloride-containing etchant. The yield curve in Figure 15.11 is measured from the etched surface with no further cleaning; it discloses the major contaminants: gold, iron (nickel?), oxygen, and carbon. The contaminants are identified by applying equations (15.1) and (15.3). The amount of any given contaminant can, by a straightforward procedure, be determined by comparing the area of the yield peak with the yield from a known amount of the same or (since the Rutherford scattering cross section is known to be proportional to the square of the atomic number of the scattering atom) another known material; usually the normal yield from the substrate is used for this purpose.

Channeling is applied to such an analysis of surface impurities with the purpose of increasing the detection sensitivity of light atoms. According to equations (15.1) and (15.3), all impurities lighter than the atoms of the thick

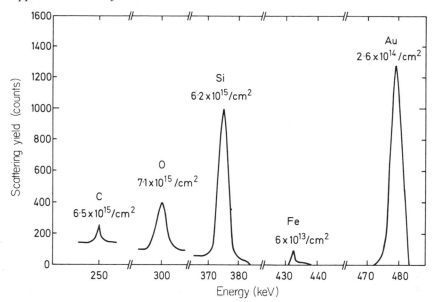

Figure 15.11. Measured channeled yield from an etched silicon surface contaminated with carbon, oxygen, iron, and gold. 0·5 MeV He ions along the ⟨110⟩ axis. Surface plane: (111). Normal yield = 1650 counts

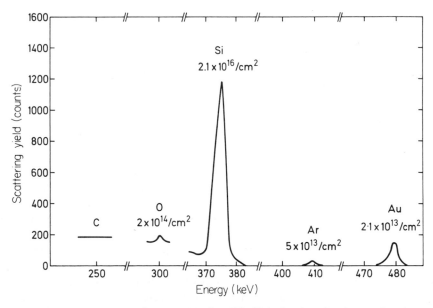

Figure 15.12. Same as Figure 15.11 after 500 eV Ar ion bombardment cleaning

substrate (silicon in Figure 15.11) appear as peaks superimposed upon a background of ions scattered in the substrate. Channeling suppresses this background—in the case of silicon by a factor of approximately 35—and thus increases the sensitivity. A further reduction of the background from the substrate is obtained with the so-called double-alignment technique[13] in which both the incident and the outgoing ion trajectories are aligned with major crystal axes.

The widths of the silicon and oxygen peaks in Figure 15.11 are larger than the instrumental resolution ΔE; this is because the silicon atoms and the oxygen atoms (at least part of them) are distributed in an approximately 25 Å-thick oxide film. It is tempting to extract the stoichiometry of the oxide from the curve, but a reliable result requires a more detailed knowledge of the conditions at the oxide/silicon interface than is available today; any disorder in the first atomic layers of the silicon immediately below the oxide will increase the silicon peak on the yield curve.

The width of the gold peak is also larger than ΔE. A possible explanation of this observation could be that the gold is distributed in the oxide. A more likely explanation, however, is that the gold is deposited as approximately 30 Å-thick islands. The latter explanation is supported by a slight difference between the width of the silicon peak and that of the gold peak; furthermore, at larger coverages, the silicon peak is observed to be slightly displaced towards lower energies, as would be expected if the gold is located on top of the oxide.

In Figure 15.12 the channeled-yield curve after the crystal had been cleaned by means of a 500 eV argon bombardment ($2 \mu A/cm^2$ for 30 min) is shown. Traces of gold and oxygen are still to be found on the surface, and a small fraction of the argon ions is incorporated in the crystal surface. The silicon peak is larger than in Figure 15.11, i.e. the ion bombardment has damaged a layer which is thicker than the original oxide layer. From the area and width of the peak it can be deduced that the damaged layer is approximately 40 Å thick and is damaged to such a degree that the original crystal structure is completely lost.

An example of the study of surface reconstruction is shown in Figure 15.13. The curve in (a) is the channeled-yield curve for gold also shown in Figure 15.10 and discussed in Section 15.3. The curve in (b) is the channeled yield measured after the gold-crystal surface had been covered with a few monolayers of epitaxially grown silver. The gold-surface peak from the silver-covered crystal is significantly smaller than that from the uncovered crystal, indicating that the original anomalous gold-surface structure has, at least partly, recovered the normal gold structure by the silver deposition.

If the gold surface had completely recovered the normal bulk structure and the silver had a perfect epitaxial growth in a homogeneous layer, then all gold atoms would be shadowed by the silver atoms, i.e. only the silver surface would appear in the channeled yield. From Figure 15.13(b) it is apparent that this is not so; both the silver peak and the gold peak appear. How far this is due to some island formation of the silver—which is indicated by the width of the silver peak—or to disorder in the silver/gold interface is at present an open question.

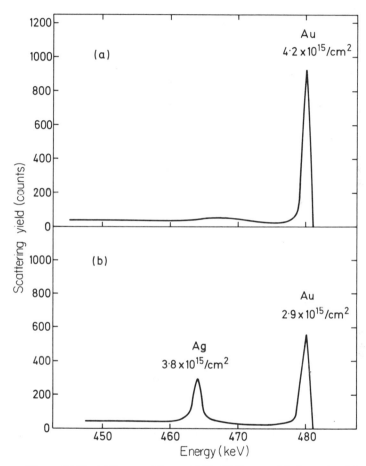

Figure 15.13. (a) Same as Figure 15.10; (b) Same as (a) after epitaxial deposition of silver on the surface

It should perhaps be mentioned that in further measurements the scattering yield from the silver showed a directional dependence almost as pronounced as that of the surface of a bulk silver crystal. This example demonstrates an important quality of the method: it is possible simultaneously to study the surface (the silver) and the deeper layers (the gold).

15.5 SUMMARY

In Section 15.2 a simple analysis of the motion of fast protons or He ions in the surface region of a single crystal indicated that the application of a combination of Rutherford scattering and channeling offers an interesting experimental tool for surface investigations. In Section 15.3 it was demonstrated that it is possible to construct experimental equipment utilizing this method, and the main expectations of Section 15.2 were verified by means of this apparatus. Finally, a few examples were described in Section 15.4.

Although the method is still under development, some of its advantages have already been established:

(i) The process upon which the method is based, namely Rutherford scattering of fast protons and He ions, is extremely well understood, in contrast, for example, to the scattering of slow electrons, which is the basic process in LEED and Auger spectroscopy.

(ii) It is possible to perform a quantitative elemental analysis of both the surface material and deeper material. For light contaminants, the sensitivity is approximately one tenth of a monolayer, i.e. somewhat poorer than the sensitivity of, for example, Auger spectroscopy. For heavy contaminants, the sensitivity is approximately one thousandth of a monolayer, i.e. comparable to Auger spectroscopy. The main advantage of the scattering method is that it gives a quantitative analysis. In most cases, quantivative measurements are very uncertain or impossible to perform by Auger spectroscopy.

(iii) It is possible to locate surface atoms relative to the substrate structure, i.e. surface disorder, surface expansion, and surface vibrations can be studied. In this application, the main advantage of the scattering method over, for example, LEED is its ability to distinguish between substrate atoms and adsorbed atoms. A further advantage is the information about the structure of the substrate below the surface, which is obtained simultaneously with the information about the surface.

It is the author's belief that these advantages are sufficient to make the scattering method a valuable tool for surface measurements and justify further efforts to develop the method and to establish its full potential—and limitations.

References

1. Lindhard, J., *Dan. Vid. Selsk., Mat. Fys. Medd.*, **34**, No. 14 (1965).
2. Bøgh, E. and Uggerhøj, E., *Nucl. Instrum. and Meth.*, **38**, 216 (1965). Also reference 13.
3. Brown, A. B., Snyder, C. W., Fowler, W. A. and Lauritsen, C. C., *Phys. Rev.*, **82**, 177 (1951).
4. Rubin, S., 'Ion Scattering Methods' Part I, Section D-1, Chapter 41 in *Treatise on Analytical Chemistry*, Ed. Kolthoff and Elving, New York, Interscience (1959).

5. Gyulai, J. and Mayer, J. W., *J. appl. Phys.*, **42**, 3578 (1971).
6. Mitchell, I. V., Kamoshida, M. and Mayer, J. W., *J. appl. Phys.*, **42**, 4378 (1971).
7. Hvalgaard, J. O., Andersen, S. L. and Olsen, T., *phys. stat. sol.* (a), **5**, K83 (1971).
8. Morgan, D. V. and Bøgh, E., *Surf. Sci.*, **32**, 278 (1972).
9. Lindhard, J., *Dan. Vid. Selsk.*, *Mat. Fys. Medd.*, **36**, No. 10 (1968).
10. Barrett, J. H., *Phys. Rev.*, B, **3**, 1527 (1971).
11. Bøgh, E., *Rad. Effect*, **7**, 115 (1971).
12. Bøgh, E., *Phys. Rev. Lett.*, **19**, 61 (1967).
13. Bøgh, E., *Can. J. Phys.*, **46**, 653 (1968).

CHAPTER XVI
Applications to Semiconductor Technology

J. W. MAYER

16.1 INTRODUCTION

The performance of semiconductor devices depends critically on the distribution of dopant species within the first few micrometres of the surface. An emitter in a bipolar transistor may contain the equivalent of a few monolayers of dopant atoms such as arsenic distributed over one or two micrometres in depth. In microelectronic circuit applications, epitaxial growth of high-quality single-crystal films is assuming a key role. In compound semiconductors, the stoichiometry of the surface layer and its dependence on preparation techniques are important factors in device technology. These are representative areas of semiconductor technology that can be investigated with channeling effect measurements utilizing energetic (100 keV to several MeV) particles such as He or H ions. It has already been demonstrated that backscattering techniques have direct application to semiconductor technology. Here, we consider how much further insight can be gained from the addition of channeling effect measurements.

Ion implantation has been demonstrated as a practical technique in the fabrication of semiconductor devices. Channeling of the implanted atoms leads to increased penetration which can be utilized to obtain special dopant distributions in a semiconductor. This combination of ion implantation and channeling has advantages in certain device applications.

The previous chapters dealt with the phenomenon of channeling and the type of information on disorder and dopant location that could be obtained. In this chapter we will consider channeling from the viewpoint of semiconductor technology; that is, application of channeling techniques to gain further understanding of processing methods used in fabricating semiconductor devices. To avoid repetition, we will draw heavily on discussions in previous chapters and will only attempt to illustrate some of the applications by specific examples. Further, we will not treat in any detail the analysis of ion implanted samples. This subject has been covered in detail in conferences, books, and in earlier chapters. However, the application of

453

channeled implantation techniques will be discussed because of potential applications in device fabrication. The subject of analysis of surface stoichiometry has been covered by Bøgh in the previous chapter.

At the outset it should be emphasized that the present techniques utilized in channeling effect analysis have limitations in their ability to detect small concentrations of either dopant atoms or lattice disorder. For example, a workable lower limit is of the order of 10^{18} to 10^{19} dopant atoms/cm^3 for high mass impurities such as arsenic or antimony in silicon. This focuses attention on high dopant concentrations typical of those found in diffused or implanted layers in semiconductors. The properties of heavily doped semiconductors are not well understood and channeling effect measurements may play a key role in the analysis.

In many aspects of problems encountered in semiconductor technology, channeling effect measurements do not provide a complete picture. It is important to compare the results of other measurements, such as electrical behaviour, with lattice location or disorder evaluation. In ion implantation studies, for example, it has long been known that it is a necessary but not sufficient condition to have dopant atoms on substitutional sites for full electrical activity. Other effects, such as radiation-damage induced effects, can lead to reductions in carrier concentrations.

16.2 LATTICE LOCATION OF DOPANT SPECIES

One of the most direct applications of channeling effect measurements to semiconductor materials is the determination of the lattice site location of dopant species. The general philosophy of this approach has been discussed in detail by Davies in Chapter 13. Backscattering measurements with MeV He ions can be used to determine the concentration and depth distribution of dopant species. These are important considerations in determining dopant solubility and diffusion parameters. However, for comparison with electrical measurements, one should determine the amount on substitutional sites. In the field of semiconductor technology, one clear-cut area that can be investigated is the heavily doped regions typified by emitters in bipolar transitors. Arsenic, because of its high solubility ($\gtrsim 1\cdot5 \times 10^{21}$ atoms/cm^3) in silicon, is an ideal candidate and is also used extensively in transistor structures.[1]

The first investigations of arsenic-diffused silicon was carried out by Chou *et al.*[2] to determine the number of electrons in relation to the number of arsenic atoms on substitutional sites. Chou[3] found that the carrier concentration near the surface was an order of magnitude lower than the substitutional arsenic concentration for samples doped to levels of $1\cdot5 \times 10^{21}$ arsenic atoms/cm^3. Analysis of the data indicated that arsenic complexes were formed and led to the decrease of the number of electrons per arsenic atom.

Recently Schwenker *et al.*[4] have measured changes in sheet resistivity of arsenic-diffused samples following a sequence of heat treatments. They found a large increase in sheet resistivity following 500–900 °C heat treatments. For example, the sheet resistance increased by nearly a factor of 3 following anneal for 120 hours at 750°C for a silicon sample doped to 1.5×10^{21} arsenic atoms/cm³. Channeling effect measurements were made by Haskell *et al.*[5] on the samples supplied by the IBM group. Figure 16.1 shows random

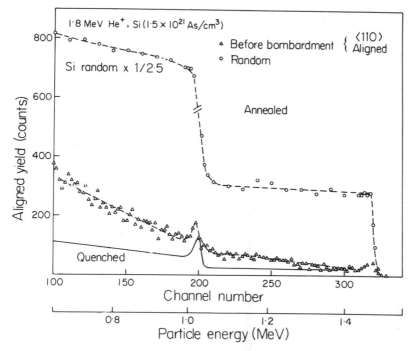

Figure 16.1. Random and $\langle 110 \rangle$ aligned backscattering energy spectra for 1.8 MeV He⁺ ions on silicon with a concentration of 1.5×10^{21} As atoms/cm³. The annealed sample (Δ) was heated at 750 °C for 120 hours. (Taken from Haskell *et al.*)[5]

and aligned spectra for quenched and annealed samples. It can be seen that the aligned yield for arsenic in the quenched sample is relatively small indicating that more than 90% of the arsenic atoms are on lattice sites. In the annealed sample the arsenic yield is comparable to that in quenched samples near the surface, but increased by about 15–20% at depths of 0.9 μm. Most of this increase in the aligned arsenic yield is due to the increase in dechanneling rate (as indicated in the increased dechanneling in the silicon signal). The increased dechanneling rate may be due to the presence of extended defects (dislocation loops, dark spots) which have been noted[4] in transmission electron microscope investigations of arsenic-doped samples processed in a similar fashion. Even excluding the effect of dechanneling (i.e. assuming the increase in the arsenic yield is caused solely by nonsubstitutional arsenic atoms), the magnitude of the increase is smaller than that predicted from sheet resistivity measurements. Thus, the changes in electrical behaviour cannot be associated entirely with nonsubstitutional arsenic atoms. Thus arsenic complexes, if formed, involve substitutional arsenic atoms. Similar effects were observed in tellurium-doped gallium arsenide in which the free carrier concentration in annealed samples was only 50% of the substitution tellurium component.[6]

Phosphorus-diffused emitters in silicon have been found to have high concentrations of dislocations and other lattice defects.[7,8] Simultaneous diffusion of arsenic and phosphorus results in a marked decrease in the number of lattice defects.[9] Fujimoto *et al.*[10] investigated lattice location and concentration distributions by use of channeling and radioactive analysis techniques. They found that a large percentage of the arsenic atoms near the surface were located off substitutional sites. It was suggested that the expansion of the silicon lattice due to nonsubstitutional arsenic atoms counteracts the contraction due to substitutional phosphorus atoms, and hence reduces the generation of dislocations.

Gold is commonly used to reduce carrier lifetime in silicon devices where fast transient response is desired. The solubility[11] of gold in lightly doped silicon is about $10^{17}/cm^3$; a doping level which is almost too low to be investigated with backscattering techniques. However, one aspect of heavily doped semiconductors is the fact that the solubility of other dopant species can be enhanced.[12] In silicon diffused with phosphorus, Chou[3] found a gold concentration substantially above the maximum equilibrium solubility in intrinsic silicon. Channeling effect measurements indicated that the gold was predominantly on substitutional sites and strongly supports the concept of gold–phosphorus ion pairing in which both the phosphorus and gold are on substitutional sites.

Channeling effect measurements played a large role in understanding implantation processes.[13] One example is the analysis of boron implanted silicon. Hall effect measurements[14] indicated reverse annealing effects (a decrease in the number of holes per cm^2 with an increase in anneal temperature) and a requirement for high anneal temperatures (800 to 900 °C) to achieve full electrical activity where the number of acceptors was comparable to the number of implanted boron atoms. This behaviour seemed strikingly different from the anneal characteristics of other dopant species implanted in silicon. North and Gibson[15] determined the substitutional fraction of boron as a function of anneal temperature and compared this behaviour with the number of holes as found by Hall effect measurements. They found a strong correlation between the number of holes per cm^2 and the number of boron atoms on substitutional sites. In the channeling effect measurements, (p, α) reactions were used to evaluate the lattice location of boron as backscattering techniques require that the mass of the dopant species be higher than that of the semiconductor host atoms. The cross section for this reaction is sufficiently large for boron to be detected easily at levels as low as $10^{14}/cm^2$ (less than $\frac{1}{10}$ monolayer).

Channeling effect techniques have become more demanding than the simple aligned and random spectra used in early analyses of implantation phenomena.[13] In the analysis of lattice location it is necessary to measure the orientation dependence (angular scan) of the yield from the impurity atoms as well as the yields along different axial directions. Obtaining angular scans adds significantly to the time required in analysis, but it is required because of flux peaking effects and possible atom locations that are displaced slightly from regular lattice site positions. Recent experiments indicate that double alignment may also be required in these cases. Double alignment, although time consuming, is a powerful technique. In an example[16] cited by Davies, the aligned yield for bismuth in silicon was 15 % of the random yield, yet a detailed analysis of the shape of the angular distribution indicated that about 50 % of the bismuth atoms were displaced slightly off lattice sites. Double alignment techniques were used to confirm this interpretation.

As the sophistication of the channeling technique increases, the integrated dose of He ions required for analysis increases leading to increased amounts of lattice disorder. There are also changes in the lattice positions of the dopant atoms during analysis owing to the bombardment with MeV particles. These additional complications are discussed in Section 16.6.

16.3 LATTICE DISORDER

Evaluation of lattice disorder in semiconductor crystals is another immediate application of channeling effect techniques to semiconductor technology. One straightforward example is the analysis of the effect of various surface preparations on the amount of lattice disorder. Figure 16.2 shows random and aligned backscattering yield for a gallium-arsenide crystal subjected to mechanical polishing and various etching treatments.[17] The yield (curve b) for the surface following a mechanical polish is markedly higher than that obtained after chemical etching (curve c). A one minute etch reduces the height of the surface peak by about a factor of eight, indicating a comparable decrease in surface disorder. In these data, curve b for example, one notes the presence of both a surface peak and a level at lower energies caused by dechanneling. For these cases where a highly disordered layer is formed at the surface, the dechanneling calculations discussed by Eisen in Chapter 14 are applicable. In fact, the data of Rimini *et al.*[18] indicate that the amount of

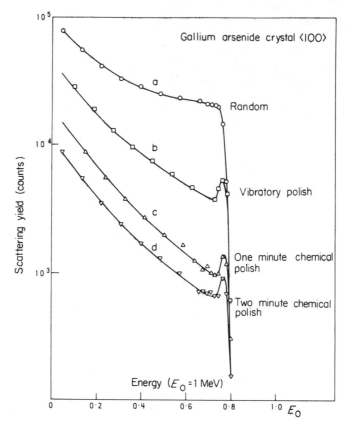

Figure 16.2. Random and $\langle 110 \rangle$ aligned backscattering energy spectra for 1.0 He$^+$ ions incident on GaAs subjected to various surface preparations. (J. L. Whitton, unpublished)

dechanneling caused by an amorphous layer can be calculated with a fair degree of confidence.

The analysis procedure is much less certain in the general case where there is a relatively large concentration of defects distributed in depth. All the effects presented by Eisen should be considered, such as flux peaking, determination of the depth scale for varied amounts of dechanneling, and the contribution from atoms slightly displaced off lattice sites. In fact, it was shown in Chapter 14 that slightly displaced lattice atoms have a relatively smaller contribution to backscattering than to dechanneling. Under these conditions it is difficult to make a simple calculation of the amount of dechanneling from the backscattering yield.

In spite of these difficulties, channeling effect measurements have been used in the determination of lattice disorder in semiconductor samples.

One of the most exciting applications that is receiving increased attention is the evaluation of epitaxial layers.[19,20] These layers as used in semiconductor device technology are often in the thickness range from 1 to 10 micrometres— thicknesses that are ideally suited to backscattering measurements. Figure 16.3 (upper drawing) shows random and aligned backscattering spectra for

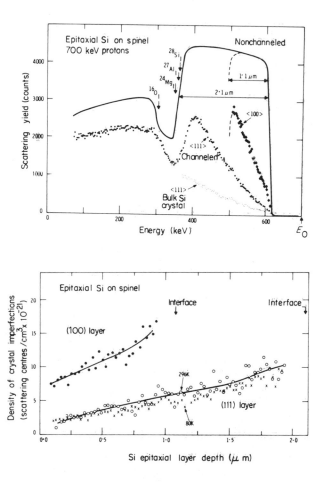

Figure 16.3. Upper drawing; random and aligned back-scattering energy spectra for 700 keV protons incident on two thicknesses of silicon epitaxially grown on spinel. The points are $\langle 111 \rangle$ aligned reference spectrum from a bulk silicon crystal. Lower drawing: density of crystal imperfections in the 1·1 μm thick layer (●) and for the 2·1 μm thick layer (○) calculated from spectra in the upper figure. (From Picraux[19])

700 keV protons incident on two thicknesses of epitaxial silicon grown on spinel ($Al_2O_3 \cdot MgO$) substrates.[19] It is clear from the figure that the aligned spectrum for both films is considerably higher than that from bulk silicon crystals. The aligned yields suggest that the amount of disorder in the thinner sample is considerably higher than that for the 2 μm sample.

To obtain quantitative information on the depth distribution of the crystal imperfections, Picraux used dechanneling calculations essentially similar to that outlined by Eisen. In this case the scattering distribution given by Keil *et al.*[21] rather than that of Meyer[22] was used. However, for this amount of disorder the two treatments give comparable results.[23] In the analysis it was assumed that the defects gave rise to both backscattering signals and dechanneling; that is, there were no slightly displaced atoms that would contribute primarily to dechanneling. The results of this type of analysis are shown in Figure 16.3 (lower drawing). The increased disorder in the thinner layer is apparent. For both layers, crystal quality improves significantly with increasing distance from the interface.

There are several inherent difficulties in this type of analysis of epitaxial layers. The first is that there is no convenient (disorder-free) reference point in the epitaxial region to check the calculated amount of dechanneling. In a sample such as shown in Figure 16.2 where the disorder is confined to regions near the surface, a self-consistent check of the dechanneling calculations can be made to ensure that the dechanneled (random) component of the beam rises to the aligned yield below the disordered layer. This procedure has been discussed by Eisen in Chapter 14. In an epitaxial layer, the disorder extends throughout the sample and an estimate of the amount of dechanneling can be obtained from the aligned spectrum of the underlying crystal substrate. In certain cases, such as homoepitaxy, the underlying crystal substrate has a high degree of order with crystal axes aligned with those in the epitaxial layer and hence provides a convenient reference point. However in other cases, such as illustrated in Figure 16.3 (upper drawing), the ratio of aligned to random yields in the substrate (spinel) is higher than that in the epitaxial layer near the interface. Even here, however, the situation is such that a self-consistent check of the analysis procedure can be made. It is possible to obtain an evaluation of the dechanneling calculation by removing a known thickness of the epitaxial layer. Re-measurement of the aligned yield and recalculation of the disorder distribution should provide a match between the amounts of disorder at equivalent depths. Picraux and Wagner[24] have applied this etching technique to the analysis of epitaxial layers. They find that close agreement can be obtained between the original analysis and the striped layers. In addition, estimates of the aligned and random stopping powers can be obtained from analysis of the energy positions of interface edges.[23,24]

Figure 16.4 shows a transmission electron micrograph of a silicon epitaxial layer similar to the one analysed in Figure 16.3. This illustrates a second

Figure 16.4. Transmission electron micrograph of a $\langle 100 \rangle$ oriented silicon epitaxial layer similar to those analyzed in Figure 16.3. Stacking faults appear to be the major defect. (From Picraux and Thomas[24])

problem that can arise when the epitaxial layers do not contain a uniform distribution of defects across the lateral dimensions of the analysis beam spot. In general, however, the defects are distributed uniformly and one has a good average of the defect concentration. This information is often of great utility in assessing the general quality of an epitaxial layer. Channeling effect measurements do not give information on the nature of the defects but only on an average concentration of imperfections. It is possible that the amount of dechanneling and the effective scattering yield will differ for various defect species. Other techniques such as transmission electron microscopy are required to determine the detailed nature of the defects.

The channeling technique does have a major application in providing guidance to the crystal grower who is investigating new epitaxial growth techniques.[25] A third problem encountered in analysis of epitaxial layers relates more to semiconductor device technology. Channeling measurements are not sensitive enough to determine the small amounts of disorder of interest in device development. For example, the amount of disorder indicated in Figure 16.3 is too high to enable these films to be useful in device structures.

After one achieves a sufficiently ordered epitaxial layer so that the defect concentration is below the sensitivity limit of channeling measurements, then one can consider use of the layers in device applications.

In spite of these difficulties in analysis, there appear to be a large number of unexplored areas involving layered structures on semiconductors where channeling can be applied. They range from the general topic of hetero-junction formation to specialized topics such as solid-phase epitaxial regrowth of germanium.[26,27]

16.4 METAL LAYERS ON SEMICONDUCTORS

In the previous sections and chapters most of the emphasis on the use of channeling techniques was directed toward impurities and lattice disorder. It is of course true that in the fabrication of a semiconductor device the starting place is the depth distribution and lattice location of dopant species. However, there is a critical problem associated with the connection of the device to the outside world, i.e. the formation of contacts. In integrated circuit technology, contacts to the semiconductor are generally achieved by metal evaporation followed by thermal processing. For the case of aluminium evaporated layers on silicon, commonly used for ohmic contacts, thermal processing causes a small percentage of silicon to dissolve in the aluminium.[28] For other metals, such as palladium, platinum and titanium, thermal process-ing creates silicide layers which act as either ohmic contacts or Schottky barriers.[29–31] Recently, backscattering techniques (without the use of channeling effects) have been used to investigate the low temperature migration of silicon into metals and formation of silicon–metal com-pounds.[32–34] In this case, backscattering of MeV He ions provides a direct measurement of the composition of the silicide and also a measure of the growth rate.

Channeling effect measurements can also be used in the evaluation of thermal processing effects on metal layers deposited on semiconductors. It is of interest to determine lattice disorder in the underlying crystal and the existence of preferred orientation effects in the silicide layers or in the alloyed region. For example, the alloying behaviour of gold and gold–germanium on gallium arsenide was investigated by backscattering and channeling effect measurements to determine the amount of lattice disorder and distribution of gold in gallium arsenide.[35] In another case, the origin of n-type behaviour in indium alloyed silicon was found to be associated with the presence of phosphorus introduced during processing rather than that of interstitial indium.[36]

The group at the Hughes Research Laboratories have used channeling and backscattering to investigate the interaction between metal films and disordered layers of silicon produced by ion implantation.[37,38] They found that amorphous regions of silicon covered by aluminium films reordered at temperatures significantly below those required in the absence of aluminium.[37] The enhanced low temperature reordering of the damaged layer is simultaneously accompanied by a migration of silicon through the aluminium. In that work, channeling methods provided a direct method to investigate annealing characteristics of a disordered semiconductor region underneath a metal film.

Another application of channeling to contact formation is the investigation of preferred orientation in the reacted metal–semiconductor layers. Figure 16.5 shows the results of analysis[38] of a 350 Å palladium layer on ⟨111⟩ silicon after anneal at 200 °C. The ratio of the yields of the palladium to silicon in the silicide layer indicate that Pd_2Si is formed. In the aligned spectra the palladium yield has decreased, indicating that there are channeling effects present, i.e. that there is some preferred orientation of the Pd_2Si film. This orientation effect is in agreement with X-ray diffraction results.[30,31] It should be noted that, for analysis of orientation effects in the silicide layers, a simpler technique would be to use the film exposure technique discussed by Barrett in Chapter 11.

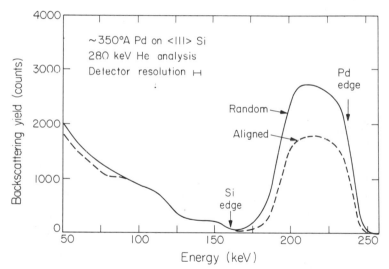

Figure 16.5. Random and ⟨111⟩ aligned backscattering spectra for 280 keV He^+ incident on a 350 Å palladium film on silicon. The silicon was implanted with 2×10^{15} phosphorus ions/cm² at 800 keV and the structure annealed at 200 °C. (From Lee *et al.*[38])

16.5 SURFACE ANALYSIS

Backscattering techniques, without recourse to channeling, have important capabilities in trace analysis of surface impurities. The first application[40] to semiconductor technology was the identification of silicon surface contamination from hydrofluoric acid solutions containing gold and copper in concentrations ranging from 0·1 to 100 parts per million. The sensitivity to high mass impurities is quite good ranging from 0·01 to 0·1 monolayers.[34] For example, Buck and Wheatley[41] have found the sensitivity to contaminants on silicon surfaces using a 100 keV He$^+$ ion beam to be 10^{12} Au atoms/cm^2 and 10^{13} Fe atoms/cm^2.

Channeling effects provide a pronounced improvement in the ability to analyse surface layers in that the contribution from the underlying crystal is reduced.[42] This permits more accurate analysis of the contribution from low mass constituents in the surface layers. For example, Morgan and Bøgh[43] have applied backscattering and channeling effect measurements to determine the stoichiometry of the surface region of GaAs and its dependence on preparation technique. Also, Poate *et al.*[44] have studied the chemical composition of two types of protective oxide coatings on single crystal GaP. In both these cases, channeling techniques were required in determination of the surface composition. This subject is treated in detail by Bøgh in the previous chapter.

16.6 BEAM EFFECTS

Davies has pointed out in Chapter 13 that bombardment of crystalline materials with MeV helium ions is not a gentle process. There are changes in the composition of surface layers, introduction of lattice disorder, and movement of dopant atoms off lattice sites. These changes in the properties of the sample must be considered in the analysis procedure.

Beam effects have been noted in channeling measurements with 1·8 MeV helium ions in arsenic-doped silicon samples.[5,39] After bombardment in a random direction, a significant fraction (up to 40%) of the arsenic atoms are found off substitutional sites. The increase in the ⟨110⟩ aligned yield (the ratio between aligned and random arsenic yields) as a function of random bombardment dose is shown in Figure 16.6 for several arsenic concentrations. The initial increase in the yield is proportional to the random dose and then reaches a saturation value in all the samples investigated. The data in Figure 16.6 give yield ratios against dose; in terms of absolute numbers the initial increase in yield is independent of arsenic concentration with about 10^{19} arsenic atoms/cm^3 moved off lattice sites per 100 μC/cm^2 of 1·8 MeV helium ions. At lower beam energies, the effect is even more pronounced.[45]

For channeling measurements, the samples receive a pre-analysis dose when they are aligned so that a major crystallographic axis is parallel to the incident beam. A typical alignment dose is about 10^3 μC/cm^2 for a beam spot of about 2 mm^2. For arsenic doped samples, this dose is sufficient to cause the aligned arsenic yield to reach its saturation value. To minimize

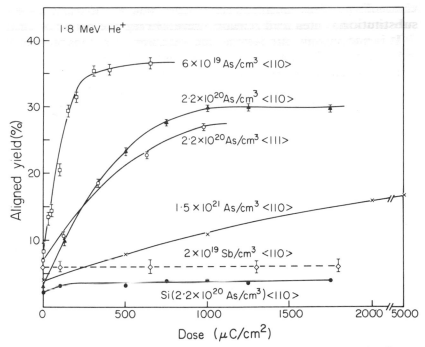

Figure 16.6. Ratio of ⟨110⟩ aligned to random yield for arsenic atoms in silicon vs. bombardment dose of 1·8 MeV He ions. Initial increase in yield indicates that approximately 10^{19} As atoms/cm³ are moved off substitutional sites per 100 μC/cm². (From Haskell *et al.*[5])

effects caused by the analysis beam, it is necessary to move the beam spot to an unbombarded region before taking the aligned spectrum. This procedure has been used for example, by Picraux *et al.*,[16] in analysis of bismuth implanted silicon.

Beam induced movement of impurity atoms off lattice sites has been observed in boron-diffused[46] and thallium-implanted silicon[47] as well as arsenic-doped silicon. The specific details of the behaviour seem somewhat different in each case. However, in all cases the motion of the dopant atoms off lattice sites is attributed to beam-created defects, such as vacancies or interstitials, which migrate to the site of the dopant atoms. The movement of the dopant atoms off lattice sites is not found for all dopant species. For example, in antimony-implanted or diffused silicon no change in the antimony aligned yield has been found with bombardment.[5]

These results indicate that an observation of an apparent nonsubstitutional lattice location of the dopant atom requires caution in interpretation. This off-lattice site location may be due to beam effects (or to flux peaking). One can recast slightly the statement made by Davies in Chapter 13;

channeling effect measurements are best for dopant atoms that are on substitutional sites and remain there during the period of analysis.

It is not merely the lattice site location of a dopant species that may be affected by analysis beam, the composition of the surface layer may also be changed. This has been investigated by Buck and Wheatley[41] in their studies of solid surfaces using 100 keV helium and hydrogen ion beams. Figures 16.7 illustrates the growth of the silicon surface peak as a function of ion dose. (The accumulated ion doses are indicated beside the curves.) Initially the surface peak represented about 5×10^{15} silicon atoms/cm^2, most of them

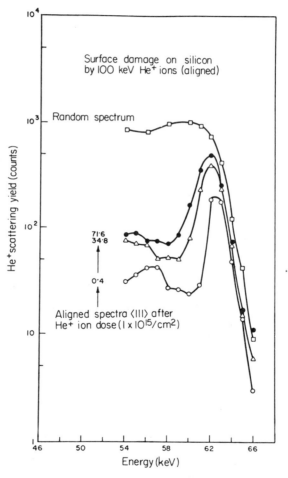

Figure 16.7. Random and $\langle 111 \rangle$ aligned backscattering spectra for 100 keV He$^+$ ions incident on silicon. The accumulated ion doses in units of 1×10^{15}/cm^2 are indicated beside the curves (Buck and Wheatley[40])

probably in an oxide film on the surface. The area under the peak grew to $1 \cdot 6 \times 10^{16}/\text{cm}^2$ for an accumulated dose of 7×10^{16} helium ions/cm^2. In attempting analysis of surface composition, then, it is generally necessary to make a series of analyses such as shown in Figure 16.7 to ensure that there are not major changes in the surface composition during analysis.

The analysis beam can cause an increase in lattice disorder during analysis. However, the opposite effect can also be noted when dealing with samples containing lattice disorder such as ion implanted semiconductors. It was found by Westmoreland et al.[48] that the analysis beam caused a decrease in the disorder peak in boron implanted samples. This effect was sufficiently severe that the beam spot was moved periodically during the analysis runs. In this case the beam apparently caused an enhanced annealing effect.

16.7 ION IMPLANTATION

There has been extensive development of ion implantation as a technique for processing semiconductor devices. Some of the advantages of the technique lie in accurate and reproducible introduction of a given number of dopant atoms per cm^2 and effective control of the depth distribution (profile) of the implanted dopants. Channeling can provide a means of altering the profile to give a deeper penetration of the implanted species and a change in the shape of the distribution. The advantages in the use of channeling to produce voltage-variable capacitors have been presented by Moline and Foxhall.[49]

The general features of channeled range profiles are discussed by Whitton in Chapter 8. Crystal orientation, target and incident ion type affect the maximum range R_{max} to which incoming particles can penetrate. In many cases the value R_{max} is well defined for fixed conditions and good agreement has been achieved in experiments at different laboratories. However, the shape of the profile is influenced by target temperature, crystal disorder (including surface oxide layers) and bombardment dose; all of these affect the dechanneling rate and hence the number of implanted dopant atoms that reach a given penetration depth. General trends can be predicted but specific profile shapes must be determined experimentally.

Phosphorus implantations in silicon provide an interesting example. In this chapter, we present only carrier concentration profiles because of the correlation with device performance. Range distributions of channeled phosphorus ions in silicon have been given by Dearnaley et al.[50] Figure 16.8 shows carrier concentration profiles resulting from 600 keV ^{31}P channeling in $\langle 100 \rangle$, $\langle 110 \rangle$, and $\langle 111 \rangle$ Si.[51] Distinct peaks are noted for the $\langle 110 \rangle$ and $\langle 111 \rangle$ profiles. The maximum penetration for $\langle 110 \rangle$ orientation, the most open channeling direction, is greater than that for the $\langle 111 \rangle$ or $\langle 100 \rangle$ orientation. The maximum range for channeled phosphorus implants is significantly greater than the random range.[51,52] In some cases this increase

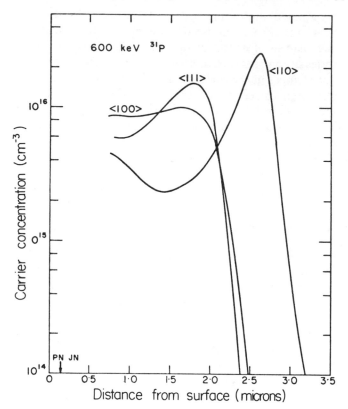

Figure 16.8. Carrier concentration profiles determined by differential capacitance techniques for nominal dose of 1·5 × 10^{12} phosphorus ions/cm² channeled along ⟨100⟩, ⟨110⟩ and ⟨111⟩ directions in silicon. For 'microns' read 'micrometres'. (From Reddi and Sansbury[51])

in penetration depth is of key importance in device performance. Note that deep dopant profiles can also be achieved by high energy random implants or by a drive-in diffusion.

To achieve reproducible channeled implants care must be taken with beam-to-sample orientation and surface preparation. In the latter case the presence of a thin oxide layer can often cause severe degradation of the dopant profile. Figure 16.9(a) shows 300 keV phosphorus implants in silicon as a function of misorientation off the ⟨111⟩ axis.[53] The critical angle for channeling is expected to be 3·1° for 300 keV phosphorus along this axis. The beam fraction which channels deeply in the crystal is only slightly affected by a one-half degree misorientation of the crystal. Further misorientation causes a marked reduction in the channel fraction of the implanted atoms.

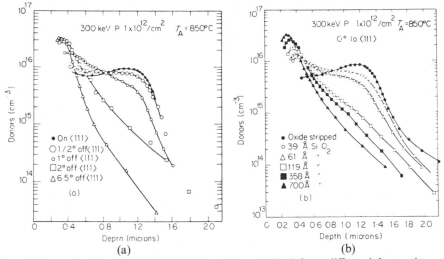

Figure 16.9. Carrier concentration profiles determined from differential capacitance for 300 keV phosphorus ions incident on $\langle 111 \rangle$ Si as a function of (a) misorientation off the $\langle 111 \rangle$ axis and (b) surface oxide thickness. The implanted layers were annealed at 850 °C. (From Moline and Reutlinger[53])

For the 100 to 300 keV phosphorus ions, the profiles are significantly altered by 1° misorientation but only slightly affected by half-degree misorientation.

The effect of surface oxide thickness on the profile of 300 keV phosphorous ions is shown in Figure 16.9(b). Samples were prepared with various thicknesses of thermally grown silicon dioxide. The data indicate that 39 Å of oxide can produce a significant effect on the implantation profile at 300 keV. For a given oxide layer thickness, the dechanneling becomes more severe at lower energies. The dechanneling is due to an increase in beam divergence caused by scattering events in the oxide layer. The implantation profile for a given misorientation can be compared directly with the profile for a certain thickness of oxide layer. Most of the phosphorus ions become dechanneled for a misorientation of about 2°, this is equivalent to 120 Å of silicon dioxide at 300 keV and 40 Å of oxide at 100 keV.

To recast this in terms of silicon and gallium arsenide, natural oxide films of about 25 Å thickness are found after exposure of the crystal surfaces to air for 24 hours. These films will cause severe dechanneling for P or S ions at energies below 100 keV. Consequently, great care must be taken in surface preparation if well channeled profiles are to be obtained for these species.

Higher mass projectiles such as arsenic or selenium exhibit a more sensitive dechanneling dependence on oxide layer thickness. For example, 300 keV arsenic ions channeled in the $\langle 111 \rangle$ direction in silicon are approximately a factor of 2 more sensitive to the oxide layer than phosphorus ions.[53] For arsenic implantations in silicon the surface oxide layer will play a dominant role in determining the dopant distribution. For lighter mass impurities such as boron the difference between random and channeled implantation penetration depths is not as great. The work of Seidel[54] has indicated that the channeled $\langle 111 \rangle$ distribution show maxima only 1·9–1·5 times deeper than the random ranges for 80 to 300 keV boron implantations. This difference in penetration may not be great enough for many device application to warrant the extra care required to make well characterized implantations.

In summary then, the present data suggest that the optimum species for channeled implants in silicon, germanium or gallium arsenide may lie in the range between magnesium and sulphur. Higher-mass dopant species exhibit a strong dependence upon surface preparation (oxide thickness) and lower-mass dopant species will not exhibit a sufficiently greater penetration to warrant the effort for channeled conditions. Of course, this presents too simple a picture. One can always anticipate device parameters such that a dechanneling profile may offer sufficient advantages to warrant application. It is too early to anticipate all device requirements.

References

1. See, for example, Ghosh, H. N., Oberai, A. S., Chang, J. J. and Vora, M. B., *IBM J. Res. Dev.*, **15**, 457 (1971).
2. Chou, S., Davidson, L. A. and Gibbons, J. F., *Appl. Phys. Lett.*, **17**, 23 (1970).
3. Chou, S., *Ph.D. Dissertation*, Stanford University (1971).
4. Schwenker, R. O., Pan, E. S. and Lever, R. F., *J. appl. Phys.*, **42**, 3195 (1971).
5. Haskell, J., Rimini, E. and Mayer, J. W., *J. appl. Phys.*, **43**, 3425 (1972).
6. Mitchell, I. V., Mayer, J. W., Kung, J. K. and Spitzer, W. G., *J. appl. Phys.*, **42**, 3982 (1971).
7. Joshi, M. L. and Wilhelm, F., *J. Electrochem. Soc.*, **112**, 184 (1965).
8. Dash, S. and Joshi, M. L., *IBM J. Res. Dev.*, **14**, 453 (1970).
9. Watanabe, M., Tokyo Shibaura Electric Co., unpublished.
10. Fujimoto, F., Komaki, K., Watanabe, M. and Yonezawa, T. *Appl. Phys. Lett.*, **20**, 248 (1972).
11. Bullis, W. M., *Solid St. Electron.*, **9**, 143 (1966).
12. Fistul, V. I., *Heavily Doped Semiconductors*, New York, Plenum Press (1969).
13. Mayer, J. W., Erikson, L. and Davies, J. A., *Ion Implantation in Semiconductors*, New York, Academic Press (1970).
14. Clark, A. H. and Manchester, K. E., *Trans AIME*, **242**, 1173 (1968).
15. North, J. C. and Gibson, W. M., *Appl. Phys. Lett.*, **16**, 126 (1970).
16. Picraux, S. T., Brown, W. L. and Gibson, W. M., *Phys. Rev.*, B, **6**, 1382 (1972).
17. Whitton, J. L., Chalk River Nuclear Laboratories, unpublished.
18. Rimini, E., Lugujjo, E. and Mayer, J. W., *Phys. Rev.* B, **6**, 718 (1972).
19. Picraux, S. T., *Appl. Phys. Lett.*, **20**, 91 (1972).
20. Itoh, T. and Nakamura, T., *Rad. Effects*, **9**, 1 (1971).
21. Keil, E., Zeitler, E. and Zinn, W., *Z. für Naturforsch.*, **15a**, 1031 (1960).
22. Meyer, L., *phys. stat. sol.*, (b), **44**, 253 (1971).

23. Picraux, S. T., Sandia Laboratories, private communication.
24. Picraux, S. T. and Thomas, G. J., *J. appl. Phys.*, to be published.
25. Harris, J. L., North American-Rockwell Science Center, unpublished.
26. Ottaviani, G., Marrello, V., Mayer, J. W., Nicolet, M-A. and Caywood, J. M., *Appl. Phys. Lett.*, **20**, 323 (1972).
27. Marrello, V., Caywood, J., Mayer, J. W. and Nicolet, M-A., *phys. stat. solidi.*, (a) **13**, 531 (1972).
28. McCaldin, J. O. and Sankur, H., *Appl. Phys. Lett.*, **19**, 524 (1971).
29. Lepselter, M. P. and Andrews, J. M., in *Ohmic Contacts to Semiconductors*, Ed. B. Schwartz, New York, The Electrochemical Society, p. 159, (1969).
30. Kircher, C. J., *Solid State Electron.*, **14**, 507 (1971).
31. Drobek, J., Sun, R. C. and Tisone, T. C., *phys. stat. sol.* (a), **8**, 243 (1971).
32. Hiraki, A., Nicolet, M-A. and Mayer, J. W., *Appl. Phys. Lett.*, **18**, 178 (1971).
33. Bower, R. W. and Mayer, J. W., *Appl. Phys. Lett.*, **20**, 359 (1972).
34. Nicolet, M-A., Mayer, J. W. and Mitchell, I. V., *Science*, **177**, 841 (1972).
35. Gyulai, J., Mayer, J. W., Rodriquez, V., Yu, A. Y. C. and Gopen, H. J., *J. appl. Phys.*, **42**, 3578 (1971).
36. McCaldin, J. O. and Mayer, J. W., *Appl. Phys. Lett.*, **17**, 365 (1970).
37. Lee, D. H., Hart, R. R. and Marsh, O. J., *Appl. Phys. Lett.*, **20**, 73 (1972).
38. Lee, D. H., Hart, R. R., Kiewit, D. A. and Marsh, O. J., *J. Appl. Phys.*, submitted.
39. Rimini, E., Haskell, J. and Mayer, J. W., *Appl. Phys. Lett.*, **20**, 234 (1972).
40. Thompson, D. A., Barber, H. D. and MacKintosh, W. D., *Appl. Phys. Lett.*, **14**, 102 (1969).
41. Buck, T. M. and Wheatley, G. H., *Surf. Sci.*, **33**, 35 (1972).
42. Mitchell, I. V., Kamoshida, M. and Mayer, J. W., *J. appl. Phys.*, **42**, 4378 (1971).
43. Morgan, D. V. and Bøgh, E., *Surf. Sci.*, **32**, 278 (1972).
44. Poate, J. M., Buck, T. M. and Schwartz, B., *J. Phys. Chem. Solids* to be published.
45. Eisen, F. H., North-American Science Center, unpublished.
46. Davies, J. A., Denhartog, J., Erikson, L. and Mayer, J. W., *Can. J. Phys.*, **45**, 4053 (1967).
47. Fladda, G., Mazzoldi, P., Rimini, E., Sigurd, D. and Erikson, L., *Rad. Effects*, **1**, 243 (1969).
48. Westmoreland, J. E., Mayer, J. W., Eisen, F. H. and Welch, B., *Rad. Effects*, **6**, 161 (1970).
49. Moline, R. A. and Foxhall, G. F., *IEEE Trans.* **ED-19**, No. 2, 267 (1972).
50. Dearnaley, G., Wilkins, M. A., Goode, P. D., Freeman, J. H. and Gard, G. A., *Atomic Collision Phenomena in Solids*, Ed. D. W. Palmer, M. W. Thompson and P. D. Townsend, Amsterdam, North Holland, p. 633 (1970).
51. Reddi, V. G. K. and Sansbury, J. D., *Appl. Phys. Lett.*, **20**, 30 (1972).
52. Moline, R. A., *J. appl. Phys.*, **42**, 3553 (1971).
53. Moline, R. A. and Reutlinger, G. W., *Proc. of 2nd Int. Conf. on Ion Implantation in Semiconductors*, Ed. I. Ruge and J. Graul, Berlin, Springer-Verlag, p. 58 (1971).
54. Seidel, T. E., *ibid.*, p. 47.

Author Index

 Channeling: Theory, Observation and Applications

Subject Index